T0280290

# Introduction to Hierarchical Bayesian Modeling for Ecological Data

# CHAPMAN & HALL/CRC APPLIED ENVIRONMENTAL STATISTICS

Series Editor

**Richard Smith**

University of North Carolina

U.S.A.

## *Published Titles*

Michael E. Ginevan and Douglas E. Splitstone, **Statistical Tools for Environmental Quality**

Timothy G. Gregoire and Harry T. Valentine, **Sampling Strategies for Natural Resources and the Environment**

Daniel Mandallaz, **Sampling Techniques for Forest Inventory**

Bryan F. J. Manly, **Statistics for Environmental Science and Management, Second Edition**

Steven P. Millard and Nagaraj K. Neerchal, **Environmental Statistics with S Plus**

Song S. Qian, **Environmental and Ecological Statistics with R**

Wayne L. Myers and Ganapati P. Patil, **Statistical Geoinformatics for Human Environment Interface**

Éric Parent and Étienne Rivot, **Introduction to Hierarchical Bayesian Modeling for Ecological Data**

Chapman & Hall/CRC
Applied Environmental Statistics

# Introduction to Hierarchical Bayesian Modeling for Ecological Data

Éric Parent
Étienne Rivot

CRC Press is an imprint of the
Taylor & Francis Group, an informa business
A CHAPMAN & HALL BOOK

CRC Press
Taylor & Francis Group
6000 Broken Sound Parkway NW, Suite 300
Boca Raton, FL 33487-2742

First issued in paperback 2020



Version Date: 20120531

ISBN 13: 978-0-367-57671-4 (pbk)
ISBN 13: 978-1-58488-919-9 (hbk)



**Library of Congress Cataloging-in-Publication Data**

Parent, E. (Eric), 1957-
Introduction to hierarchical Bayesian modeling for ecological data / Eric Parent, Etienne Rivot.
p. cm. -- (Chapman & Hall/CRC applied environmental statistics)
"A CRC title."
Includes bibliographical references and index.
ISBN 978-1-58488-919-9 (hardcover : alk. paper)
1. Ecology--Statistical methods. 2. Bayesian statistical decision theory. I. Rivot, Etienne, 1974- II. Title.

QH541.15.S72P37 2012
577.072'7--dc23 2012015479

**Visit the Taylor & Francis Web site at**
**http://www.taylorandfrancis.com**

**and the CRC Press Web site at**
**http://www.crcpress.com**

# Contents

# *List of Figures*

# List of Tables

# Foreword

*Statistics? Not for us!* Surprisingly, this is still what can sometimes be heard when discussing statistics with scientists from applied disciplines. How is it that such people disregard the science of doubt? No doubt that practitioners possess a deep understanding of phenomenological behavior of the system under study. They also have an acute feeling of its variability and its complexity. They have elaborated a mental image of the phenomenon, with operational shortcuts and permanent reformulations. Dare we say: the seed for a *model* is already set in their minds. At the same time, they never blindly trust model outputs and agree that statistics is necessary to depict things that happened in the past and to make projections about what may occur in the future. However, they also often complain that the basic statistical procedures are too simple and not flexible enough to address real problems they have to struggle with, and one has to acknowledge that instruction manuals for more advanced statistical tools are rarely funny readings! The present book might not be exactly a funny reading, but we hope that it could contribute to make the first steps of statistical modeling and inference more accessible, and tear-free, to a wide community of practitioners.

In Ecology, interestingly, complex issues have often been the hotbed of development of innovative statistical methodology. Statistical Ecology, a branch of ecological sciences, has proved incredibly productive in the last thirty years. However, temptation remains to apply *ready-to-use* statistical recipes, and frustration can be heavy when the data and the problem at hand do not fit exactly into the mold of the classical statistical toolboxes. But, to sacrifice the model realism for the sake of statistical tractability, or to defer to the sole mathematician the inference of the more complex model formulation, is that the unavoidable dilemma? In both cases, the analyst is deprived from the exciting intellectual adventures of model making!

This book was written because we believe that Hierarchical Bayesian Modeling is like a golden key to be given to scientists in Ecology. This avenue of thought should highly contribute to free their creativity in designing statistical models of their own.

In this book, we first recall that the principles of Bayesian statistics are not difficult to learn and can be intuitively understood. Bayesian statistical inference is merely learning from data by updating prior probabilistic judgments into posterior beliefs. Bayesian statistical modeling offers an intuitive avenue to put structures on data, to test ideas, to investigate several competing hypotheses and to assess degrees of confidence of predictions.

We also illustrate how conditional reasoning is a way to dismantle a complex reality into more understandable pieces. The construction of complex models is achieved by assembling basic elements linked via conditional probabilities that form the core of the hierarchical model structure. As conditional reasoning is intimately linked with Bayesian thinking, considering hierarchical models within the Bayesian setting offers a unified coherent framework for modeling, estimation and prediction.

Key ideas of this book have emerged through discussions with colleagues and PhD students. Étienne Prévost spent a whole research career working on salmons and Bayes, and many chapters are inspired by our collaborations. We are grateful that he allowed us to present in this book a lot of examples and data from his work. For the Scorff databases, field data collection has been carried out under his supervision by the personnel from the Moulin des Princes experimental station. Many thanks to the team of Fisheries and Oceans Canada in Moncton, where it all began. In 2005, we set up there a two-week Bayesian school with our friend Étienne Prévost. Gerald Chaput, Hugues Benoit and their colleagues suggested we write this book. We heard again the same suggestions during the 2006 and 2007 doctoral sessions that we organized for the Marine Exploited Ecosystems Research Unit (UMR EME) in Sète, France, to promote Bayesian thinking for modeling ecological data. We are particularly grateful to Daniel Gaertner, Nicolas Bez, Jean-Marc Fromentin and Frédéric Ménard who encouraged us on the way. We would like to thank Jean-Luc Baglinière (INRA, UMR ESE, Rennes) without whom the long series of data on A. salmon populations in the Oir and Scorff Rivers would not exist, and all the staff of the Unité Expérimentale d'Écologie et d'Écotoxicologie (INRA, U3E, Rennes), in particular Nicolas Jeannot and Frédéric Marchand for their invaluable work in the field collecting the salmon trapping data in the Scorff and Oir Rivers.

There are many people who also deserve our thanks, although they did not know they were so actively influencing us and contributing to this book, such as our families, our mentors, our PhD students and also the many collaborators from our research institutions, which fostered a nurturing environment for the whole lifetime of this book project. We also had the great fortune of counting on our long-standing friendship with Jacques Bernier and Lucien Duckstein for useful discussions and careful proofreading.

Jérôme Guitton patiently helped us to master enough *html* programming language to implement our website *hbm-for-ecology.org*. This website gives many of the datasets, R and WinBUGS or OpenBUGS codes that we used to derive inference and figures in the book.

Éric Parent and Étienne Rivot
Paris, France

# Part I

# Basic blocks of Bayesian modeling and inference for ecological data

# Chapter 1

## Bayesian hierarchical models in statistical ecology


### Summary

The Salmon life cycle and the biomass production model exemplify the key idea of this book: Hierarchical Bayesian Modeling (HBM) is a Directed Acyclic Graph (DAG) modeling technique with the capacity to cope with high-dimensional complex models typically needed for ecological inferences and predictions ([64]; [65]; [67]; [75]; [313]).

HBM works through conditional decomposition of high-dimension problems into a series of probabilistically linked simpler substructures. HBM enables to exploit diverse sources of information to derive inferences from large numbers of latent variables and parameters that describe complex relationships while keeping as close as possible to the basic phenomena.

Based on these two motivating case studies from fisheries sciences, we detail the three basic layers of hierarchical statistical models ([75]; [76]; [312]; [313]):

1. A data level that specifies the probability distribution of the observables at hand given the parameters and the underlying processes;

2. A latent process level depicting the various hidden ecological mechanisms that make sense of the data;

3. A parameter level identifying the fixed quantities that would be sufficient, were they known, to mimic the behavior of the system and to produce new data statistically similar to the ones already collected.

HBM stands out as an approach that can accommodate complex systems in a fully consistent framework and can represent a much broader class of models than the classical statistical methods from ready-to-use toolboxes. Finally, we use graphical modeling techniques as a guiding thread to announce the contents of the book's chapters, starting with simple stochastic structures (elementary blocks) in Part I that will progressively be assembled to lead to the more elaborate hierarchical models of Part II.

## 1.1 Challenges for statistical ecology

### 1.1.1 The three steps of quantitative modeling

Quantitative modeling is typically motivated by knowledge and action: *i)* modeling seeks to improve our understanding of a phenomenon (*cognitive objective*); *ii)* many applied ecological questions require models as a tool to derive predictions (*applied objectives*) (see Fig. 1.1).

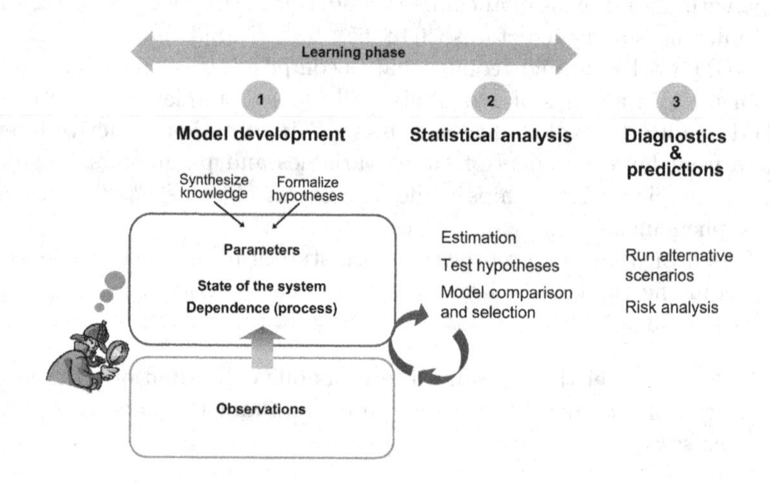


**FIGURE 1.1**: The three steps of quantitative modeling. Steps 1 and 2 correspond to the learning phase. The little character on the left side is inspired from the Ecological Detective in the book of R. Hilborn and M. Mangel ([136]).

1. *Step 1.* Propose a tentative model for the process under consideration.

Models are used to synthesize knowledge and speculate about the system of interest. But modeling is also governed by the objectives of the study and some subjectivity from the analyst ([236]). For instance, plants or animals in ecosystems can be represented one by one by dynamically interacting rule-based agents (well-documented examples of individual-based models for ecology can be found in [125] and [122]). In such models, local heterogeneity among entities is built in for studying, by simulation, the global scale consequences of a particular trait of ecological interest, such as body size or some behavioral characteristics such as an alteration in reproductive strategy. On the contrary, in a population model ([164]), like the biomass production model depicted in Fig. 1.2, the characteristics of the population are averaged together and the model attempts to depict the evolution of the population as a whole. Knowledge is generally understood as *structural quantitative* knowledge with mathematical equations that formalize hypotheses and quantify interactions between variables. Models rely on *i)* key variables of interest with respect to the objectives of the modeling enterprise, and *ii)* the main cause-to-effects interactions between these variables which are built from a *deductive reasoning.* Interactions mimic either deterministic or stochastic cause-to-effect relationships. Consider as an example, a $\pi = 40\%$ survival rate between two stages $A$ and $B$ of some population with size $N_A$ and $N_B$ like the juvenile to smolt transition in the salmon life cycle of Fig. 1.4; a deterministic model would state that the population at stage $B$ is 40% of the one at stage $A$ while a stochastic model would say that population size $N_B$ is some random variable, distributed as a Binomial with order $N_A$ and probability 0.4 (mathematical layouts for the Binomial distribution are given in Chapters 2 and 4). Because natural systems are often poorly understood, there may be alternative competing hypotheses. Equations generally involve unknown quantities (*e.g.*, unknown parameters or hidden system states) that must be estimated: for instance the survival rate $\pi$ from state $A$ to state $B$ in the previous example might be left unknown and one would expect this quantity to be estimated from the available data.

Knowledge also involves *qualitative* knowledge or *expertise* ([223]). From past experience when dealing with a similar situation in another location, use of proxies or a long life understanding of the biological phenomenon, scientists may sometimes specify valuable best guesses and credible intervals for unknown model quantities. For instance, referring to the previous example, some probabilistic judgment could be made a priori for the unknown survival rate $\pi$

such as, "considering the number of potential predators and the length of juveniles, the probability of survival $\pi$ is most likely between 0.25 and 0.5, unlikely below 0.20, but extremely rarely above 0.8."

2. *Step 2.* Learning from observations.

   In the second step, one seeks to learn from the data at hand, *i.e.*, to use the data to update the prior knowledge that has been coded in Step 1. This is often referred to as *confronting the model to the data* or more simply as *model fitting* or *inference.* Objectives are *i)* to compare the credibility of alternative competing hypotheses with respect to the observations; *ii)* to estimate the unknown quantities of the model and the associated uncertainty conditionally upon the data at hand. This step relies heavily on statistical science to infer pattern and causality from the data. By comparison with the deductive way of thinking that works with cause-to-effect relationships, inference relies on *inductive reasoning* and promotes a reverse way of thinking, working on effect-to-cause relationships. Depending on the field of application, it is sometimes termed *inverse modeling* (engineering), *data assimilation* (geosciences), *statistical learning* (neural networks), among others.

3. *Step 3.* Use the model in a deductive way as a decision tool.

   In this third step, the model can be used in a deductive way as a tool for prediction. This generally takes the form of simulations designed to explore the response of the system under different scenarios. Typically, models for renewable resource assessment are used to establish a diagnostic about past evolution and the present state of the resource, to estimate key management reference points such as the sustainable level of exploitation, and to predict the potential evolution of the resource under several scenarios for future exploitation or evolution of the environment. As an example of choice, the 2000 special report of the International Panel on Climate Change of Working Group III includes a summary for policy makers that describes long-term greenhouse gas emissions depending on different future storylines with contrasted demographic, socioeconomic, technological and environmental developments.

   However, as models become more complex, they require more parameters and they inevitably become more sensitive to parameter values. Before using the model in a predictive mode, a critical issue for the Ecological Detective ([136]) is to have a precise and rigorous quantification of the uncertainties in the different components of the model ([133]; [184]), not least because the general adoption of the precautionary principles and precautionary approaches to management of ecological systems ([67]; [106]) requires an assessment of the risks that will occur as a result of the different scenarios. Risk can be defined as the probability distribution of the consequences of an undesirable event, (*e.g.*, irreversible environmental damage). As such, risks are inevitable consequences of uncertainty. If there is no uncertainty, the concept of risk is irrelevant because the probability of any event degenerates to either 0 or 1.

Here, the first two steps have been isolated for the purpose of clarity. But this is somewhat artificial as they are practically embedded within an iterative process in which the confrontation of the proposed model to the available data (Step 2) may lead to reconsider the hypotheses that were made in Step 1.

The relative importance of each of the three steps may vary depending upon the modeling objectives. Step 1 will receive primary attention for fundamental ecological studies. If the management purpose is of primary interest, more modeling effort will be devoted to Step 3.

Throughout this book, we will exemplify how the Bayesian setting enables one to combine these three steps (model design, model fitting and predictive simulations) within a single unified and rigorous statistical framework. A deeper discussion on the Bayesian setting is postponed to Section 1.3.2.

### 1.1.2 Models are intrinsically complex, highly uncertain and partially observed

The growing interest in solving ecological problems has provided a new impetus for the development of complex models, which raised statistical challenges for quantitative modeling analyses. Complexity in ecological models arises from two main reasons: the motivation to get enough realism and the need to account for various sources of uncertainty.

#### 1.1.2.1 Increasing model realism and dimension

Improving the realism of ecological models inherently results in an increase of model dimension and complexity. Various components are now increasingly used in recent ecological models, such as dynamic or spatial components (*e.g.*, spatial heterogeneity, migration patterns), and multiple interactions forms (*e.g.*, nonlinear intra-specific density dependence, inter-specific interactions) often combined with the effects of covariates to trigger the main influences of the environment (*e.g.*, changing forcing conditions as a climate under global warming).

#### 1.1.2.2 Accounting for different sources of variability and uncertainty

The field of environmental sciences is becoming increasingly aware of the importance of accurately accounting for the multiple sources of variability and uncertainty when modeling environmental processes and making forecasts. This development is motivated in part by the desire to provide an accurate picture of the state of knowledge of ecosystems, and to be able to better assess the quality of predictions and decisions resulting from the models. Models are functional caricatures of (complex) natural systems which are not easily observed. Uncertainties in ecological models are inherent to our incomplete knowledge, due to both the difficulty to gather exhaustive observation data on natural systems and to our relatively poor ability to propose mechanistic models with good predictive power. Uncertainty is hence a rather difficult notion because it encompasses the elements that fluctuate due to unpredictable or unaccessible varying factors (uncertainty by essence) and those that are just partially known (uncertainty by ignorance).

In ecology, uncertainties stem from three main sources (see [63], [133], and [173], but also [252] for a general discussion about uncertainty): *i)* model errors; *ii)* process stochasticity; *iii)* observation errors.

1. Model errors

   As any model is a functional simplification of a real process, it thus provides an imperfect and a potentially misleading representation of how the real system works. Model errors (also referred to as structural errors) reflect our ignorance about the very nature of the system. Model error may of course yield major consequences if the model is used for forecasting (in particular when extrapolating beyond the range of the data that have been collected). The role of hypotheses is to restrict the number of plausible models. By convention, a zero probability is put on the models that do not belong to the set delimited by the hypotheses. The model error term is to be quantified only for the members of this remaining set.

2. Process stochasticity

   Models with deterministic process are useful to learn from the interaction of complex processes. However, when confronting models with data, models must account for unpredictable variability or stochasticity that cannot be explained by deterministic processes. Process stochasticity can take different forms of which the most common are demographic stochasticity (due to random differences between individuals or statistical groups of individuals with the same characteristics) or environmental stochasticity (due to unpredictable random variations of parameters) ([100]).

3. Observation or measurement errors

   Observation or measurement errors stem inherently from the impossibility to accurately and exhaustively observe Nature. They are the consequence of imperfection in the data records. Field observations are rarely issued from an optimized sample scheme, and the expense of data collection restrict the collection of as much data as might be desirable. Data are often incomplete, and the amount of sampling and measurement errors resulting from either sampling, measurement or estimation mistakes is often large and unknown. In the words of J. Schnute ([272]), "*counting fish is like counting trees except they are invisible and they move.*"Hence, the Ecological Detective in Figure 1.1 has to cope with indirect, noisy and incomplete observations which have to be put in coherence with some hypotheses about the hidden process that underlines the observed phenomenon.

In statistical ecology, uncertainty due to model errors generally represents the major source of uncertainty ([252]; [277]). Model weighting or model averaging offers some tools to handle model errors ([139]; [182]; [287]). In statistical models, uncertainty due to process stochasticity is generally irreducible, in the sense that it does not decline when the sample size increases. The uncertainty due to estimation errors embed uncertainty stemming from measurement errors, and results from imprecision (quantified by statistical inference) when estimating the unknown of the systems from the available data. In contrast with the uncertainty due to process stochasticity, estimation errors can be reduced by improving the data collection procedure. In theory, they decline asymptotically when more data become available (only under the stringent hypotheses of a perfect model and a stationary phenomenon).

### 1.1.3 Embedding ecological models within a statistical approach

As these different sources of errors and variability are conceptually and quantitatively different, modern statistical ecology seeks to propose quantitative tools to separate out the uncertainty due to the process stochasticity to the one due to the observation sampling process. This requires flexible approaches to fusing models with data, approaches that can accommodate uncertainties in the way ecological processes operate and the way we observe them.

### 1.1.4 Motivating examples

#### The Biomass production model

Biomass production models ([137]; [244]) have a long history in quantitative fisheries sciences and continue to prove useful in stock assessment under a data-poor environment. Although they are based on a crude simplification of the fish population dynamics, they remain a tool of choice for providing fisheries management advice in situations in which:

- There are insufficient age- and size-composition data to permit the use of assessment methods based on age- or size-structured models;
- One is mostly interested in estimating the maximum sustainable yield ($C_{MSY}$), and its associated biomass ($B_{MSY}$) (mathematical details about $MSY$ calculations are postponed to Chapter 7, see Fig. 7.5);
- It is worthwhile predicting the biomass levels and surplus production for the future and sometimes in the past, to impute missing data records. The latter is particularly important to test the performance of alternative management scenarios when efforts are being made to control the level of catches and promote a sustainable harvest policy.

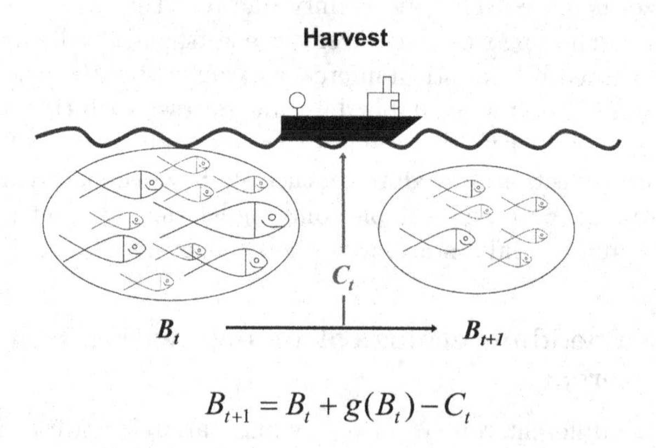


**FIGURE 1.2**: Biomass dynamics of an exploited fish stock.

The motivating example of the Biomass production model illustrates the three complementary steps (Fig. 1.1).

1. *Step 1.* Propose a tentative model for the process under concern.

   The dynamic biomass production model is a crude but useful simplification of the dynamics of a harvested fish population. The key variable of interest is the total *biomass* of the population at each time. The dynamics is modeled in discrete time, most often on a year-to-year basis. The biomass at the beginning of time step $t+1$, denoted $B_{t+1}$, is obtained from $B_t$ through a rather simple balance equation:

   $$B_{t+1} = B_t + g(B_t) - c_t \tag{1.1}$$

   where $c_t$ is the observed harvest between $t$ and $t+1$ and $g(B_t)$ is the *production function.* It quantifies the balance between recruitment (arrival of new individuals in the stock biomass), growth (weight), natural mortality, and possibly emigration-immigration. The most classic choice for the production function is the *logistic* one with two parameters, the population growth rate $r$ and the carrying capacity $K$, which are generally hypothesized constant over time (see also Chapter 11 for more details):

   $$g(B_t) = r \cdot B_t \cdot (1 - \frac{B_t}{K}) \tag{1.2}$$

   A Log-Normal random noise term is generally added to capture the biological variability due to (unpredictable) environmental variations:

   $$\begin{cases} B_{t+1} = (B_t + g(B_t) - c_t) \cdot e^{\epsilon_{t+1}} \\ \varepsilon_{t+1} \overset{iid}{\sim} Normal(0, \sigma^2) \end{cases} \tag{1.3}$$

   with the $\sim$ sign meaning that $\epsilon_{t+1}$ , the logarithm of the perturbation, is a normally distributed $N(0, \sigma^2)$ random term standing for the environmental noise. Mathematics for the Normal Distribution will be developed in Chapters 3 and 6.

2. *Step 2.* Learning from observations.

   Let $t = 1, ..., n$ denote the epochs for which observations are available. Available data in that case typically consist of a series of observed catches $c_1, ..., c_n$ and of abundance indices $i_1, ..., i_n$, the latter are often assumed proportional to the current biomass $i_t = q \cdot B_t$, $\forall t \in \{1, ..., n\}$.

   Abundance indices can be of different nature. They can be derived from commercial data such as commercial catches per unit of fishing effort, or from a scientific survey such as scientific trawling or acoustic surveys. The abundance indices $i_{t=1:n}$ are often assumed proportional to the current biomass $i_t = q \cdot B_t$, $\forall t \in \{1, ..., n\}$ with a catchability parameter $q$, hypothesized constant over time. A common, although simplifying assumption is that observed catch and abundance index of each year $t$ are respectively related to the unobserved true catch and unobserved biomass through stochastic observation models:

$$\begin{cases} c_t = C_t \cdot e^{\omega_{1,t}} \\ i_t = q \cdot B_t \cdot e^{\omega_{2,t}} \end{cases} \tag{1.4}$$

with $\omega_{1,t}$ and $\omega_{2,t}$ normally distributed $N(0, {\tau_{1,o}}^2)$ and $N(0, {\tau_{2,o}}^2)$ random terms describing the uncertainty in the observed catches and abundance indices due to measurement and sampling error (observation error).

Fisheries scientists would typically rely on the available knowledge (expertise and past data) and observation (data at hand) to provide answers to the following questions of interest: *i*) Is the logistic growth function appropriate or does another form of production function fit the observed data better? *ii*) What are the credible values for the parameters $(r, K)$ and the associated uncertainty? Can the growth rate $r$ and the carrying capacity $K$ be elicited from some probabilistic prior judgmental expertise? *iii*) What are the credible values for the historical trajectory of the biomass level, say $B_1, ..., B_n$ and what is the level of the Biomass depletion over the time series $\frac{B_n}{B_1}$?

Figure 1.3 relies on a graphical representation based on different layers to illustrate the conceptual difference between the stochastic process for the biomass dynamics and the sampling model for the observations. Statistical inference aims at using the information contained in the data $(i_t, c_t)_{t=1:n}$ to learn about the unknown biomass dynamics.

3. *Step 3.* Use the model in a deductive way as a decision tool.

   Quantities that interest the natural resources manager, such as the management reference points related to long-term equilibrium, can be directly derived from the parameters $(r, K)$ of Eq. (1.2):

$$\begin{cases} C_{MSY} = \dfrac{r \cdot K}{4} \\ B_{MSY} = \dfrac{K}{2} \end{cases} \tag{1.5}$$

   What are credible values for the management reference points $C_{MSY}$ and $B_{MSY}$ and their associated uncertainty? Do they match

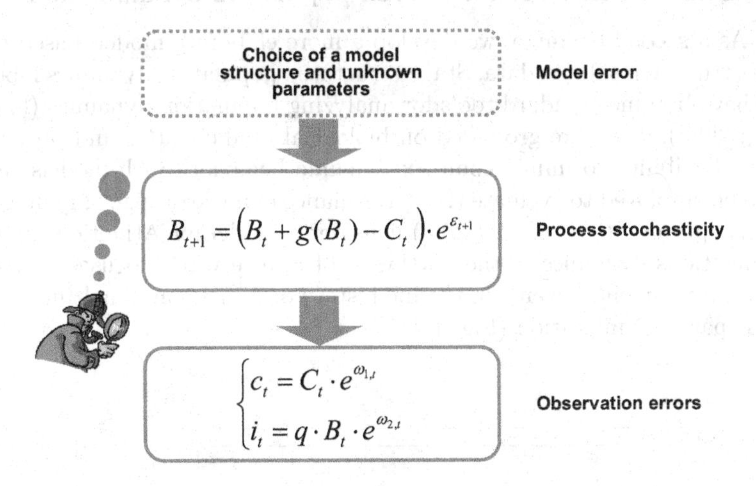


**FIGURE 1.3**: Biomass production model. Hierarchical representation to explicitly disentangle the stochastic model for the hidden dynamics from the sampling model for the observations. The arrow from the hidden dynamic to the observations represents the sampling process giving rise to observations $(i_t, c_t)_{t=1:n}$.

the common sense of fishery scientists? Fisheries scientists may also be interested in deriving predictions of future trajectories of the biomass and catches over time $t = n+1, ..., n+k$ under alternative *management scenarios.* Simulations typically aim at comparing different harvest control rules, *e.g.*, different fishing mortalities $F_{n+1}, ..., F_{n+k}$ leading to a series of catches $C_{n+1} = F_{n+1}B_{n+1}, ..., C_{n+k} = F_{n+k}B_{n+k}$. Implementation uncertainties can also be modeled at this stage to evaluate the consequences of the unavoidable discrepancies between a recommended management policy and its actual implementation in operational practice. Predictive simulations are also of interest to help implement experimental designs. Suppose for instance that, due to budget limitations, only half of the scientific campaigns previously planned in the next ten years will be undertaken. When and how should the measurements be taken to make the best of the future information?

### Atlantic salmon stage-structured population dynamic model

As a second example, we develop a more elaborate model, based on age-structured salmon data. Stage-structured population dynamics models have become standard tools for analyzing population dynamics ([50]; [51]; [300]). They are grounded on biological fundamentals and provide some flexibility to mimic complex dynamic behavior. Such models can also be employed to evaluate the performance of a wide range of management options. Rivot *et al.* ([259]) consider a model for Atlantic salmon populations dynamics in the northwest of France which focuses on two of the fundamental events in the life history of A. salmon: Smoltification and spawning migration (Fig. 1.4).

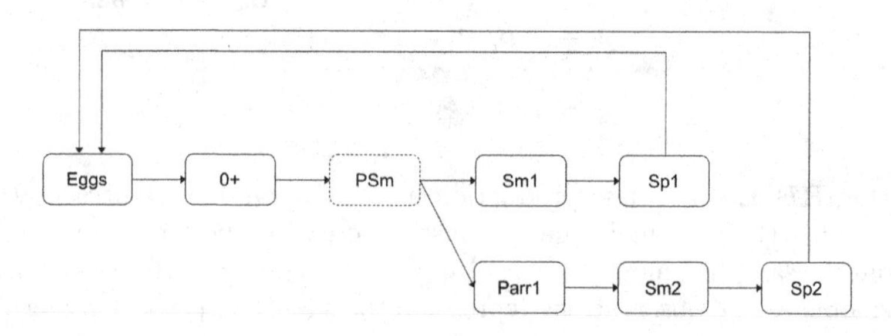


**FIGURE 1.4**: Simplified Atlantic salmon life cycle model with only one sea-age class. Spawners $Sp1$ return and spawn one year before spawners $Sp2$ of the same cohort.

Spawning occurs in the river in late fall and early winter and most spawners die shortly after their first reproduction. Eggs hatch in the gravel and alevins emerge during the next spring. After emergence, young-of-the-year salmon (denoted 0+ in Fig. 1.4) spend one or several years in the river before undergoing smoltification in the spring. At this time, they migrate downstream to the sea as smolt. Adults return to their home river for spawning after one or several years at sea. In France, a major portion of the juveniles become smolts at one year of river age (1+ Smolt, denoted $Sm1$ in Fig. 1.4), in the spring following their emergence as young-of-the-year salmon, leaving behind the smaller juveniles ($Parr1$) to spend an additional year in the river before seaward migration (2+ Smolt, denoted $Sm2$ in Fig. 1.4). Fish from the two smolt age classes return as spawners after one or two years at sea. Other possible life histories (early maturation in freshwater before seaward migration, adults spending more than 2 winters at sea, spawners surviving after spawning) are rare and only have a negligible influence on population dynamics.

In this motivating example, we will consider a simplified life cycle model by forgetting the variability of sea age classes and consider that fish from the two smolt age classes return as spawners after only one year at sea as shown in Fig. 1.4 ($Sp1$ and $Sp2$, respectively). Two additional simplifying hypotheses are made: *i*) We suppose that the life history is not an inheritable characteristics; *ii*) We use a common life cycle for male and female fish.

Below we illustrate what could be the three complementary steps of the quantitative modeling approach (Fig. 1.1) applied to this A. salmon population dynamics model.

1. *Step 1.* Propose a tentative model for the process under consideration.

   Hereafter a mathematical representation of the life cycle model is proposed in Fig. 1.4. The model is built on a discrete, yearly basis time step. The number of spawners, eggs, young-of-the-year, pre-smolts, 1+ smolts and Parr 1 and 2+ smolts at each time step $t$ are denoted $Sp_t$ ($Sp_t = Sp1_t + Sp2_t$), $W_t$, $0{+}_t$, $PSm_t$, $Sm1_t$, $P1_t$ and $Sm2_t$, respectively.

   (a) *Spawners → Eggs.* The number of eggs spawned by the adults returning in year $t$, $W_t$, is modelled as a deterministic function of the number of spawners, the proportion of females $p_f$ and of the mean fecundity of these females denoted $fec$, both considered as known and constant over time:

   $$W_t = Sp_t \cdot p_f \cdot fec \tag{1.6}$$

   (b) *Eggs → 0+ juveniles.* A density-dependent process is used to model the freshwater production of juveniles resulting from the reproduction of the spawners returning in year $t$. Density dependence is modelled by the widely used dome-shaped Ricker curve with unknown parameters $(\alpha, \beta)$. Environmental variability renders the stock-recruitment process stochastic. Again, this is classically introduced via independent and identically distributed LogNormal errors:

   $$\begin{cases} 0{+}_{t+1} = \alpha \cdot W_t \cdot e^{-\beta \cdot W_t} \cdot e^{\epsilon_t} \\ \varepsilon_t \overset{iid}{\sim} N(0, \sigma^2) \end{cases} \tag{1.7}$$

   As in Eq. (1.3), $\epsilon_t$ is a normally distributed $N(0, \sigma^2)$ random term standing for the environmental noise (see also Chapter 7).

(c) *0+ juveniles → Smolts.* The young-of-the-year $0+_{t+1}$ will then survive to the next spring of year $t+2$ as pre-smolts $PSm_{t+2}$, with probability $\gamma_{0+}$ (considered as invariant over time). Survival can be modeled as a Binomial process that captures the demographic stochasticity:

$$PSm_{t+2} \sim Binomial(0+_{t+1}, \gamma_{0+}) \tag{1.8}$$

In each cohort, a proportion $\theta_{Sm1}$ (also assumed time invariant) of the pre-smolts will migrate as 1+ Smolts, the remaining part will stay one additional year as 1+ Parrs. Life history choice for smoltification is also commonly modeled using a Binomial distribution:

$$Sm1_{t+2} \sim Binomial(PSm_{t+2}, \theta_{Sm1}) \tag{1.9}$$

1+ Parrs will survive (with a survival rate $\gamma_{Parr1}$) and will migrate as 2+ Smolts. Demographic stochasticity in the survival of resident parrs is also modelled using a Binomial distribution:

$$Sm2_{t+3} \sim Binomial(Parr1_{t+2}, \gamma_{Parr1}) \tag{1.10}$$

(d) *Smolts → Returning spawners.* Hypothesizing a strict homing of adults to their native stream, the smolt-to-spawner transition can be modeled as the result of Binomial processes with $\gamma_{Sm}$ the survival probability at sea, considered as invariant over time and between the two smolt age classes:

$$\begin{cases} Sp1_{t+3} \sim Binomial(Sm1_{t+2}, \gamma_{Sm}) \\ Sp2_{t+4} \sim Binomial(Sm2_{t+3}, \gamma_{Sm}) \end{cases} \tag{1.11}$$

The spawners returning in year $t+3$ and $t+4$ will contribute to the reproduction of year $t+3$ and $t+4$, respectively.

2. *Step 2.* Learning from observations.

All parameters $(\alpha, \beta, \sigma, \gamma_{0+}, \theta_{Sm1}, \gamma_{Parr1}, \gamma_{Sm})$ in the A. Salmon stage-structured population dynamic model are usually unknown. They could ideally be estimated from time series of observations of the number of fish in each of the development stages. However, these numbers are generally not directly observable. For instance, the true number of 1+ Smolt migrating downstream each year $t$

is unknown. Typically, only partial knowledge is available through a trapping experiment. The available observations will then consist of a number of smolts caught in a downstream trap, denoted $C_{Sm1,t}$. Assuming a standard Binomial trapping experiment with trap efficiency $\pi_{Sm}$ (considered constant), such an observation process follows a Binomial distribution:

$$C_{Sm1,t} \sim Binomial(Sm1_t, \pi_{Sm}) \tag{1.12}$$

Similar Binomial counting processes with eventually different capture efficiencies can be connected at stages $Sm2$, Spawners $Sp1$ and $Sp$ and 0+ juveniles, thus providing information on the dynamics of all development stages but accounting for the sampling uncertainty. Statistical inferences will consist of using this information to learn about the A. salmon population dynamics. In other words, we want a probabilistic judgment concerning the various stages and parameters given the data. Chapter 11 is devoted to this analysis.

3. *Step 3.* Use the model in a deductive way as a decision tool.

   Although the recreational fishery (angling) is often the main source of salmon exploitation, associated harvest rates could be rather high (near 50%) and have different harvest control rules (limited fishing periods, quotas, etc.) which can be used to regulate exploitation. Such a population dynamic model can be a useful tool to assess the performance of different management strategies. This will be developed extensively in Chapter 12.

---

## 1.2 Conditional reasoning, graphs and hierarchical models

### 1.2.1 Use conditional probability distributions to model probabilistic transitions

Conditional models are networks of components (variables, also called nodes in graph theory jargon), but many of the nodes are unknown. Those variables are linked by deterministic or probabilistic conditional dependencies ([152]). In this book, conditional probability distributions are used to mimic uncertainty and stochastic influences. We will use the Gelfand's ([115]) bracket notation for probability distributions.

Let

$$[V_1|\ V_2]$$

denotes the probability of event $V_1$ given the event $V_2$ has occurred. It states that, given $V_1$, the issue $V_2$ is uncertain and described by probabilistic bets. The bracket notation will be used indifferently for probability distributions of discrete variables and for probability density functions (pdf) of continuous variables.

**Example (continued): A. salmon population dynamic model**

The A. salmon stage-structured model can be viewed as a conditional model. Following our bracket notation above, the *Eggs → 0+ juveniles* transition (Eq. (1.7)) can be written as a conditional probability distribution under the form $[0+_{t+1}|W_t, \alpha, \beta, \sigma]$. Here, the conditional pdf represents Log-Normal environmental variability.

All mechanisms of the *0+ juveniles → Smolts* transitions were modeled as Binomial distributions that can be written as conditional probability distributions $[PSm_{t+2}|0+_{t+1}, \gamma_{0+}]$, $[Sm1_{t+2}|PSm_{t+2}, \theta_{Sm1}]$ and $[Sm2_{t+3}|Parr1_{t+2}, \gamma_{Parr1}]$. Here, the conditional pdf represents demographic stochasticity.

The Binomial smolt-to-spawner survival transitions can also be written as conditional probability distributions $[Sp1_{t+3}|Sm1_{t+2}, \gamma_{Sm}]$ and $[Sp2_{t+4}|Sm2_{t+3}, \gamma_{Sm}]$.

### 1.2.2 Graphical models

Graphical models, which are often called *Bayesian networks* or *Bayes nets*, are a useful metaphor for such conditional models ([73]). The use of graphical models significantly improves the modeling process in two ways: *i)* They are useful in the two tasks of statistical modeling ([285]), deductive way of thinking when designing the model structure (cause-to-effect) and inference when seeking to learn about the unknowns (effect-to-cause); *ii)* They help visualize the factorization of complexity; subsequently, they proved a valuable tool for catalyzing interactions between experts during the process of model building.

Figure 1.5 sketches the possible basic conditional relationships between three variables $V_1, V_2, V_3$ (to get all configurations, consider all permutations since the ordering $1, 2, 3$ does not matter).

In graphical models, conditional reasoning is efficiently summarized and pictured by means of ellipses, boxes and arrows. Graphical conventions help to explain the statuses of the variables under the inferential context, and the conditional dependencies and independencies between the variables: *i)* Ellipses are reserved for variables that are defined by a

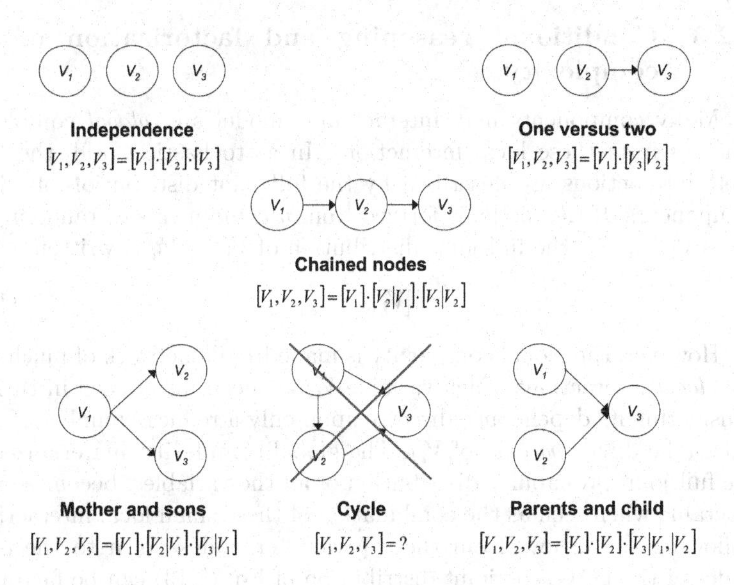


**FIGURE 1.5**: Some acyclic conditional links between three variables.

probability distribution. Nonobserved variables are represented as white ellipses, and observed variables (*data*) as shaded ellipses. An ellipse with no parents is often called a *parameter*; *ii)* Shaded rectangular boxes can be used to represent fixed constant or covariates which may *physically* help to understand the structure of the model; *iii)* Arrows are used to model dependencies between the variables. Solid line arrows represent stochastic transitions between variables, and dotted line arrows represent logical transition (deterministic). The direction of the arrows follows the line of reasoning the modeling phase ([215]).

In these graphs, no cycle is allowed, so that nodes are unambiguously defined conditionally to their parents, following the arrows. The very stringent conditions for the existence of a joint distribution $[V_1, V_2]$, when the two conditionals $[V_1|V_2]$ and $[V_2|V_1]$ are given (*i.e.*, there exists a cycle in the previous graph), are detailed in [8]. Graphical models conveniently designed without cycle via successive conditioning links are called Directed Acyclic Graphs (DAG). They have proved to be valuable tools for many ecological studies ([34]; [127]; [167]; [171]; [188]; [203]; [239]; [255]; [257]; [259]; [303]; [304]).

### 1.2.3 Conditional reasoning and factorization of the complexity

Many components may interact in a model and *global* complexity emerges from these local interactions. In a stochastic world, the complete interactions are described by the full joint distribution of all the components. If the model is formed from $n$ components or random variables $V_1, ..., V_n$ , the full joint distribution of $V_1, ..., V_n$ is written as:

$$[V_1, ..., V_n] \tag{1.13}$$

However, the global complexity is made from a network of much simpler *local* interactions. Each variable $V_i$ has been introduced in turn by considering its dependence directly upon only a reduced number of variables (the direct *parents* of $V_i$). The global complexity, materialized by the full joint probability distribution of all the variables, becomes easily tractable when seen as the combination of these small local interactions. Following the arrows from these parameter nodes down to the other nodes of the DAG, the joint distribution of Eq. (1.13) can be factorized into more simple parent-child interactions:

$$[V_1, ..., V_n] = \prod_{i=1}^{n} [V_i | pa(V_i)] \tag{1.14}$$

The *parameters* have been previously defined as variables without parents; hence, they cannot be defined as conditional distributions in Eq. (1.14) and we must make the convention that $pa(V_i)$ is the null set for the $i$'s corresponding to parameters. Distributions for these entry nodes of the graph are typically called *a priori* distributions or *priors*. It is worth noting that the interpretation of probabilistic conditional models combines naturally with the Bayesian paradigm that quantifies uncertainty by means of probabilistic judgments ([26]; [82]; [153]; [268]; [299]). As random variables generated by priors, parameters naturally appear as white ellipses in the DAG and the corresponding terms in Eq. (1.14) collapse to $[V^* | pa(V^*)] = [V^*]$, and Eq. (1.14) writes:

$$[V_1, ..., V_n] = [V^*] \times \prod_{V_i \neq V^*} [V_i | pa(V_i)] \tag{1.15}$$

Finally, one will merely find the various elementary conditional configurations illustrated in Fig. 1.5 as the small pieces of interaction issued from such decomposition (Eq. (1.15)) of the joint distribution.

Constants and forcing conditions introduced into the analysis as deterministic covariates have no probabilistic status (conversely to the random variables $V_1, ..., V_n$) and should not in principle be considered in

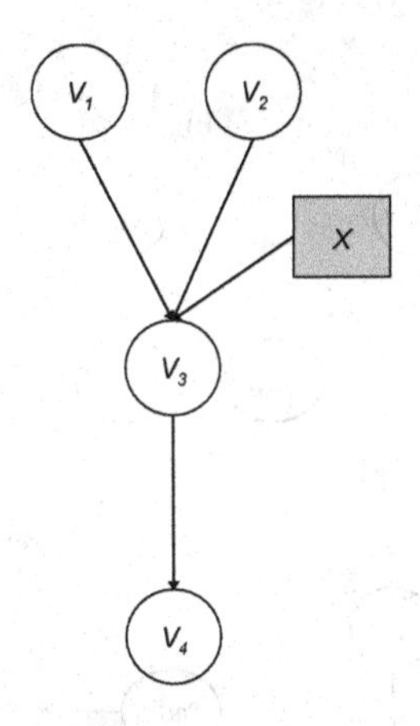


**FIGURE 1.6**: A simple example of DAG with four variables and one covariate. $V_1$ and $V_2$ have no parents and are considered as parameters with independent unconditional distributions $[V_1]$ and $[V_2]$. The full joint distribution has been designed according to the factorization $[V_1, V_2, V_3, V_4|X] = [V_1] \times [V_2] \times [V_3|V_1, V_2; X] \times [V_4|V_3]$.

probabilistic computations. However, it is often useful to make them appear in a DAG in order to precisely detail the model structure. Yet, note that we used *shaded* rectangular *boxes* instead of *ellipses* as graphical conventions to emphasize their particular status. Once they appear in the DAG, the full joint distribution of the model is implicitly conditioned by the constants and covariates which are part of the deterministic model structure. As far as the bracket notation is concerned, they will be subsequently mentioned in the conditioning side of a probability distribution when necessary (but one may also wish to separate them by a semicolon from the random variables). Figure 1.6 provides a very simple example of conditional model with four variables $(V_1, V_2, V_3, V_4)$ (in which $V_1$ and $V_2$ are parameters) and one covariate $X$.

**Example (continued): A. salmon population dynamic model**

Figure 1.7 gives simple examples of DAG for the different demographic transitions that characterize the A. salmon life cycle, and provides the DAG for the full life cycle model running for one cohort originated from eggs spawned at year $t$. The DAG is built by connecting

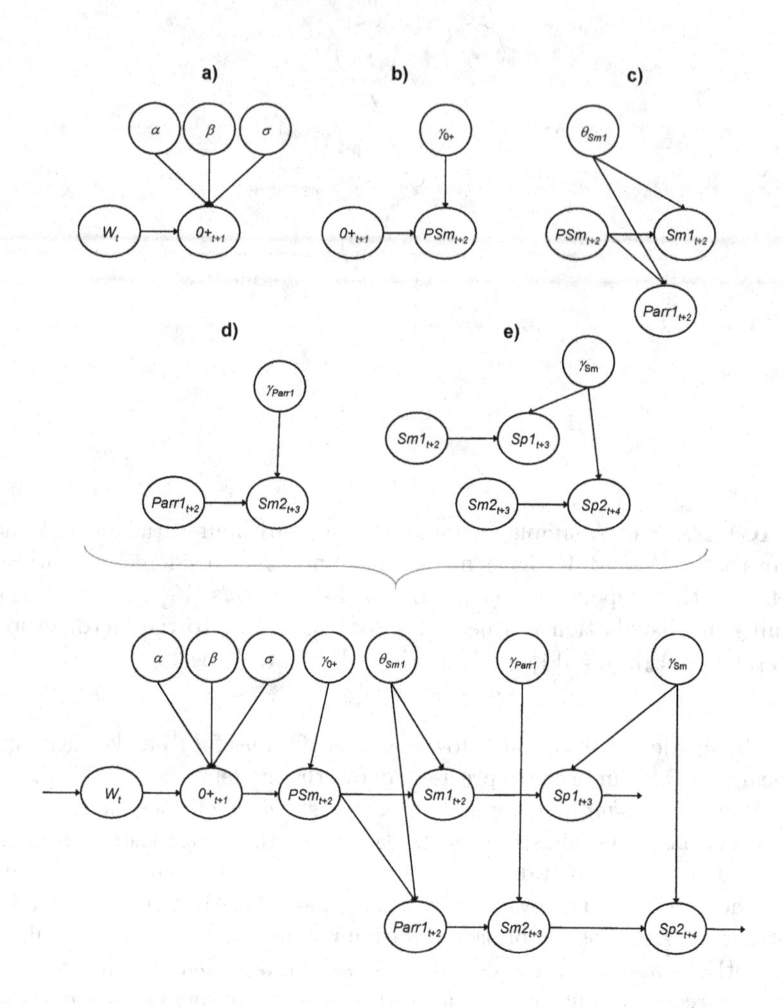


**FIGURE 1.7**: Step-by-step construction of a complex model by articulating blocks to describe the full A. salmon life cycle model. Upper panel: (a) Ricker stock-recruitment relationship with environmental noise; (b) Survival from 0+ juveniles to pre-smolts; (c,d) Smoltification transitions with demographic stochasticity; (e) Post-smolts marine survival rates. Bottom panel: DAG for the entire life cycle describing the dynamic of one cohort originated as eggs spawned at year $t$.

the sub-modules that mimic the density-dependent eggs → 0+ juveniles survival (Ricker relationship), the 0+ juveniles → pre-Smolts survival, the smoltification phase and the marine survival of post-smolts. This illustrates how models can be built step-by-step, by first building some bricks independently that will be connected together in a second step.

The basic properties of DAGs simply derive from standard probability theory. Consider for instance the set of all variables in Fig. 1.7. $(\alpha, \beta, \sigma, W_t, \gamma_{0+}, \theta_{Sm1})$ have no parents; the parents of $0+_{t+1}$ are $W_t$ and $(\alpha, \beta, \sigma)$; the parents of $Sm1_{t+2}$ are $PSm_{t+2}$ and $\theta_{Sm1}$ and so on for the ongoing of each cohort. The full joint distribution of all unknown variables for one cohort originated from eggs spawned at year $t$, denoted $J_t$, can be factorized relying on probability theory. Following the local dependencies in the DAG (Fig. 1.7), and assuming that all parameters are independent, this factorization of the joint probability distribution reads:

$$\begin{aligned}
J_t =& [\alpha] \times [\beta] \times [\sigma] \times [W_t] \times [\gamma_{0+}] \times [\theta_{Sm1}] \times [\gamma_{Parr1}] \times [\gamma_{Sm}] \\
&\times [0+_{t+1} | W_t, \alpha, \beta, \sigma] \times [PSm_{t+2} | \gamma_{0+}, 0+_{t+1}] \\
&\times [Sm1_{t+2}, Parr1_{t+2} | PSm_{t+2}, \theta_{Sm1}] \\
&\times [Sm2_{t+3} | Parr1_{t+2}, \gamma_{Parr1}] \\
&\times [Sp1_{t+3} | Sm1_{t+2}, \gamma_{Sm}] \times [Sp2_{t+4} | Sm2_{t+3}, \gamma_{Sm}]
\end{aligned} \tag{1.16}$$

### 1.2.4 Embedding observation processes into a probabilistic conditional model

Following the above section, the DAG is a graphical representation of probabilistic cause-to-effect relationships. Looking at the graph by following the direction of the arrows emphasizes the child-parent dependence (and eventually conditional independence) of the variables. This corresponds to the modeling way of thinking. The DAG depicts the tentative structure that has been proposed for the model. The mathematical (probabilistic) result is the factorization of the full joint distribution as in Eq. (1.16).

Such a probabilistic conditional model can be further developed by including probabilistic observation process. Indeed, observation processes are rarely perfect, especially in ecological sciences. Sampling and measurement errors may occur in many experimental protocols, and are also modeled via conditional probability distributions. For instance, a variable $V_i$ in a model can be an observable, and the conditional distribution $[V_i | pa(V_i)]$ stands for the *observation process* or a *sampling process*, that is the stochastic mechanism that leads to an observation. The DAG depicts the structure that is proposed for both the process and the observation model. As far as probability is concerned, the observation process is simply additional (observed) variables fully integrated into the conditional probability model.

### 1.2.5 Hierarchical modeling: Separating out process stochasticity from observation errors

Variables in a complex graphical model that include both a process for state variables and an observation process, can advantageously be classified into three categories to emphasize the so-called *Hierarchical Modeling* structure.

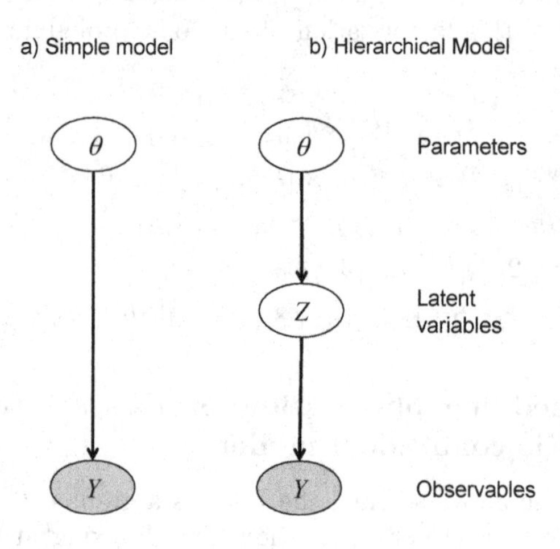


**FIGURE 1.8**: Directed acyclic graph (DAG) of a prototype Hierarchical Model.

Nodes without any parent, nonobservable nodes with parents, and observed nodes are denoted $\theta$, $Z$, and $Y$, respectively. The nonobservable nodes with parents, denoted $Z$, are also called *latent variables.*

The conventional notation $(\theta, Z, Y)$ splits the DAG variables into three layers as depicted in Fig. 1.8 where, conversely to DAG, we consider each layer as a block. Using the generic set of variables from the previous section $V_1, ..., V_n$, let us for convenience sake suppose that the $p$ first nodes $V_1, ..., V_p$ are not observables, and that the last $n-p$ nodes $V_{p+1}, ..., V_n$ are observables. We hence have distinguished $Y$, a block notation for all the *observables* in the graph (those previously termed

$V_{p+1}, ..., V_n$) and the other nonobserved quantities $V_1, ..., V_p$=$(\theta, Z)$ that are unknown. Among these unknown quantities, we further make the distinction between the parameters (suppose for convenient notation that the $k$ first variables $V_1, ..., V_k$ are nodes without parent) that are entry points to the graph and are denoted $\theta$ and the $p - k$ inner ones (the remaining $V_{k+1}, ..., V_p$) that are denoted $Z$. Parameters $\theta$ are often called *state of nature* while latent variables $Z$ are commonly named *state of the system.* Such a three-layers structure as depicted by Fig. 1.8 is the core of Bayesian Hierarchical Modeling (HBM).

Keeping in mind the groupings $\theta = (V_1, ..., V_k)$, $Z = (V_{k+1}, ..., V_p)$, $Y = (V_{p+1}, ..., V_n)$, and assuming that parameters have an unconditional prior probability distribution $[\theta]$, the full joint distribution of the HBM may be factorized as in Eq. (1.17):

$$[\theta, Z, Y] = [\theta] \times [Z|\theta]\ times[Y|\theta, Z] \tag{1.17}$$

Equation 1.18 offers a more explicit interpretation of the factorization of a hierarchical model, which is illustrated in Figs. 1.8 and 1.12:

$$\begin{aligned}[Parameters,\ Process,\ Observables] = \\ &[Parameters] \\ &\times [Process|Parameters] \\ &\times [Observables|Process,\ Parameters]\end{aligned} \tag{1.18}$$

It is worth noting that $[Z|\theta]$ and $[Y|Z, \theta]$ are compact representations of eventually highly complex probabilistic state and observation processes. These process and observation equations might be constructed from the combination of many more simple local interactions, which are themselves reorganized through parent-child conditional dependencies using Eq. (1.14).

### Example (continued): A. salmon population dynamic model

In the A. Salmon stage-structured population dynamic model, the number of fish in each of the development stage are unknown but are observed indirectly through capture experiments during migration in downstream (for smolts) and upstream (for adults) traps and through electrofishing experiments for freshwater resident 0+ juveniles. As described previously, those capture experiments can be modeled by Binomial sampling probability distributions, with probabilities of capture denoted $\pi_{Sm}$, $\pi_{Sp}$, and $\pi_{0+}$, and considered constant for all years $t$:

$$\begin{cases} C_{0+,t+1} \sim Binomial(0+_{t+1}, \pi_{0+}) \\ C_{Sm1,t+2} \sim Binomial(Sm1_{t+2}, \pi_{Sm}) \\ C_{Sm2,t+3} \sim Binomial(Sm1_{t+3}, \pi_{Sm}) \\ C_{Sp1,t+3} \sim Binomial(Sp1_{t+3}, \pi_{Sp}) \\ C_{Sp2,t+4} \sim Binomial(Sp2_{t+4}, \pi_{Sp}) \end{cases} \tag{1.19}$$

Equation (1.19) brings an additional block to be linked to the previous DAG. Figure 1.9 extends Fig. 1.7 where Binomial counting processes have been connected at stages 0+ juveniles, smolts and spawners. Following our graphical conventions, the corresponding ellipses for the Binomial catches in the graph are shaded if the catches are observed, and the catchabilities, considered known, appear in shaded boxes.

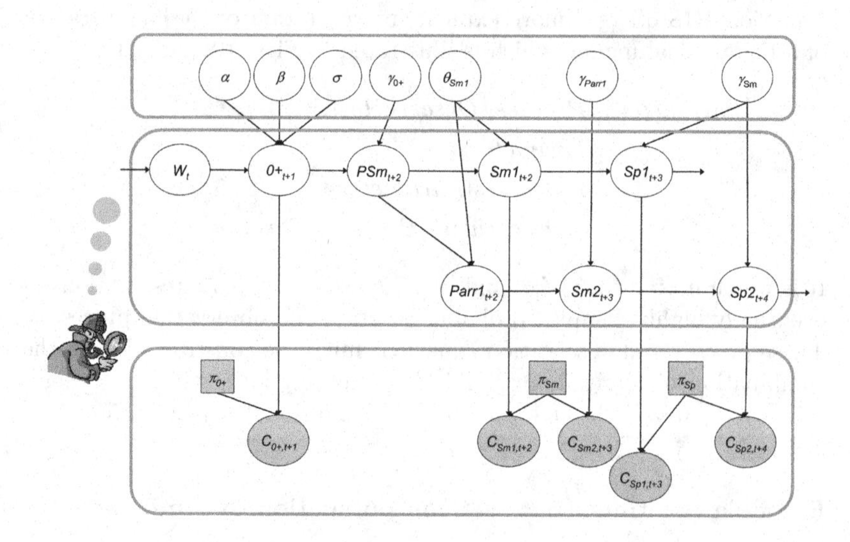


**FIGURE 1.9**: DAG describing the A. salmon life cycle model with Binomial observation process. Parameters ($\theta$ following our notations) appear in the upper layer. The middle layer describes the dynamics of hidden states ($[Z|\theta]$). The observation model ($[Y|Z,\theta]$) is presented in the bottom layer.

Keeping in mind our block notation $(\theta, Z, Y)$ for a generic hierarchical model, the vector of all the *parameters* $\theta$ would correspond to the vector of transition probabilities between each stage (survival and mortality ratios, Ricker parameters, etc.). The *latent states* $Z$ (the other nodes in between parameters and observables) would be mainly the number of individuals in the possible stages for all time indices, and the observations $y$ will be the catches:

- Nodes without parents $\theta = (\alpha, \beta, \sigma, W_t, \gamma_{O+}, \gamma_{Parr1}, \theta_{Sm1}, \gamma_{Sm})$
- Nonshaded nodes with parents $Z = (O+_{t+1}, PSm_{t+2}, Sm1_{t+2}, Parr1_{t+2}, Sm2_{t+3}, Sp1_{t+3}, Sp2_{t+4})$
- Shaded observed nodes $Y = (c_{O+_{t+1}}, c_{Sm1_{t+2}}, c_{Sm2_{t+3}}, c_{Sp1_{t+3}}, c_{Sp2_{t+4}})$.

*Note*: In this example, all the catchabilities $\pi$ are considered known. But the $\pi$'s could also be estimated. This is easily performed by expanding the observation process to include capture-mark-recapture experiments designed to provide information on both capture efficiencies $\pi$ and total numbers (for details, see [255], [257] and [259]).

---

## 1.3 Bayesian inferences on hierarchical models

### 1.3.1 The intuition: What if a node is observed?

The main intuition for *Bayesian statistical learning* (in other words *Bayesian inference*) can be advantageously presented via fluxes of information on a DAG. What happens when an observation becomes actually available? Let us, for convenience, suppose that the observable nodes Y (*e.g.*, the $n-p$ last nodes in the graph $V_{p+1}, ..., V_n$, have been *observed* and take the value $y$. Following our graphical conventions, the corresponding ellipses in the graph then become shaded. Once the phenomenon $Y$ has been observed (note we are using latin capital letters for the random phenomenon and lowercase ones for data), this certainly impacts the probability distributions of all the other nodes in the graph. When looking at the flow of information in the DAG, the information contained in $Y = y$ will *propagate* in the reverse direction of the arrows. An *effect* was observed: How will the distribution of the different *parents* of $y$ be modified? This is the *inferential* way of thinking. In other words, and following our block notations $(\theta, Z, Y)$, observable nodes in the block $Y$ are no longer random but they have been set to fixed observed values $Y = y$, and the range of possible values of the other nodes $(\theta, Z) = (V_1, ..., V_p)$ in the graph will subsequently be restrained to be *probabilistically compatible* with the observations $y$. The distribution $[\theta, Z, Y]$ does not matter any longer; we are now interested in the conditional distribution:

$$[theta, Z|Y = y] \tag{1.20}$$

namely the distribution describing what we do not know (the nonshaded nodes in the ellipses of the graph) given what we have seen (the shaded ellipses). In the Bayesian rationale, $[\theta, Z|Y = y]$ stands for the *a posteriori* distribution, *i.e.*, the joint posterior distribution of all unknowns updated by the information conveyed by the recorded occurrence $Y = y$.

For instance, Fig. 1.10 illustrates how the information contained in the observed variables $V_4 = v_4$ will propagate through the graph (following the inverse sense of the arrows) to update the distribution of variables $V_1, V_2, V_3$ into the so-called posterior $[V_1, V_2, V_3|V_4 = v_4]$.

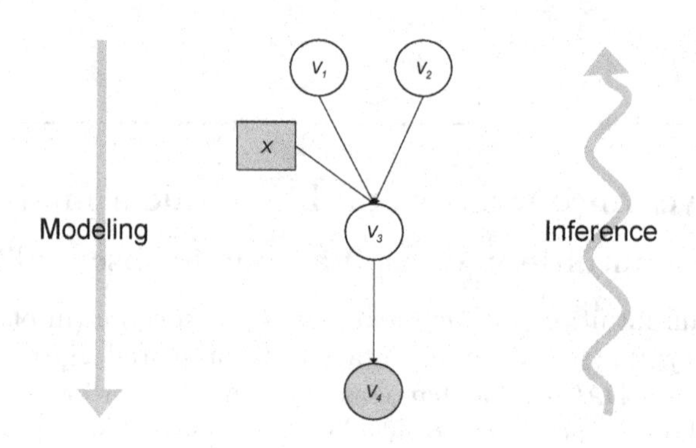


**FIGURE 1.10**: What happens when the variable $V_4$ is observed? The information will propagate through the graph following the inverse sense of the arrows to update the distribution of variables $V_1, V_2, V_3$ into the posterior $[V_1, V_2, V_3|V_4 = v_4]$.

### 1.3.2 Getting the Bayesian posterior distribution

The previous section provided an *intuitive* explanation of how the information brought by the observations are used to update the probability distribution of all unknowns in a probabilistic model. Clearly, such an updating only makes sense in the Bayesian paradigm. According to Reverend Thomas Bayes in his famous *post mortem* paper [20], communicated to the Royal Society by his friend Richard Price: *"The probability of any event is the ratio between the value at which an expectation depending on the happening of the event ought to be computed, and the chance of the thing expected upon it's happening."* In modern words (see [82] or [268]), the probability of the event $E$ is the price at which you would buy or sell a bet that rewards you one currency unit if event $E$ happens, and nothing if it fails. Conversely to the frequentist definition of probability, the limiting fraction of positive outcomes in a repetitive experiment of Jacques Bernoulli (1654-1705), which restricts probability to observable events only, the Bayesian definition *directly* quantifies the degree of uncertainty of a scientific judgment.

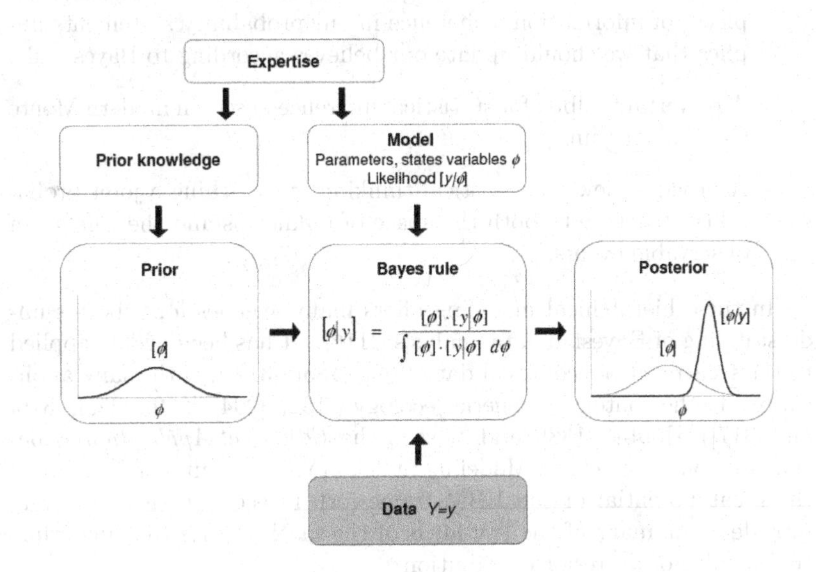


**FIGURE 1.11**: The Bayes rule: an information processor. The idea of Bayesian updating is represented here with data and unknowns denoted by $y$ and $\phi$, respectively.

Bayesian inference (see Fig. 1.11) has been extensively discussed in several classical books such as [26], [36] or [117] or [153]. The long story of struggles between Bayesian scientists and frequentist ones is brightly depicted in [201], putting forward the prominent role of Bayes' rule in turning points of scientific history. Funny enough, it seems that the many revivals of the often overlooked Bayesian reasoning stemmed from scientists in applied fields, mostly engineering. In ecology too, the Bayesian controversy, still a matter of serious discussion in the previous century (see for instance [92]), is now becoming out of fashion. Yet, frequentist arguments like the ones from [88] or [174] remain challenging because they point out where Bayesian answers have to be developed and popularized so as to conduct more convincing statistical analyses.

Although there are many frequentist methods to fit hierarchical models ([87]), Bayesian reasoning offers a convenient probabilistic rationale to derive inference on hierarchical models. Following [64], we adopt in this book the Bayesian approach for the following theoretical and practical advantages:

- An intuitive comprehension of the uncertain world in terms of predictive probability statements, which seems rather natural to many life scientists. We are reasoning conditionally and, when facing new pieces of information, coherence in our probability statements implies that we should update our believes according to Bayes' rule;

- A powerful toolbox for statistical inference based on modern Monte Carlo algorithms;

- A broader view on statistical thinking by attaching a joint probability structure to both the space of unknowns and the algebra of observable events.

In turn, hierarchical modeling offers many avenues for a better understanding of Bayesian data analysis ([116]). It has been widely applied to the treatment of ecological data ([64]; [183]; [312]), with many applications in the context of fisheries ecology ([132]; [204]; [239]; [255]; [316] and [317]). Hobbs ([138]) and most of the *Ecological Applications'* special issue on Hierarchical Modeling in 2009 (Vol. 19, Number 3) detailed the great potential of the HBM framework in ecology. Here, we only provide a summary of the key ideas of the mathematics that underline the calculus of posterior distributions.

The full joint distribution $[\theta, Z, Y]$ can be factorized as in Eq. (1.17)

$$[\theta, Z, Y] = [\theta] \times [Z|\theta] \times [Y|\theta, Z]$$

but also in the following form by making the reverse conditioning:

$$[\theta, Z, Y] = [Y] \times [\theta, Z|Y] \tag{1.21}$$

Once the observable $Y$ have been observed, some data $Y = y$ are available, and combining Eqs. (1.17) and (1.21) yields the Bayes formula:

$$[\theta, Z|Y = y] = \frac{[Y = y, Z, \theta]}{[Y = y]} \tag{1.22}$$

with

$$[Y = y] = \iint_{z,\theta} [Y = y|\theta, z] \times [z, \theta] dz d\theta \tag{1.23}$$

Using the hierarchical factorization of $[Y = y, Z, \theta]$ in Eq. (1.17) one obtains the practical way to get to the Bayes rule:

$$[\theta, Z|Y = y] = \frac{[\theta] \times [Z|\theta] \times [Y = y|\theta, Z]}{[Y = y]} \tag{1.24}$$

The denominator of Eq. (1.24) detailed in Eq. (1.23) is called the *predictive prior pdf.* It encodes the credibility degrees that can be rationally given to any possible value of the observable $y$ by integrating out the prior uncertainty about the unknown parameter $\theta$ and the (unknown) latent variable $Z$. When searching to specify the posterior distributions $[\theta, Z|Y = y], [Z|Y = y]$ and $[\theta|Y = y]$, the quantity $[Y = y]$ is to be considered as a constant because it is not a function of $\theta$ nor of $Z$. Consequently, the full joint posterior distribution (Eq. (1.24)) is often written as proportional to the factorization (Eq. (1.17)):

$$[\theta, Z|Y = y] \propto [\theta] \times [Z|\theta] \times [Y = y|\theta, Z] \tag{1.25}$$

The joint conditional distribution $[\theta, Z|Y = y]$ in Eq. (1.25) is the *posterior* joint posterior distribution of the unknowns. This is the cornerstone of Bayesian inference of hierarchical models. It quantifies the credibility levels of the state of nature $\theta$ and the state of the latent variable $Z$ given that the experimental result $y$ has occurred.

Let us now comment on some of the key features appearing in Eq. (1.24):

1. It is helpful to understand the Bayes formula as an information processor. The prior knowledge about the unknowns $(\theta, Z)$ is updated into a posterior knowledge $[\theta, Z|Y = y]$ thanks to the data $y$ that have been conveyed through the model by the complete likelihood $[Y = y|\theta, Z]$. As in Fig. 1.11, this formula is commonly presented as $[\phi|Y = y] = \frac{[Y=y|\phi] \times [\phi]}{\int_\phi [Y=y|\phi] d\phi}$, simply writing $\phi$ for the couple $(\theta, Z)$ of unknowns.

   From a rather philosophical point of view, updating the prior knowledge $[\phi]$ into a posterior distribution $[\phi|Y = y]$ combines two different natures of uncertainty. One may say that $\phi$ in the prior distribution $[\phi]$ is a random quantity by *ignorance* whereas the data $y$ is the realization of the variable $Y$ that is random by *essence* (according to $[Y = y|\phi]$).

2. $[\theta]$ is the prior distribution of the components in parameter vector $\theta$. Setting the *a priori* distribution $[\theta]$ is also modeling according to the Bayesian school of thought, with the delicate task of summarizing the knowledge available to the analyst about the unknown $\theta$ prior to having observed $y$. Note that in the context of a DAG, the joint prior distribution for the vector $\theta$ with $k$ components may itself be a network that also factorizes as $[\theta_1, ..., \theta_k] = \prod[\theta_i|pa(\theta_i)]$. Some nodes will be taken as parameters *stricto sensu* (*i.e.*, nodes without parents), and others will be defined conditionally upon the already defined ones, following a prior model structure.

3. We call $[\theta, Z] = [Z|\theta] \times [\theta]$ the joint prior knowledge for the unknowns. It combines expert knowledge (encoded in $[\theta]$, see below) and phenomenological knowledge ($[Z|\theta]$ which encodes how to go from state of nature $\theta$ to hidden system states $Z$).

4. $[Y|\theta, Z]$ defines the *complete likelihood.* Specifying likelihoods is the essential task of modeling. It amounts to describing with mathematical objects (probability distributions and deterministic functions) how we go from the *causes*, *i.e.*, the unknown quantity of interest $\theta$, to the nonobservable phenomenon of interest $Z$ (the *latent layer*), which in turn will give rise to the observable *consequences* $Y$. In these most important tasks, the analyst introduces theoretical hypotheses about the nature of the processes giving rise to the observed data and the latent variables. The *likelihood* in the classical sense of the term goes directly from the parameters to the tangible quantities, which is:

$$[Y = y|\theta] = \int_z [Y = y|\theta, Z] \times [Z = z|\theta]dz$$

   In the above expression, the hidden state variables $Z$ are integrated out. Indeed from a pure likelihood point of view, the possible values taken by the hidden part of the system do not matter. Latent variables might be considered as statistical nuisance quantities since they appear mostly as mental intermediate tools that help going from the causes $\theta$ to the observed consequences $y$. They are nevertheless meaningful in terms of model design and interpretation.

5. Suppose we were interested in new values $Y^{new}$ produced by the model given the already observed data $y$. This time, the degree of credibility $[Y^{new}|Y = y]$ that can be rationally given to any possible value of the observable $Y^{new}$ is obtained by integrating

out the posterior uncertainty against the unknown parameter $\theta$ :

$$[Y^{new}|Y=y] = \int_{\theta}\int_{z}[Y^{new},\theta,z|Y=y]d\theta dz$$
$$= \int_{\theta}\int_{z}[Y^{new}|\theta,z,Y=y]\times[\theta,z|Y=y]d\theta dz$$

Keeping in mind statistical sufficiency, *i.e.*, that once $\theta$ and $z$ are known, it is no longer necessary to keep observed values $y$ in the conditioning terms to generate new ones $Y^{new}$ :

$$[Y^{new}|y] = \int_{\theta}\int_{z}[Y^{new}|\theta,z]\times[\theta,z|Y=y]d\theta dz \tag{1.26}$$

This pdf is known as the *posterior predictive* distribution and underpins *predictive analysis* in the Bayesian setting. Equation (1.26) shows that, with a random sample from the posterior distribution $[\theta,Z|Y=y]$ at hand, generating posterior predictive values is a sequential straightforward procedure: (i) draw at random $\theta^{new},Z^{new}\sim[\theta,Z|Y=y]$, (ii) given $\theta^{new}$ and $Z^{new}$ draw at random $Y^{new}\sim[y|\theta^{new},Z^{new}]$.

### 1.3.3 Capturing the posterior distribution in practice

Equation 1.24 seems rather simple in appearance; unfortunately, it can be a hard task to obtain the posterior for the unknowns $\theta$ and $Z$ in many real case studies. Although the complete likelihood $[Y=y|\theta,Z]$ and the joint prior $[\theta,Z]=[Z|\theta]\times[\theta]$ are directly handled by the analyst, getting the posterior becomes burdensome when the unknown $\theta$ or $Z$ are defined in a high-dimensional space. In such a case, performing the multiple integrations to derive the predictive pdf $[Y=y]$ may be intractable, even numerically. The advent of the Markov chain Monte Carlo algorithms (MCMC) makes getting samples of posterior distributions technically possible for hierarchical models ([38]; [117]; [120]; [260]). The analyst can now have access to $[\theta,Z|Y=y]$ by means of a random sample of replicates $\{\theta^{(g)},Z^{(g)}\}_{g=1:G}$ of size $G$ drawn from $[\theta,Z|Y=y]$. Monte Carlo algorithms avoid the explicit computation of the Bayes formula denominator since they only need to know the distribution from which to sample, up to a constant. Some of them, like the Gibbs sampler, take advantage of the conditional independence structures encountered in the DAG of HBM to implement more efficient sampling in the parameters' space ([170]).

The free software WinBUGS ([71]; [185]; [203]; [284]) is a tool of choice for Bayesian inference when the posterior distributions contain high-dimensional system states and parameters, which is the case of HBMs for ecological data ([68]; [121]). King *et al.* ([164]) is of special interest for the practitioner because, for each case study, they provide their inference routines both in R (as a stand-alone program) and in WinBUGS. Although only Chapter 17 makes recourse to Bayesian thinking (indeed, the last chapter, WinBUGS being the ultimate tool in the hand of the analyst to give solutions to problems not solved by other conventional analyses), Zuur *et al.* ([321]) is an interesting book from the field of statistical ecology with many programs detailed in R. Kery ([160]) delivers a remarkable step by step WinBUGS initiation course, recovering parameters from simulated data that could have been met during ecological studies; as a logical sequel, real data are analyzed in depth with more elaborate model structures in [161]. Recent publications of interest for the ecologist wishing on hierarchical modeling also include [67], [181] or [265].

---

## 1.4 What can be found in this book?

What makes a statistical model especially adapted to the treatment of ecological data? Many of the classical ready-to-use statistical toolboxes have technical limitations and are not able to cope with the kind of models that the environmental scientists would like to have. Most of the classical statistical modeling toolboxes propose *data-driven* modeling approaches in which models are built under the data constraint that variables of the model are often chosen such that the data can be considered as direct realizations of the state variables. In other words, state variables and observations are collapsed into a single layer. Such an approach is frustrating because it stifles creativity. Indeed, the most interesting and stimulating models are the ones that are built from the interaction of state variables introduced because of their interest, rather than because of their direct observation. Classical statistical toolboxes have also introduced confusion in the origin of the randomness. Variability/stochasticity in the process and uncertainty arising from observations (*e.g.*, the sampling process or observation errors) are most often collapsed.

This book intends to illustrate how the HBM framework enables to go beyond these caveats. One also has to reply on Ogle ([221]) whose paper ends with the following warning: *"The primary issues that the field of ecology faces with respect to this exciting and powerful approach to ecological data analysis is adequate training and education. In the absence of sufficient training, these methods may be underutilized, misunderstood, or incorrectly applied."* We will emphasize how HBM supports multidimensional models involving complex interactions between parameters and latent variables as well as various observation equations to be connected with. In contrast with classical statistical toolboxes, both observation errors and process variability are explicitly acknowledged. The framework offers an original way to learn about the complex ecological process from various sources of data and information. In addition, when following the HBM avenue of thought, the scientist recovers the freedom to build models of his own.

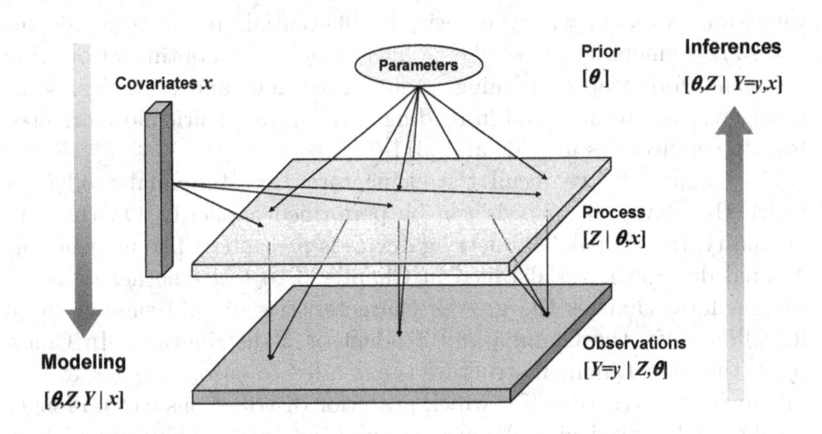


**FIGURE 1.12**: Hierarchical modeling strategy. Factorization of the complexity and Bayesian inferences.

The specific *ecological* features will be found in the observation layer of Fig. 1.12 (for instance, some peculiar data collection technique such as successive removals or capture-mark-recapture) or in its process layer (*e.g.*, the description of a complex life cycle), or in both. This book proposes a structured progression through the design of models for ecological data. We start with simple models that progressively become more complex (and more realistic) by introducing explanatory covariates and intermediate hidden states variables. When fitting the models to data, we also progressively present the various concepts and techniques of the

Bayesian paradigm, from a very practical point of view and on real case studies. Most data and programs are available from our website, including a companion booklet of exercises with research suggestions for each chapter.

## Part I: Basic blocks of Bayesian modeling and inference for ecological data

Part I refers mainly to models without a latent layer. Working with the elementary DAG depicted in Fig. 1.8-a might seem quite frustrating at first look. However, the intent is to make the reader familiar with the usual links between a set of parameters and another set of observables. Such training models have to be understood as elements to be stored for further use when assembling more elaborate models in Part II. Of course, we assume a reader with the basics in differential and integral calculus and some elementary knowledge in linear algebra and optimization. Had he or she forgotten everything about probability and statistics, some good revision can be found in reading [102] or, in a spirit more oriented toward engineers as in [151] and [311].

In Chapter 2, we recall the elementary Beta-Binomial model for which the Bayesian analysis can be performed explicitly. On that opportunity, the WinBUGS inferring device is presented. The unavoidable Normal distribution is detailed in Chapter 3 to test whether or not a salmon farm changes the growth characteristics of wild salmon (with its fellow pdfs, the Gamma and Student or T-distribution). In Chapter 4, the Beta-Binomial structure is extended to encompass larger useful stochastic structures for which posterior distributions can no longer be obtained analytically. We also complement the probabilistic toolbox of the Bayesian apprentice with the Poisson and Negative Binomial distributions. General methods of inference are presented as well as the Bayesian principles of model choice. Sticking to Binomial distributions for observables and Beta pdfs for parameters, a rather elaborate model with information conveyed by six different sources of observations is sketched in Chapter 5. In that chapter, latent variables are deterministic intermediate budget numbers for inferring Salmon population size. Chapters 6, 7 and 8 make use of explanatory covariates in linear and nonlinear models. The classical Normal linear model is given in Chapter 6 with its Bayesian treatment. Among the stock recruitment models introduced in Chapter 7, the Ricker model can be understood as a special case of the well-known linear model. Simple stock-recruitment analysis assumes that the stock (number of eggs) and the recruitment (number of juveniles) are known without errors and the SR parameters are estimated with their associated uncertainty. Further advances on model choice are also presented in that chapter. More sophisticated parameter-data links are developed in Chapter 8 where generalized linear models are exemplified on fishery data: Binomial logistic regression on smolt ages, and ordered probit model for skate presence. Additional material on model choice is also given in that chapter.

## Part II: Setting up more sophisticated hierarchical structures

In Part II, full HBM with latent layers of various nature are developed.

Chapter 9 focuses on random effect models (see Fig. 1.13). An exchangeable hierarchical structure as in Fig. 1.13 assumes that the $Z_k$ (that could be random effects for instance) are sampled from a common *regional* distribution $[Z_k|\theta]$, conditional on some unknown parameters $\theta$:

$$[Z_{1:n}, \theta | Y_{1:n} = y_{1:n}] \propto [\theta] \times \prod_{k=1}^{k=n} [Z_k|\theta] \times \prod_{k=1}^{k=n} [Y_k = y_k | Z_k, \theta] \quad (1.27)$$

with $y_k = (y_k^1, ..., y_k^{p_k})$ the vector of observations for the unit $k$. For instance $Z_k$ might characterize the density of a given species of marine invertebrates within a homogeneous fishing zone $k$ while $(y_k^1, ..., y_k^{p_k})$ will correspond to the number caught during the $p_k$ repetitions of the experiment in zone $k$.

The hierarchical structure sets the dependency between the units by expressing both similarity and heterogeneity among the $Z_k$. It allows for between-units variations among the $Z_k$. The prior distribution on the parameters $\theta$ will be updated by the observations in all units. This updating of the regional parameters allows transferring of information between units.

Examples developed in Chapter 9 ground the theory on the commonly encountered Binomial and Normal distributions. For instance, how to design a model such that parameters of the Stock-Recruitment relationships do depend on their own ecological system, but at the same time some transfer of information is allowed between neighboring rivers. Due to the historical significance of the shrinking effect ([286]), we feel free to divert from ecological issues to revisit in appendix E the baseball example detailed by Efron and Morris ([98]) as a simple introduction to hierarchical modeling.

Chapter 10 keeps on playing with *LEGO* bricks to build more and more complex hierarchical models by piling up several simple layers resulting in complex and versatile models purposely tailored to solve a scientific question. We show that hierarchical structures are fruitful to

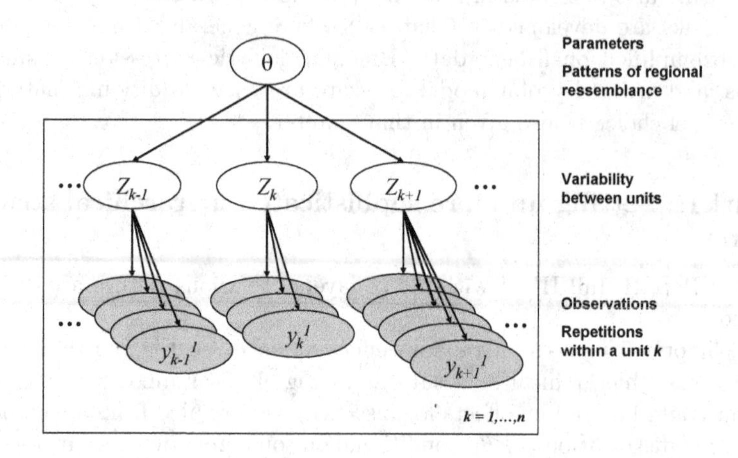


**FIGURE 1.13**: DAG with random effects in the latent layer of the HBM.

integrate multiple and various sources of data to learn from characteristics of ecological systems. The first example shows that HBM is flexible and effective for the treatment of successive removal catch data introduced in Chapter 4 to estimate A. salmon 0+ juvenile population size. We show how to develop a general model to assess the effects of temporal variations and habitat type, on two latent quantities of interest: the density of fish and the probability of capture. In the second example, we show how to consider top layers of the HBM built in the first example as a prior construction for juvenile abundance to be updated by measurements taken further in the Salmon life cycle (the smolt phase). Normal distributions are conveniently used for approximations in the log scale. The third example turns to a more intriguing sophistication where we extend the model with a level depicting a new rapid sampling technique as a possible cheap alternative to the costly successive removal procedure.

Chapter 11 is devoted to dynamic hierarchical models (see Fig. 1.14), with discrete and continuous latent states of various dimensions. Such models with a dynamic state transition in the latent layer of the hierarchical structure are commonly called state-space models.

At each time step $t$, let $Z_t$ denote the state vector, that is the vector of all the unknown quantities of the model which are time dependent. For instance the states may be the unobserved number of individuals in

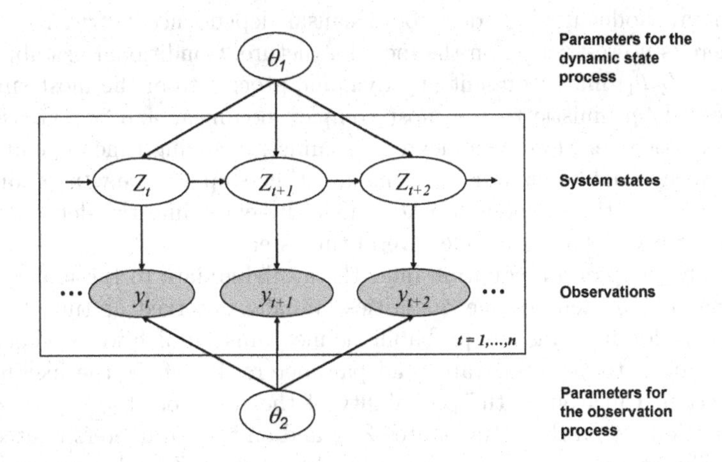


**FIGURE 1.14**: DAG for a generic state-space model with a dynamic state transition in the latent layer of the HBM.

each class of an age-structured model. $\theta = (\theta_1, \theta_2)$ denotes the vector of the parameters, that is all the time stationary unknowns involved in the dynamics ($\theta_1$) and in the observation process ($\theta_2$). Typically $\theta_1$ will be a vector of survival rates while $\theta_2$ will for instance describe catchability and measurement dispersion. Conditionally on the parameters $\theta_1$, the sequence of unknown states $\{Z\}_{t=1,\ldots,n}$ (denoted $Z_{1:n}$ in the following) is allowed to follow a multidimensional Markov chain. The transition kernel of the Markov process is defined by the dynamic process equations: The time dependence between states is introduced by successively conditioning the future states on the past and the parameters. Usually, the dependence of the current state $Z_t$ only reduces to that on the most recent state $Z_{t-1}$. For instance this is the case if $Z_{t+1}$ is the number of individuals that survive from an age class (previously $Z_t$) to the next one. Then, the state process is first-order Markovian. Thanks to conditional independence property, one can split the whole joint pdf into the product of single unit time steps. Once a prior distribution $[Z_1]$ is specified for the first state, the process equation can be written as:

$$[Z_{1:n}, \theta] = [\theta_1] \times [Z_1] \times \prod_{t=1}^{t=n-1} [Z_{t+1} | Z_t, \theta_1] \tag{1.28}$$

Such a formalization encompasses a wide class of dynamics in discrete time. Conditional dependence within the Markov chain is the most general model used to describe a causal dependence between events. There is no restriction on the model structure. Conditional distribution $[Z_{t+1}|Z_t, \theta_1]$ may represent any dynamic process from the most simple linear deterministic to the most complex nonlinear and non-Gaussian stochastic one. More complicated dynamics, involving time dependence between variables greater than one unit time step, can be written under the form of the previous Markov model by extending the definition of the state variable to include several time steps.

The observation equation links the available data to the underlying dynamics. $y_t$ denotes the quantities actually observed at time $t$. The observation model is any probabilistic mechanism which would generate data given the system states and parameters. It defines the likelihood function, which gives the probability of the series of observations $y_{1:n}$ conditionally on the actual states $Z_{1:n}$ and on the parameters related to the observation model in $\theta$. Conditionally on state $Z_t$ and parameters $\theta$, the observations $y_t$ are mutually independent, and the likelihood can be written as:

$$[Y_{1:n} = y_{1:n}|Z_{1:n}, \theta_2] = \prod_{t=1}^{t=n} [Y_t = y_t|Z_t, \theta_2] \tag{1.29}$$

This general framework enables the incorporation of data of a different nature and from different sources to perform inferences. In the simplest case, $y_t$ is a noisy measure of the state $Z_t$; however, observation $y_t$ may be related to the states of interest by a more complex stochastic process, (*e.g.*, a capture-mark-recapture model). When the state variables are not directly accessible to measurement, latent variables are introduced in $Z_t$ as intermediate steps for conditional modeling to relate observations to parameters. The dimension of the observation vector can be different between years (because of missing data or different observation processes) and can be different from the dimension of the state vector.

Following the general factorization rule of a HBM in Eq. (1.25), combining the joint prior equation (Eq. (1.28)) and the observation equation (Eq. (1.29)) yields the posterior distribution:

$$\begin{aligned}[Z_{1:n}, \theta|Y_{1:n} = y_{1:n}] \propto [\theta] \times [Z_1] \times \prod_{t=1}^{t=n-1} [Z_{t+1}|Z_t, \theta] \\ \times \prod_{t=1}^{t=n} [Y_t = y_t|Z_t, \theta]\end{aligned} \tag{1.30}$$

Surprisingly, estimation in the Bayesian setting of complex nonlinear non-Gaussian state-space models remains easily tractable. In Part II, the flexibility of the Bayesian analysis of state-space models will be illustrated using examples of the dynamic of the biomass of a fish stock, and of the salmon life cycle.

Chapter 12 deals with decision making and natural resources planning. In this closing chapter, statistical decision theory ([178]; [246]) is used as a natural extension of the Bayesian paradigm upon which we relied throughout the book.

# Chapter 2

## Introductory step into Bayesian modeling: The Beta-Binomial model


### Summary

From a basic fisheries science perspective, this chapter illustrates how:

1. A very simple Beta-Binomial model is built to mimic a mark-recapture experiment. The conditional dependencies network of this model are conveniently and efficiently summarized by means of a directed acyclic graphical representation. For this simplest DAG, only two nodes are necessary (one parameter and one observable, probabilistically linked by a very simple conditional relationship) with no latent layer involved.

2. The conceptual toolbox of probability distributions allows to build models accounting for both substantive knowledge and uncertainty, two sides of the same coin.

3. The interactions between the observed data, covariates and unknown quantities governing probability distributions can be rigorously, simply and comprehensively depicted by way of conditional reasoning.

4. Bayesian analysis offers a coherent deductive framework for making inferences about the unknown quantities (*e.g.*, parameters) of a model from observed data. It opens the door to the introduction of quantified scientific expertise about unknowns of interest beyond those already introduced through the conditional structure of the model. This expertise is combined with information brought by the data to derive relative degrees of credibility summarized by (*a posteriori*) probability distributions.

## 2.1 From a scientific question to a Bayesian analysis

The Scorff is one of the twenty rivers of Brittany (France) colonized by Atlantic salmon. Juveniles migrate out of the Scorff in April and undertake a long-range migration in the ocean. After having spent from 16 months to more than 2 years at sea, the adults come back to their native river to spawn. During their stay in the river prior to reproduction, the adults are exploited by a rod and line fishery. To avoid overexploitation and collapse of the population, the fishery is regulated by management measures (catch quotas, fishing periods, reserves, etc.). To assess the effect of these measures, the exploitation rate, *i.e.*, the proportion of the population removed by the anglers, needs to be assessed.

**FIGURE 2.1**: The Moulin des Princes trapping device on the Scorff river (Brittany, France).

But how could one measure the *exploitation rate*? It is a conceptual quantity stemming from the human brain to formalize a problem of resource management. However, it may be assessed indirectly by designing an experiment whose outcome would be, in the fishery scientist's mind, dependent on the actual exploitation rate. Based on the observed results of his experiment, the fishery scientist would seek to make some inductive statement about the exploitation rate.

A classical mark-recapture experiment can be set up to assess the exploitation rate. At the Moulin des Princes (Fig. 2.1), the scientific trapping facility (belonging to the French National Research Institute of

Agriculture) located at the river mouth, adults are sampled and tagged before being released into the river and submitted to the rod and line exploitation. Thanks to the cooperation of fishermen who report their catch, the number of tagged fish recaptured by angling is recorded. In 1999, $n = 167$ salmon were tagged at the trapping facility and $y = 16$ were recaptured by angling. What might have been the exploitation rate $\pi$? Assuming the behavior of the tagged and untagged fish issimilar with regards to the rod and line fishery, $\hat{\pi} = \frac{y}{n} = 0.096$ is certainly an intuitively appealing guess for $\pi$. But is it the correct answer? How confident are we about this value?

The reason we feel uncertain about the *guess-timate* $\frac{y}{n}$ we just calculated is that we know angling is a rather random activity. Had we perfect knowledge of the exploitation rate and of the number of tagged fish prior to the opening of the fishing season, we could still not determine exactly the number of fish that would be caught. The Bayes theorem provides the link between the uncertainty about $\pi$ knowing $y$ (and $n$), and that of $y$ knowing $\pi$ (and $n$) :

$$[\pi|y] = \frac{[y|\pi] \times [\pi]}{\int\limits_{\pi} [y|\pi] \times [\pi] d\pi} \tag{2.1}$$

The statistical language distinguishes the following elements already given in the introductory chapter:

- $Y$ is the observable (phenomenon), $y$ denotes the observation (data).
- $\pi$, the parameter of the model which is here the only *unknown*.

It is important to understand that $n$, a quantity which helps for the phenomenological explanation, is assumed to be known without any uncertainty in the study. As such, $n$ is called a *covariate*: Conversely to $y$ or $\pi$, it does not have any probability distribution attached to it. Formally speaking, it should not appear as a conditioning term in the DAG nor in the bracket notation which only deals with possibly varying quantities of the model (in a way the number $n$ is sort of embedded in the model structure). To improve readibility when necessary, we suggest to introduce covariates in the bracket notation in the conditioning term after a semicolon to indicate its different (nonrandom) status. The Bayes rule would consequently be written as

$$[\pi|y; n] = \frac{[y|\pi; n] \times [\pi]}{\int\limits_{\pi} [y|\pi; n] \times [\pi] d\pi} \tag{2.2}$$

This is a specific reformulation of the general Bayes'rule (Eq. (1.22)) with no latent variable. Provided that we can specify the two components $[y|\pi;n]$ (the *likelihood*), $[\pi]$ (the *prior*) and perform the integration $\int_{\pi}[y|\pi;n] \times [\pi]d\pi = [y|n]$, Bayes theorem tells us that we shall be able to infer our knowledge/uncertainty about $\pi$ conditionally on the known observation $y$ (and the covariate $n$).

In the sequel, we will successively present in detail how we can achieve the above three tasks. Then we will show how the three components $[y|\pi;n]$, $[\pi]$ and $[y|n]$ are combined according to Bayes theorem to get the *a posteriori* distribution of ultimate interest $[\pi|y;n]$ (or *posterior*). Finally, we propose to guide the reader through a WinBUGS practical session about Bayesian modeling and inference based on the previous case study.

---

## 2.2 What is modeling?

### 2.2.1 Model making

Remember that $n = 167$ adult salmon were marked and released into the Scorff River in 1999. The anglers catch these tagged fish with an efficiency $\pi$. Suppose we know $\pi = 0.15$. In this section, we temporally forget about the 1999 recorded catch, a number denoted hereafter by $y_{1999}$, and ask the reader to think of how many fish might be caught.

Again, no one can give a definitive answer to that question. A good guess (before actually recording $y_{1999}$) is some *uncertain* quantity $Y$ with probable values about $n \times 0.15$. A *model* is a conceptual construct to describe the phenomenon and quantify the uncertainty about the possible outcomes of the mark-recapture experiment assuming we know the number of tags available to the angling fishery and its exploitation rate. Note that we use a capital letter for the random *phenomenological* quantity $Y$ to make a clear distinction with the actual *number* observed $y_{1999}$. The model relies on two fundamental hypotheses:

- We first assume that the fate of each salmon facing the angling activity, *i.e.*, being caught or escaping, is ruled by the same *Bernoulli* mechanism. One after the other, the destiny of every individual $i$ of the population is picked by drawing at random a binary variable $X_i$ such that $X_i = 1$ with probability $\pi = 0.15$ (get caught) and $X_i = 0$ with probability $1 - \pi = 0.85$ (escape). The traditional image of a Bernoulli trial is drawing a ball from an urn with black balls in proportion $\pi$ and white balls in proportion $1-\pi$. If a black ball is drawn, the corresponding fish is caught, if it is a white ball, the fish escapes.

- Second, we consider that the $n$ tagged salmon behave independently with regard to the rod and line fishing. Under this assumption, the fate of fish $i$ ($X_i = 1$ or $X_i = 0$) brings no information on that of any other fish $j$ (*i.e.*, the possible value of $X_j$).

### 2.2.2 Discussing hypotheses

These two hypotheses have important features: they explicitly link the unknown parameter of interest $\pi$ and the covariate $n$ with the observation $y$ and they account for the randomness of the fishing success. They correspond to the most common assumption of identical and independent distribution (*iid* in the statistical jargon) for the fish behavior. It is so frequently made that it is barely mentioned or interpreted with regard to its ecological meaning. Although it makes inference about the exploitation rate considerably easier, as will be shown later, this assumption is a drastic simplification. Speaking against the identical distribution assumption for instance, the adult salmons are caught at the trapping facility over a period extending over several months and it could be argued that salmon tagged and released at the same time could exhibit similarities in their ability to escape angling. They may stay grouped together and spend a more or less prolonged period in the zone closed to any fishing located just upstream from the trap. Such behavior would violate the independence hypothesis, as the fate of a salmon tagged one day would be informative about the fate of the other fish tagged the same day. We can also wonder about a constant catchability $\pi$ of a fish during the experiment. Maybe when many fish have been captured, there remains much more space to escape from the anglers and it gets easier for the fish to escape, or the less resilient fish are captured in the beginning and it becomes harder to catch the rest of the population.

### 2.2.3 Likelihood as a probabilistic bet on the outputs given that the parameter is known

To sum up, a model is a conceptual representation of a phenomenon, relying on explicit simplifying hypotheses while still capturing essential features, with the aim of providing a quantified version of the phenomenon including its uncertainties. Here, the *iid*-Bernoulli model provides a probability distribution of all the possible outcomes $Y = \sum_{i=1}^{n} X_i$

from a recapture experiment of $n = 167$ released salmon with a fishing efficiency of $\pi = 0.15$:

$$Y|n, \pi \sim Binomial(n, \pi) \tag{2.3}$$

And we can write the well-known Binomial formula for the corresponding probability distribution

$$[Y = y|\pi; n] = \frac{n!}{(n-y)!y!}\pi^y(1-\pi)^{n-y} \tag{2.4}$$

and its moments

$$\begin{cases} \mathbb{E}(Y) = n\pi \\ \mathbb{V}(Y) = n\pi(1-\pi) \end{cases} \tag{2.5}$$

Using an *R*-like notation, we will sometimes write in the following of the book the Binomial formula 2.4 as $dbinom(y, n, \pi)$. It is worth stressing the nature of the various components of the probability distribution $[Y = y|\pi; n]$:

- $Y$ is the random variable of interest: it is unknown since it has not been observed yet. Roughly speaking, $Y$ describes all the possible outcomes of the experiment, taking values from 0 to $n$ with their *credible weights* (probabilities). In this book, such observables will be written using capital Latin letters. $y$ represents any numerical value, *i.e.*, the possible observation of the phenomenon $Y$. Lowercase Latin letters will be generally used for observations.

- $\pi$ and $n$ are on the right-hand side of the conditioning bar | of the probability distribution. In the modeling step, they are assumed to be known.

With such a model, the analyst is able to bet on the possible experimental results. Figure 2.2 gives the probability distributions of $[Y = y|\pi = 0.15; n = 14]$ and $[Y = y|\pi = 0.15; n = 167]$ as functions of $y$ (mind the different scales of the coordinates). Random draws from this distribution can generate repetitions of the observations from the phenomenon under study. The scientist can even make winning bets. Given $\pi = 0.15$ and $n = 167$, the random event $A = \{20 \leq Y \leq 30\}$ is more likely than its complementary one $\bar{A} = \{20 > Y \text{ or } Y > 30\}$. Since $\frac{[A|\pi=0.15 \quad ;n=167]}{[\bar{A}|\pi=0.15 \quad ;n=167]} = \frac{0.7677}{0.2323} \approx 3.30$, odds such as 1 against 3 could even be taken with confidence.

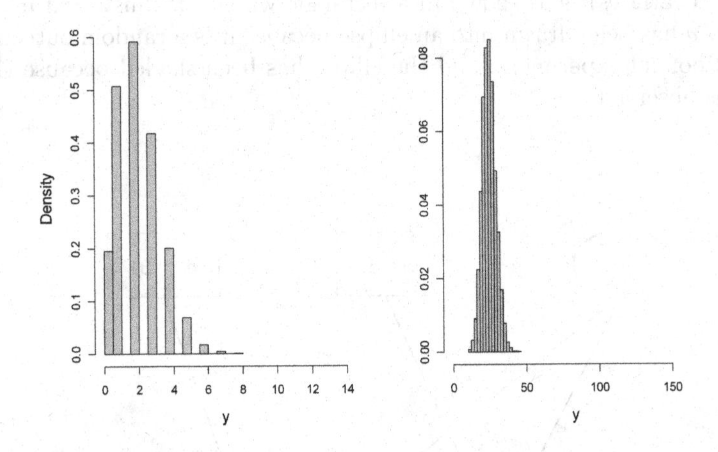


**FIGURE 2.2**: Histograms of 10000 random draws from a Binomial distribution $Y|n, \pi \sim Binomial(n, \pi)$ with $(\pi = 0.15; n = 14)$ (left) and $(\pi = 0.15; n = 167)$ (right).

## 2.3 Think conditionally and make a graphical representation

When thinking conditionally from $n$ and $\pi$ to the possible values of $Y$ (in capital letters because the random variable refers to all virtual outcomes of the Binomial draw, *i.e.*, the observable), we make a model (see Eq.( 2.4)). More generally, one can equivalently understand a model as a list of instructions that is able to output fake data with the same (hypothesized) statistical properties as the observed sample.

When thinking conditionally from $y$ (in lowercase letters because we are talking about the actually observed value of $Y$) and the complementary explanation $n$, to the unknown $\pi$, we develop an inferential point of view. The Bayesian paradigm relies on the theory of conditional probabilities to make full use of its internal mathematical coherence and to exploit the symmetric nature of modeling and inference.

Figure 2.3 draws the corresponding Directed Acyclic Graph that offers a graphical representation of the relations between variables using *nodes* and *arrows*. The direction of the arrows follows the reasoning lines of the modeling phase. Both $n$ and $\pi$ are necessary to get $y$ (according to the Binomial model). As $n$ is not uncertain (see the above discussion about covariates), it is shown in a rectangle while $\pi$ is illustrated in an ellipse. $y$ has been drawn into an ellipse because it is a random outcome of a Binomial experiment, and the ellipse has been shaded because the issue is observed.

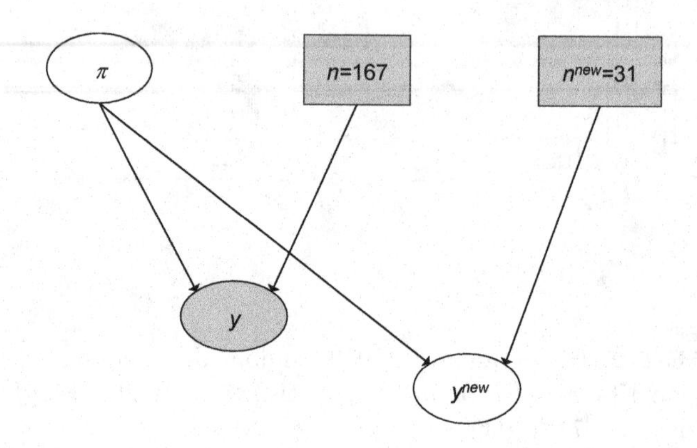


**FIGURE 2.3**: Directed Acyclic Graph (DAG) for a basic Beta-Binomial model.

Figure 2.3 also includes an additional experiment to the previous situation: $n^{new} = 31$ salmon are marked and released into the same river. We wonder how many salmon $y^{new}$ from the 31 new ones will be recaptured with the same new fishing device.

---

## 2.4 Inference is the reverse way of thinking

The next problem does not differ much, at first look, from the previous one: 167 salmon are marked and released into an isolated portion of a river. A recapture experiment is conducted using a (new) fishing device. This time we tell you that $y = y_{1999} = 16$ fish are caught, and the question now is: what can the efficiency of the fishing device be?

A rough empirical estimation of the fishing efficiency is $\frac{16}{167}$. But once again, recourse to probabilistic concepts is unavoidable to quantify the reliability of this empirical estimation $\frac{16}{167}$. The analyst will again describe the experiment by the same Binomial model leading to Eq. (2.4). Conversely to the previous situation, the observation is now recorded: $y$ is known but the fishing efficiency $\pi$ is not. The model maker simply imagines that $\pi$ has been set to a fixed value. Equation (2.4) is now to be interpreted as: If the efficiency of the new fishing device were set to a given value $\pi$, what is the probability that the experiment leads to $y = 16$ recaptured salmons? When thinking of $[Y = y|\pi; n]$ as a function of $\pi$, we refer to the *likelihood* of the model. Figure 2.4 gives the likelihood profile $[Y = 16|\pi; n = 167]$ as a function of $\pi$.

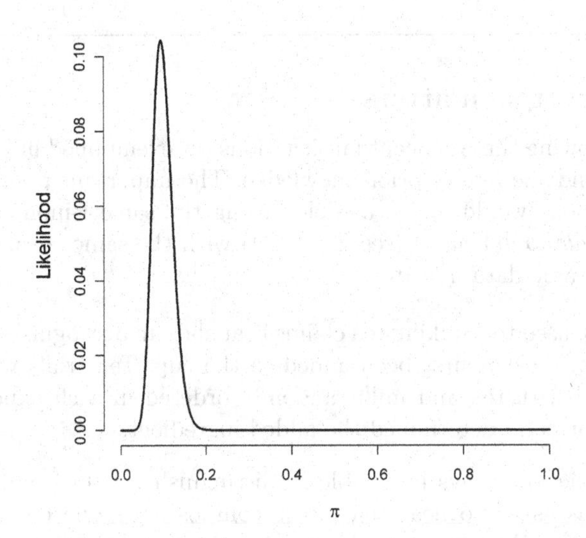


**FIGURE 2.4**: Likelihood profile for $y = 16$ and $n = 167$ as a function of $\pi$ between 0 and 1.

Making inference is describing our state of knowledge about the unknown $\pi$ once we have observed the data $y$. Bayesian inference interprets this consideration under a full probabilistic framework and focuses on the mathematical object $[\pi|y; n]$ given by Eq. (2.2).

$[\pi|y, n]$ is called the *posterior*, or the *a posteriori* probability distribution of $\pi$.

Betting about possible values of $\pi$ gives sensible interpretation to the random nature of $\pi$ in Eq. (2.2). The process began when we accepted the idea of probabilistic bets to describe the state of knowledge about the unknown $\pi$ once we have observed the data $y$. As we will work with this quantity as being observed, Fig. 2.3 shows this situation by means of a shaded ellipse. On the other hand, we wonder about possible values of $y^{new}$ which therefore has the status of a latent variable in an ellipse (unknown or missing data).

To go one step further, one has to describe the state of knowledge about the unknown $\pi$ before we observe the data $y$, *i.e.*, how would one bet about the possible values of $\pi$ under the *no data a priori* state of information? We first show in the following section that the prior state of knowledge may make an important difference even with the same data.

---

## 2.5 Expertise matters

The following three conceptual situations (as exemplified in [26]) help to understand the role of prior knowledge. The important point is that these situations would all be modeled using the same Binomial model $[Y\,|\pi, n] = dbinom(Y, n, \pi)$ (see Eq. (2.4)) with the same covariate $n = 10$ and the same data $y = 10$.

1. A lady used to drinking tea claims that she can distinguish whether milk or tea has first been added to the cup. Ten trials were conducted with tea and milk randomly ordered in each experiment. She correctly determined the added ingredient.

2. A music lover says he is able to distinguish Haydn from Mozart. He was asked to hear ten pieces composed by one or the other composer. He gave the right answer each time.

3. After a night spent drinking with colleagues, a senior statistician claimed he was a perfect forecaster. His colleagues flipped a coin ten times and each time he guessed the right outcome.

In all situations, the unknown $\pi$ refers to the skill of giving a correct answer and the same *iid* conditions can realistically hold. Assuming $\pi$ greater than $\frac{1}{2}$ means that human skill works at least as well as pure randomness. After consideration of the results in situation 2, most people would feel that $\pi$ close to 1 is more likely than $\pi$ in the neighborhood of 0.5 which is not a surprise since the man claimed he was an expert. In situation 3 conversely, the data do not help to make a convincing statement and we do not feel at ease with the discrepancy between our prior belief and the experimental results. We still believe that the coin is a fair coin and that $\pi$ is 0.5. The lack of personal experience often leaves us dubious to give *a priori* a strong statement under situation 1. Most people (except perhaps the British) would confess that they do not know and bet rather evenly on possible values of $\pi$ between 0.5 and 1, but the experimental results have somehow changed their mind in favor of English lady's skill in tea tasting.

How is it that the same Binomial models of occurrence with the same data and the same hypotheses do not lead to the same conclusions? For Bayesian statisticians, the answer is that the *a priori* state of knowledge does matter; *priors* are definitely not the same in every situation, so that, even through learning from the same data, the conclusions may remain quite different *a posteriori*. In other words, the *prior* belongs to the body of hypotheses. Changing the *prior* is changing the model structure, which may in turn impact the outcomes.

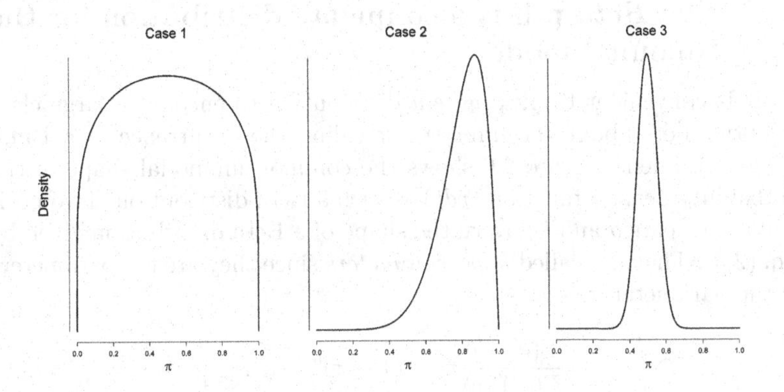


**FIGURE 2.5**: Priors matter: Three possible unscaled priors explained. Case 1: Lady drinking tea; case 2: Music lover; case 3: Flipping a coin.

## 2.6 Encoding prior knowledge

Unknown parameters are often "not as unknown" as they seem to be at first glance; even without data, all subsets of possible values do not bear the same prior belief. Unscaled representations of commonsense prior distributions of knowledge for each of the three previous situations are depicted in Fig. 2.5. In situation 1, if we doubt the skills of the English lady drinking tea with milk or milk with tea, a rather uniform prior belief is assumed for all possible values. Conversely, in situation 3, the prior belief is highly concentrated around $\pi = 0.5$. Some people would even put all the weight at 0.5 and nothing elsewhere. Some prior weight is still kept on the left side of the interval in situation 2, but most of our credibility is lent to values larger than 0.5.

In commonly encountered situations, prior information is available through local expertise about the ecosystem, a meta-analysis based on the scientific literature, past years of observations, and auxiliary information collected in the field by other practitioners.

---

## 2.7 The *Beta* pdf is a conjugate distribution for the Binomial model

It is convenient to pick a *Beta* distribution to encode (a large class of) prior bets about the unknown $\pi$ ruling the occurrence of a Binomial experiment. Figure 2.6 shows the common unimodal shape of the probability density function (*pdf*) of such a Beta distribution. Two positive coefficients $a$ and $b$ govern the shape of a Beta distribution given by Eq. (2.6). They are called *hyperparameters* since they are the parameters of the parameter $\pi$.

$$[\pi] = \frac{\Gamma(a+b)}{\Gamma(a)\Gamma(b)}\pi^{a-1}(1-\pi)^{b-1} \times 1_{[0,1]}(\pi) \qquad (2.6)$$

The function $1_{[0,1]}$ means that $1_{[0,1]}(\pi) = 1$ if $\pi$ is within the interval $[0,1]$ and 0 otherwise. Using an *R*-like notation, we will sometimes write in the following of the book the Beta formula (2.6) as $dbeta(\pi, a, b)$. Past information such as a histogram of records of success ratio with a similar fishing device can be used to empirically "fit"a Beta *pdf.* Most of the time, the analyst is happy with an estimation of the first moments or the median and a quartile. For instance, it can be shown that the first two moments of the Beta distribution are simply related to the hyperparameters:

$$\begin{cases} \mathbb{E}(\pi) = \dfrac{a}{a+b} \\ \mathbb{V}(\pi) = \dfrac{ab}{(a+b+1)(a+b)^2} = \dfrac{\mathbb{E}(\pi)(1-\mathbb{E}(\pi))}{a+b+1} \end{cases}$$

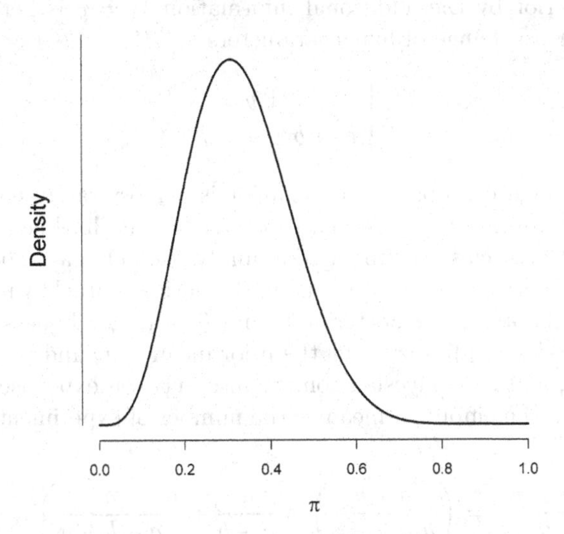


**FIGURE 2.6**: Example of a Beta($a$,$b$) *pdf* with parameters $a$=5 and $b$=10.

For given values of the moments $\mathbb{E}(\pi)$ and $\mathbb{V}(\pi)$, the hyperparameters $a$ and $b$ are simply estimated by inverting Eq. (2.7). The uniform distribution correspond to the special case $a = 1 = b$. Values of $a$ and $b$ below 1 provide $U$ shaped pdfs.

When no data are available, the expert can be directly asked questions such as: "If a hundred fish were released, what would be the most likely number of recaptured fish with the present fishing device? On what number of fish would you bet with odds 3 against 1?" Note that the mode and the quartile are also unubiquitously linked to $a$ and $b$, which can therefore be extracted with some algebra from the expert's answers.

In the special case of a Binomial likelihood with Beta *prior*, the quantity $[y]$, *i.e.*, the denominator of the Bayes formula 2.2 can be analytically found. Specifying a Beta *prior* (Eq. (2.6)) for $[\pi]$ with a Binomial likelihood, Eq. (2.2) gives the following explicit posterior:

$$[\pi|y] = \frac{\Gamma(a+b+n)}{\Gamma(a+y)\Gamma(b+n-y)}\pi^{a+y-1}(1-\pi)^{n+b_\pi-y-1} \times \mathbb{1}_{[0,1]} \qquad (2.7)$$

We notice that the *a posteriori* distribution of $\pi$ remains within the same *Beta* family as the *prior*. That is why the *Beta* family of pdf is said to be a natural *conjugate* of the Binomial likelihood. Updating the *prior* into a posterior by the additional information $Y = y$ is here encoded through a mere change of hyperparameters:

$$\begin{cases} a \to a + y \\ b \to b + n - y \end{cases} \tag{2.8}$$

This conjugate probabilistic behavior is a most rarely encountered situation, stemming from specific properties of the likelihood function (belonging to the class of exponential family, see [30], page 265 or [153], Chapter 8). Further details on this mathematical miracle will be given in the next chapters. The posterior mean $\mathbb{E}(\pi|y)$(a good guess once data have been collected) lies between the prior mean $\mathbb{E}(\pi)$ and the empirical estimate $\hat{\pi}$ , and this Bayesian compromise between expertise and data gets closer to the empirical mean as the number of experiments increase since

$$\frac{a+y}{a+b+n} = \left(\frac{a+b}{a+b+n}\right) \times \frac{a}{a+b} + \left(\frac{n}{a+b+n}\right) \times \frac{y}{n}$$

A dimension analysis of Eq. (2.8) suggests a useful interpretation of the Beta prior in terms of equivalent data. Hyperparameter $a$ corresponds to the number of prior successes in an experiment with $a + b$ virtual trials.

---

## 2.8 Bayesian inference as statistical learning

In Bayesian statistics, probabilistic bets can be made about unknowns. From a technical point of view, the explicit introduction of probability distributions for unobservable quantities (the unknowns) is the main difference between conventional statistics and the Bayesian approach that is developed in this book. In addition, this means that there are only conditional probabilities because, depending on the available knowledge, such bets may change. We denote $[\pi\,|K]$ such a probability distribution putting weight on the probable values of the unknown $\pi$ when the state of knowledge is depicted here by the notation $K$. When no data are available to convey information about the unknown, the prior state of knowledge $K$ will simply write the *a priori* distribution $[\pi\,|K = prior]$ or more simply $[\pi]$ (or *prior*). A *prior* is a probability distribution function representing the knowledge about the parameters at hand before the data are collected. When data $y$ have been observed, the state of knowledge has changed ($K = \{y\}$) and we deal with the *a posteriori* bet $[\pi \,|y]$ or *posterior*. To sum up, the Bayes rule given at Eq. (2.2) (or more generally at Eq. (1.22)) is to be interpreted as an information processor, relying on a model (here a Binomial one), that updates the knowledge about the unknown $\pi$ from the *prior* $[\pi]$ into the posterior $[\pi \,|y]$ by sort of *crunching* the data $y$. As such, Fig. 2.7 may suggest a more convincing illustration of the statistical learning mechanism than Figure 1.11 in Chapter 1.

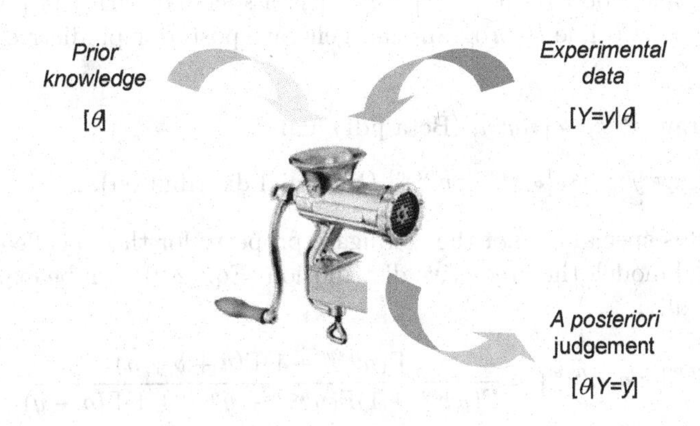


**FIGURE 2.7**: The Bayesian crank: Bayes rule as a statistical learning mechanism.

## 2.9 Bayesian inference as a statistical tool for prediction

Suppose we are told that an additional number of $n^{new} = 31$ have been marked and released in the Scorff River, with the same experimental conditions. Given that we have already seen $y = 16$ recaptured fish for $n = 167$ tagged fish, what can we predictively say about $y^{new}$? We keep on making probabilistic bets about the unknown $y^{new}$ (see

Fig. 2.3). From a technical point of view, we place a condition on the state of knowledge $\pi|y$ and simply use probability theory to evaluate the *posterior predictive* as in Eq. (1.26) from Chapter 1:

$$\begin{aligned}[y^{new}|y;n,n^{new}] &= \int_0^1 [y^{new},\pi|y;n,n^{new}]d\pi \\ &= \int_0^1 [y^{new}|\pi,y;n,n^{new}][\pi|y;n,n^{new}]d\pi \\ &= \int_0^1 [y^{new}|\pi;n^{new}][\pi|y;n]d\pi \end{aligned} \tag{2.9}$$

The likelihood term $[y^{new}|\pi;n^{new}]$ is associated with the posterior $[\pi|y;n]$. A two line R program can generate posterior predictive values $y^{new}$:

1. draw $\pi^{new} \sim [\bullet|y;n]$ (Beta pdf) and
2. draw $y^{new} \sim [\bullet|\pi^{new};n^{new}]$ (Binomial distribution).

In this special case of the conjugate property for the so-called Beta-Binomial model, the predictive distribution (Eq. (2.9)) can be expressed analytically:

$$\begin{aligned}[y^{new}|y;n,n^{new}] = &\frac{\Gamma(n^{new}+1)\Gamma(a+b+n)}{\Gamma(y^{new}+1)\Gamma(n^{new}-y^{new}+1)\Gamma(a+y)} \\ &\times \frac{\Gamma(a+y+y^{new})\Gamma(b+n-y+n^{new}-y^{new})}{\Gamma(b+n-y)\Gamma(a+b+n+n^{new})}\end{aligned}$$

This is known as the *Polya* distribution in the statistical literature ([102]): starting from an urn with $a+y$ black balls and $b+n-y$ white ones, one repeats $n^{new}$ the following scheme: when a ball is drawn, it is replaced and an additional one of the same color is put into the urn. The random variable $y^{new}$ is the number of black balls obtained at the end of the experiment. Using again *R*-like notation:

$$dPolya(y,n,a,b) = \int_0^1 dbinom(y,n,\pi) \times dbeta(\pi,a,b)d\pi$$

With this notation, the posterior predictive distribution (Eq. (2.9)) is defined as $dPolya(y^{new},n^{new},a+y,b+n-y)$. Due to a conjugate property, the Bayes formula 2.2 for the Binomial model (see Eq. (2.7)) can also be specifically written as:

$$dbeta(\pi,a+y,b+n-y) = \frac{dbinom(y,n,\pi) \times dbeta(\pi,a,b)}{dPolya(y,n,a,b)} \tag{2.10}$$

## 2.10 Asymptotic behavior of the Beta-Binomial model

In this section, we see what happens if after $n$ trials, we get $y$ successes (fish recaptured by the fishing device), with $n$ being a very large number, ideally infinite. Starting with a Beta *prior* as in Eq. (2.6) with hyperparameters $(a, b)$, Eq. (2.7) leads to:

$$\begin{aligned} \mathbb{V}ar(\pi\,|y;n) &= \frac{1}{n}\frac{n}{(a+b+n+1)}\frac{(a+y)/n}{(a+b+n)/n}\frac{(b+n-y)/n}{(a+b+n)/n} \\ &= \frac{1}{n}\frac{1}{(\frac{a+b+1}{n}+1)}\frac{(\frac{a}{n}+\frac{y}{n})}{(\frac{a+b}{n}+1)}\frac{(\frac{b}{n}+1-\frac{y}{n})}{(\frac{a+b}{n}+1)} \\ &< \frac{1}{n}\frac{1}{(0+1)}\frac{(1+1)}{(0+1)}\frac{(1+1-0)}{(0+1)} \text{ if } n > a+b \end{aligned} \tag{2.11}$$

In other words, when $n$ increases toward infinity, the posterior knowledge about the unknown $\pi$ gets more and more precise around a single value, since its variance decreases as $1/n$. As already mentioned in the previous section, the distribution (Eq. (2.6)) narrows toward the empirical mean $\frac{y}{n}$ because:

$$\mathbb{E}(\pi\,|y;n) = \frac{y}{n} : \frac{\frac{a}{y}+1}{\frac{a+b}{n}+1} \rightarrow \hat{\pi} = \frac{y}{n}$$

This points out a coherence property of Bayesian analysis: the more information brought by the data, the less uncertainty left for the unknown, whatever the *prior* might be. This result also holds for other models: an asymptotic theorem in Bayesian analysis proves that, when the *prior* does not systematically exclude values from the support of $\pi$, the larger the sample size, the more concentrated on the unknown the posterior distribution will become. This is a Bayesian version ([26]) of the *law of large numbers* that states in the frequentist paradigm that the empirical mean $\frac{y}{n}$ tends to the theoretical mean $\pi$. In addition, the posterior distributions can be asymptotically approximated by a Normal *pdf* whose variance decreases proportionally to the inverse of sample size.

## 2.11 Practical exercise: A simple Beta-Binomial model with WinBUGS

In the following, we will simply illustrate how the Beta Binomial example, which has been developed in this chapter, can be written with WinBUGS. The aim is to estimate the probability of success $\pi$ of a Binomial process knowing the number of trials $n$ and the number of success $y$. We will illustrate how the DAG in Figure 2.3 is simply translated into a WinBUGS code, and show some figures which can be easily derived from $R$ to illustrate posterior inferences and sensitivity to the choice of *prior*.

The Beta-Binomial model of Figure 2.3 is merely declared within the WinBUGS language by the following lines of code:

```
♯Model
model
{
π ~ dbeta(1,1)
y ~ dbin(n,π)
y_new ~ dbin(n_new,π)
}

♯Data
list(n = 167, y = 16, n_new = 31)
♯Inits
list(π = 0.15, y_new = 3)
```

- A Uniform *prior* on the parameter $\pi$ is specified by the line $\pi \sim dbeta(1,1)$, whereas the line $y\_new \sim dbin(n\_new, \pi)$ specifies the likelihood. It is worth stressing that although $\pi$ and $y$ are of very different nature ($\pi$ is random with a prior distribution and $y$ is observed), both appear at the left-hand side of a sign $\sim$). $\pi \sim dbeta(1,1)$ encodes that $\pi$ is distributed *a priori* in a Beta distribution. $y\_new \sim dbin(n\_new, \pi)$ means that $y$ is to be considered as a random issue of a Binomial experiment (but $y$ is fixed in the dataset);
- $list(n = 167, y = 16, n\_new = 31)$ is the dataset;
- $list(\pi = 0.15, y\_new = 3)$ contains the initialization of the MCMC

sampling process. MCMC for Markov chains Monte Carlo are the main numerical methods in Bayesian computation, and references are given at the end of the chapter. WinBUGS is a so-called MCMC sampler (*i.e.*, based on Markov chain Monte Carlo algorithms) that outputs a sample from the posterior distribution of all the model unknowns: the $i^{th}$ draw is random conditionally on the $i-1^{th}$ draw. The Markov chain needs to be initialized by specifying the starting point for $\pi = 0.15$.

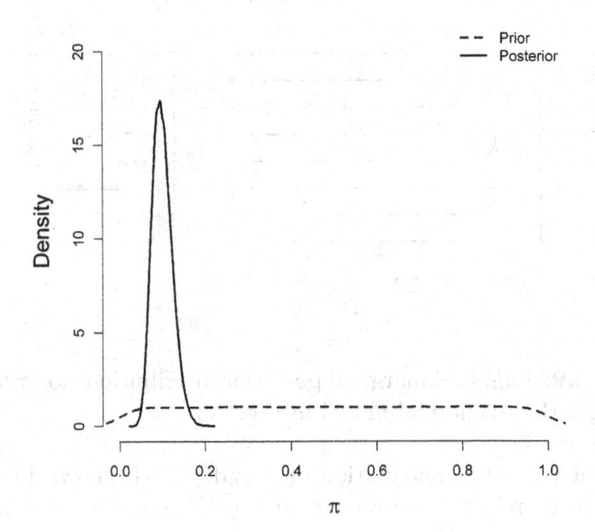


**FIGURE 2.8**: Prior and posterior distributions for the parameter $\pi$ obtained from the simple Beta-Binomial model in Figure 2.3. The prior (dotted line) is $Beta(1,1)$. The posterior (solid line) is $Beta(17,152)$.

After the convergence of the MCMC chains to their ergodic distribution has been checked (WinBUGS offers some ready-to-use tools to check convergence; other tests can be done using the *coda* or *boa* packages of R), the posterior distribution of $\pi$ is approximated by the MCMC sample. Figure 2.8 shows the posterior pdf which is estimated from a 10,000 MCMC sample, together with the Beta(1,1) prior. We invite the reader to verify that the estimated posterior is a Beta($\alpha'$,$\beta'$) distribution with updated parameters $\alpha' = 1+16 = 17$ and $\beta' = 1+167-16 = 152$

(see Eq. (2.8)). Thanks to the large sample size ($n = 167, y = 16$), the posterior shows considerable concentration in comparison with the prior.

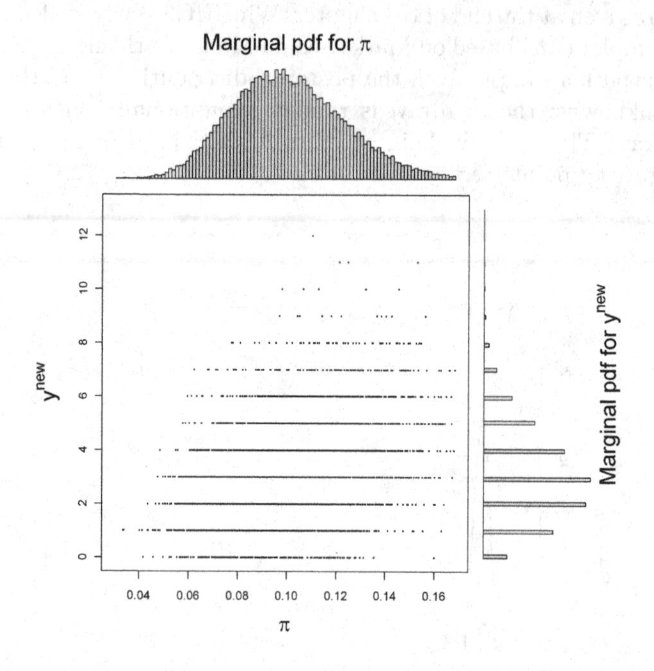


**FIGURE 2.9**: Joint and marginal posterior distributions for $\pi$ and $y^{new}$ in the simple Beta-Binomial model in Fig. 2.3.

The joint posterior distribution of $\pi$ and $y^{new}$ is shown in Fig. 2.9. The positive correlation between $\pi$ and $y^{new}$ is easily explained: the greater the fishing efficiency $\pi$, the greater the number of future recaptured fish $y^{new}$ among the $n^{new}$ marked ones.

To conclude this first step in WinBUGS, we propose a simple experiment to illustrate the sensitivity of the posterior distribution to the *prior* and to the amount of data conveyed by the dataset. We will keep on working with the same Beta-Binomial model, but change the prior distribution and the dataset.

Three alternative prior distributions are tested:

- $\pi \sim Beta(1, 1)$ (little informative, uniform)
- $\pi \sim Beta(9, 1)$ (informative, optimistic: favoring high values of $\pi$)
- $\pi \sim Beta(1, 9)$ (informative, pessimistic)

These three priors are combined with Binomial likelihood using three

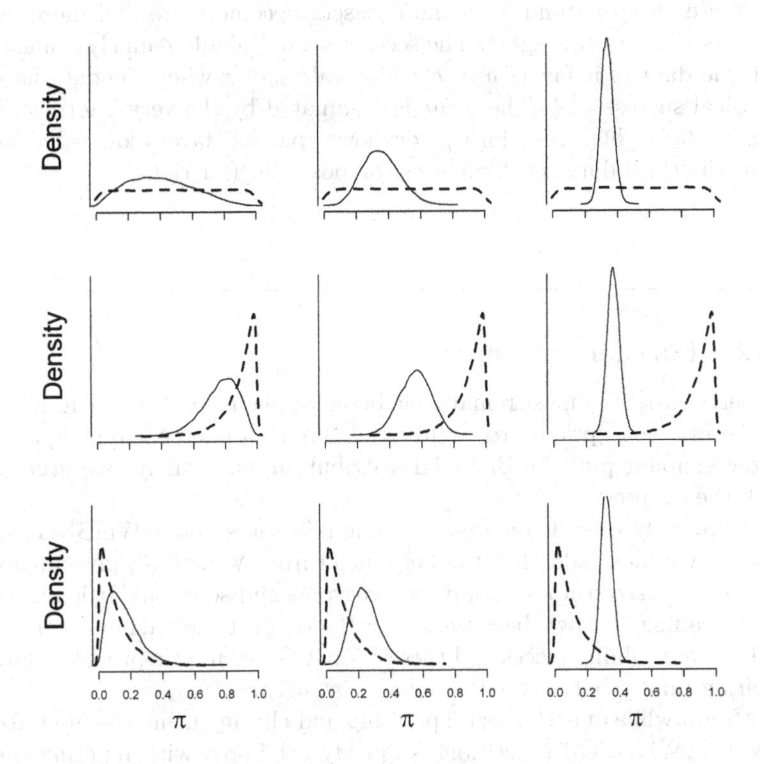


**FIGURE 2.10**: Sensitivity of the Beta posterior pdf to prior and data. Prior are shown as dotted lines, posterior are shown in solid lines. Upper pannel: Uniform(0,1); Middle pannel: Beta(9,1). Lower pannel: Beta(1,9). Dataset differs by column. Left column: $n = 3$, $y = 1$. Middle column: $n = 15$, $y = 5$. Right column: $n = 150$, $y = 50$.

different datasets, which all have the same empirical ratio of success set to $\frac{1}{3}$ but with a different sample sizes:

- $n = 3, y = 1$ (weakly informative)
- $n = 15, y = 5$ (informative)
- $n = 150, y = 50$ (strongly informative)

Figure 2.10 shows the results obtained for the nine different estimations performed independently by crossing the three different Beta prior distributions with the three alternative datasets. The figure shows that with a noninformative *prior* (upper pannels), the posterior gets more and more sharp around $\frac{1}{3}$ as the datasets become more and more informative (left to the right). The second *prior* (middle pannel) conflicts with the data as it favors *a priori* high values of $\pi$ when the data have empirical success $= \frac{1}{3}$. The *prior* is dominated by the very informative dataset (far right). The third *prior* (lower pannel) favors low values of $\pi$ but the third data set dominates the posterior (far right).

## 2.12 Further references

Berry ([31]) is a most remarkable book for beginners because it delivers the first concepts in probability and Bayesian inference relying only on the simplest pdf, the Binomial distribution, with many case studies in all the chapters.

We already provide on page 33 some references about WinBUGS in ecology. We must add that the help menu from WinBUGS gives many examples of case studies with data, programs and solutions. It is highly worth reading through because in the HBM spirit, new models can be made by assembling pieces and these WinBUGS examples provide many inspiring (and proof checked) stochastic structures.

After a while of enthusiastic pointing and clicking using the interface provided, WinBUGS practitioners quickly get bored with repeting the same actions for their statistical analyses. Luckily, the statistical analyses can be conducted from *R* using add-on packages. Chapter 5 of [160] is a detailed session on how to run WinBUGS from *R* (and back) via the package R2WinBUGS (see also [288]).

In the practicals of the companion site of this book (*hbm-for-ecology.org*), we also call the package Brugs ([292]) that works in connection with OpenBUGS, an open source companion version of WinBUGS.

Gamerman ([109]) and Robert and Casella ([260]) are reference textbooks to understand the Monte Carlo algorithms used to sample probability distributions known up to a constant as the posterior pdf given by Eq. (1.25). At an intermediate level, the main ideas for posterior sampling are developed in Box ([35]), Brooks ([37]), Chib and Greenberg ([56]), Kass *et al.* ([156]), and Tierney ([296]). For readers with only basic familiarity with probability, very accessible presentations of MCMC algorithms can be found in Hoff ([140]) Chapters 6 and 10 or in Chapter 10 of Kadane ([153]). The statistical revolution as termed by Brooks ([38]) is still going on and the Bayesian toolbox is being continuingly improved by a flourishing research stream of advanced probabilistic methods for inference ([9]; [21]; [44]; [95]; [120]).

# Chapter 3

## The basic Normal model: A Bayesian comparison of means


### Summary

On a simple fish sampling experiment, we illustrate how to take advantage of the mathematical properties of conjugate distributions. After the Beta-binomial distributions presented in Chapter 2, we now enrich the toolbox of the ecological detective with the Gamma-Normal model. No recourse to MCMC computation is needed here. We derive closed-form expression of the posterior distribution of the difference between the mean lengths of two groups of fish. Rather than focusing on the complexity of the model itself, we point out the easy analytical derivation of posterior pdfs.


## 3.1 Motivating example: does the salmon farm's pollutants influence the growth of juveniles?

In September 2000, the French National Institute for Agronomical Research (INRA, Rennes, France) conducted a large survey to evaluate the number of salmon juveniles in the Scorff River. The Scorff River, located in Brittany, has long been colonized by wild salmon [225]. "Juveniles" are fish that have stayed either one year or two years in the river before getting ready for their sea migration. The two age classes are easily separated according to the number of rings on their scales. As the main part of the catch consists of one-year-old juveniles, to keep this example simple we will not bother with the very few fish that have spent two years in the river. Samples of one-year-old juveniles were regularly taken on 38 spots sites along the river, by electrofishing, using the same duration and with the same protocol at every site (see details about the sampling method in Chapter 4). The size (fork length) and weight of the captured individuals (such as the one in Fig. 3.1) are recorded in Table 3.1.

**FIGURE 3.1**: 0+ A. salmon juveniles caught by electrofishing on the Scorff River.

In 2000, an issue was raised by the installation of a Salmon fish farm on this river: does the aquaculture activity perturb the growth of the natural wild salmon? By comparing records of wild juveniles upstream and downstream of the fish farm, can we help to answer the question?

More precisely, Table 3.1 gives the lengths (in $mm$) of all fish captured downstream the site of the sampling farm, and upstream of the fish farm. One denotes $n^d = 21, n^f = 27, n^u = 12$ the number of fish captured at these three locations of interest. Index $d$, $f$ and $u$ denote the position of the sites, downstream, immediate proximity and upstream the fish farm, respectively. Let $y_i^s$ be the length of the $i$ juvenile captured at site $s$ ($s = d, f, u$), and $\mathbf{y}^s = (y_1^s, ..., y_{n^s}^s)$ the vector of all fish measured at site $s$. The full dataset is written as $\mathbf{y} = \{\mathbf{y}^d, \mathbf{y}^f, \mathbf{y}^u\}$.

## 3.2 A Normal model for the fish length

Fish length is subject to many sources of variation. Some may be systematic: if we suspect the fish farm to release nutrients in the river,

| | | | | | | | | |
|---|---|---|---|---|---|---|---|---|
| | 112 | 110 | 117 | 103 | 137 | 103 | 131 | 130 |
| Downstream | 120 | 116 | 117 | 116 | 104 | 136 | 108 | |
| | 130 | 131 | 122 | 116 | 102 | | | |
| | 112 | 114 | 131 | 92 | 105 | 126 | 85 | 111 |
| Fish farm | 110 | 131 | 101 | 128 | 124 | 76 | 102 | 119 |
| | 124 | 129 | 124 | 100 | 129 | 119 | | |
| | 128 | 90 | 116 | 94 | 111 | | | |
| Upstream | 98 | 121 | 108 | 100 | 107 | 93 | 123 | 101 |
| | 98 | 108 | 102 | 103 | | | | |

**TABLE 3.1**: Length (fork length in mm) of one-year-old juvenile salmons sampled by INRA, downstream, on the site and upstream of the fish farm at Pont Callec, on the Scorff River, in September 2000.

we would expect the wild fish downstream to be systematically bigger than the ones upstream. Other variations are random due to natural (*i.e.*, uncontrolled) genetic or environmental conditions. Even if the odds in favor of such an event are likely to be fairly low, one cannot exclude that all the bigger juveniles were captured upstream the salmon farm, or the other way around, only by chance.

Histograms of salmon length (sampled by electrofishing all along the river in September 2000) exhibit random variability (see Fig. 3.2). Recourse is currently made to the Normal model to mimic the fish length distribution. First, the Normal model is parsimonious: only one location parameter $\mu$ and one dispersion parameter $\sigma$ are required. It is equivalent to write that $Y \sim Normal(\mu, \sigma^2)$ or to say that $Y = \mu + \varepsilon$, $\varepsilon \sim Normal(0, \sigma^2)$, making clear that the phenomenon under study, $Y$, is the combination of a systematic effect $\mu$ plus a random Normal and centered perturbation $\varepsilon$. Second, the Normal structure is sometimes adopted due to the central limit theorem that says that the sum of independent random noises of the same possible magnitude (whatever their distribution) becomes asymptotically Normal.

The natural variability of fish length in each site may stem from such an addition of many environmental random noises. If we drew three histograms, one for each subsample of Table 3.1, the well-known bell shaped curve would not appear with any striking evidence due to the limited size of the subsamples: consequently, the choice of the model is left to the scientists but beware! All the subsequent results rely on this model as long as no sensitivity analysis is performed.

In this example, six parameters $(\mu^u, \sigma^u, \mu^f, \sigma^f, \mu^d, \sigma^d)$ could be required to describe the unknown state of nature (we keep the indices $u, f$ and $d$ to denote the three portions of the river). The simplifying hypothesis $\sigma = \sigma^u = \sigma^f = \sigma^d$ is added, reducing the dimension of the state of

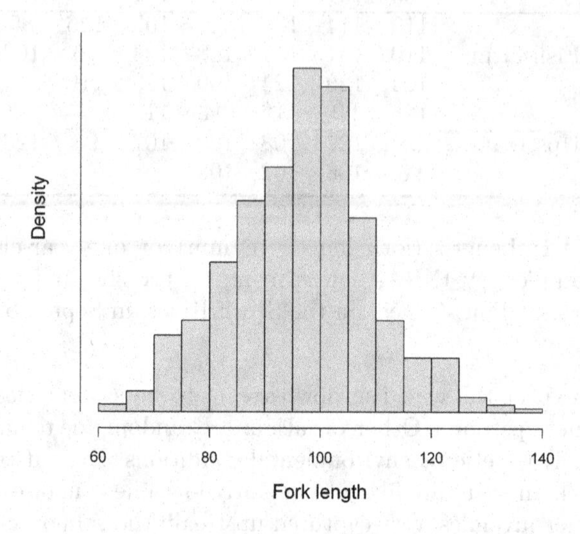


**FIGURE 3.2**: Histogram of the fork lengths from juvenile salmon caught by electrofishing in the Scorff River during September 2000.

nature from six to four components $\theta = (\mu^u, \mu^f, \mu^d, \sigma)$. The ecological meaning of this hypothesis is that the three sites share the same random conditions (and therefore produce the same natural variability between salmon) but they only differ by the mean length of the animals that can be captured at each spot (systematic effect due to the spot). With this model, one could simulate the length $Y^s$ of a juvenile picked at random in the river at site $s$:

$$Y^s \,|\mu^s, \sigma \sim Normal(\mu^s, \sigma^2)$$

In this chapter, the reader is required to blow the dust, if any, from his probabilistic toolbox. We recall the expression of the Normal pdf, that we also called, using *R*-like notations, $dnorm(y, \mu^s, \sigma)$:

$$[Y^s = y\,|\mu^s, \sigma] = \frac{1}{\sigma\sqrt{2\pi}} \times \exp\left(-\frac{1}{2\sigma^2}(y - \mu^s)^2\right) \tag{3.1}$$

All random variables $Y^u$ and $Y^f$ and $Y^d$ are defined similarly from Eq. (3.1) (after changing the indices). Finally we state that $Y^u$, $Y^f$ and $Y^d$ are independent assuming the parameters $\theta = (\mu^u, \mu^f, \mu^d, \sigma)$ are known: there is no statistical correlation between the various measurements because data collection is made at random at each spot.

Finally, the probability distribution of the sampled data (also named the *likelihood* when considered as a function of the parameters) can be written as the product of pdf of Normal random variables:

$$
\begin{aligned}
[\mathbf{y} \,|\theta] &= [\mathbf{y}^d, \mathbf{y}^f, \mathbf{y}^u \,|\mu^u, \mu^f, \mu^d, \sigma] \\
&= [\mathbf{y}^u \,|\mu^u, \sigma] \times [\mathbf{y}^f \,|\mu^f, \sigma] \times [\mathbf{y}^d \,|\mu^d, \sigma] \\
&= \left(\frac{1}{\sigma\sqrt{2\pi}}\right)^{n^d+n^f+n^u} \times \exp\left(-\frac{1}{2\sigma^2}\sum_{j=1}^{n^u}(y_j^u - \mu^u)^2\right) \\
&\times \exp\left(-\frac{1}{2\sigma^2}\sum_{j=1}^{n^f}(y_j^f - \mu^f)^2\right) \\
&\times \exp\left(-\frac{1}{2\sigma^2}\sum_{j=1}^{n^d}(y_j^d - \mu^d)^2\right)
\end{aligned}
\tag{3.2}
$$

---

## 3.3 Normal-Gamma as conjugate models to encode expertise

### 3.3.1 Encoding prior knowledge on $\mu$

Although fish can theoretically take any possible length according to Eq. (3.2), the researcher of the French National Research Institute of Agronomy possesses some prior knowledge that does not rely on the data collected around the fish farm only. As he has been working for a long time on Salmon from Brittany, he won't bet *a priori* equivalently on any possible value $\mu$ for the population of one-year-old juvenile on the river from which the data were recorded. By proposing various values for $\mu$, one can weigh the relative credibility that is granted to each proposed value and sketch the *a priori* pdf $[\mu]$. A comprehensive approach for *elicitation*, *i.e.*, the process of extracting expert knowledge about some unknown quantity and formulating that uncertain judgment as a probability distribution can be found in OHagan *et al.* ([223]). Another way to encode the prior knowledge is to look for data stemming from the same (or some similar) phenomenon. Luckily, on the same river, the same year, there were 35 other spots not related to the fish farm. With the 35 averaged lengths, we can draw a histogram (see Fig. 3.3) which gives a good idea of a prior distribution of the mean $\mu$. As a first approach, this empirical distribution can be approximated by a parametric pdf. With regard to the shape of the histogram of the mean estimates, a simple idea is to take again a Normal pdf, but this time for $[\mu]$:

$$[\mu] = \frac{1}{s_\mu\sqrt{2\pi}} \times \exp\left(-\frac{1}{2s_\mu^2}(\mu - m_\mu)^2\right) \tag{3.3}$$

When centered on $m_\mu = 100$ with a standard deviation $s_\mu = 10$, such a Normal pdf fits well enough this prior knowledge of $\mu$ (see Fig. 3.3).

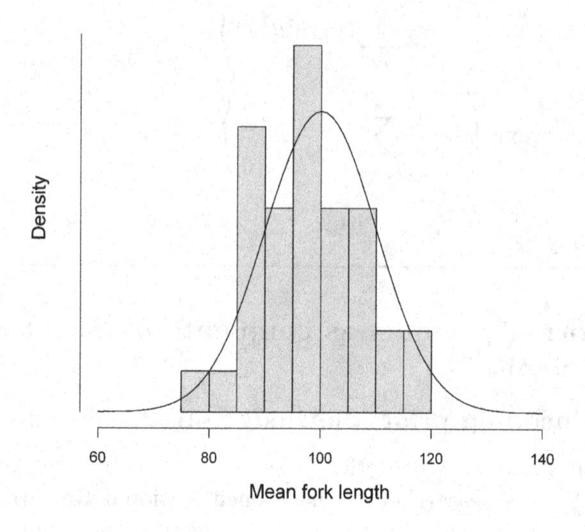


**FIGURE 3.3**: *A priori* knowledge about the mean length of salmon juveniles of the Scorff River. A Normal distribution with mean 100 and standard deviation = 10 was found to fit reasonably well the empirical distribution (solid line).

As in the previous chapter, $m_\mu$ and $s_\mu$ are called the *hyperparameters* (of the prior pdf of the parameter of interest $\mu$).

### 3.3.2 Encoding prior knowledge on $\sigma$

Figure 3.4 shows a similar result for the empirical variances of the 35 same samples. The histogram of these variance estimates could be used to get a prior for $\sigma^2$. More specifically, we work with the precision (*i.e.*, the inverse of the variance, $\sigma^{-2}$) and introduce the Gamma distribution, parameterized by two coefficients $a$ and $b$ that we fit to the empirical variances obtained with the 35 sites.

$$\sigma^{-2}|a_\sigma, b_\sigma \sim Gamma(a_\sigma, b_\sigma)$$

A good fit is obtained by adjusting the hyperparameters, $a_\sigma = 3.4$ and $b_\sigma = 250$. The Gamma *a priori* distribution for $\sigma^{-2}$ has the following pdf (a new one to remember! $dgamma(\sigma^{-2}, a, b)$ using *R*-like notations), with moments expressed as a function of the hyperparameters:

$$\begin{cases} [\sigma^{-2}] = \dfrac{b_\sigma^{a_\sigma}}{\Gamma(a_\sigma)}(\sigma^{-2})^{a_\sigma - 1} \exp\left(-b_\sigma \times \sigma^{-2}\right) \\ \mathbb{E}(\sigma^{-2}) = \dfrac{a_\sigma}{b_\sigma} = 1.36 \times 10^{-2} \\ \mathbb{V}(\sigma^{-2}) = \dfrac{a_\sigma}{b_\sigma^2} = \dfrac{\mathbb{E}(\sigma^{-2})}{b_\sigma} = 5.44 \times 10^{-5} \end{cases} \tag{3.4}$$

Working with the precision $\sigma^{-2}$ instead of the standard deviation $\sigma$ or the variance $\sigma^2$ is a matter of mathematical convenience that will be justified in the next section.

### 3.3.3 Bayesian updating of the mean, a quick approximate solution

Suppose we knew $\sigma$, either because we do take for granted that the sample variance $\sigma^2 \approx \frac{250}{3.4} = 73.5$ is precisely estimated from the data (and check later if the results are sensitive to this estimation) or because we work for the moment *as if* $\sigma$ were known for probabilistic computations in a sort of mind experiment. We express the prior variance of the mean as a fraction $\nu_\mu$ of the sample mean: $s_\mu^2 = \frac{\sigma^2}{\nu_\mu}$, *i.e.*, $\nu_\mu = 0.735$. In other words, the prior information on $\mu$ is worth a 0.735 virtual sample size (a little less than one datum). The joint distribution of data upstream and downstream of the fish farm $\mathbf{y} = (\mathbf{y}^u, \mathbf{y}^d)$ and the unknown means $\theta = (\mu^u, \mu^d)$ is written as:

$$\begin{aligned} [\mathbf{y}, \theta] = & \left(\frac{1}{\sigma\sqrt{2\pi}}\right)^{n^d + n^u} \left(\frac{\nu_\mu^{0.5}}{\sigma\sqrt{2\pi}}\right)^2 \\ & \times \exp\left(-\frac{1}{2\sigma^2}\left(\sum_{j=1}^{n^u}\left(y_j^u - \mu^u\right)^2 + \nu_\mu(\mu^u - m_\mu)^2\right)\right) \\ & \times \exp\left(-\frac{1}{2\sigma^2}\left(\sum_{j=1}^{n^d}\left(y_j^d - \mu^d\right)^2 + \nu_\mu(\mu^d - m_\mu)^2\right)\right) \end{aligned} \tag{3.5}$$

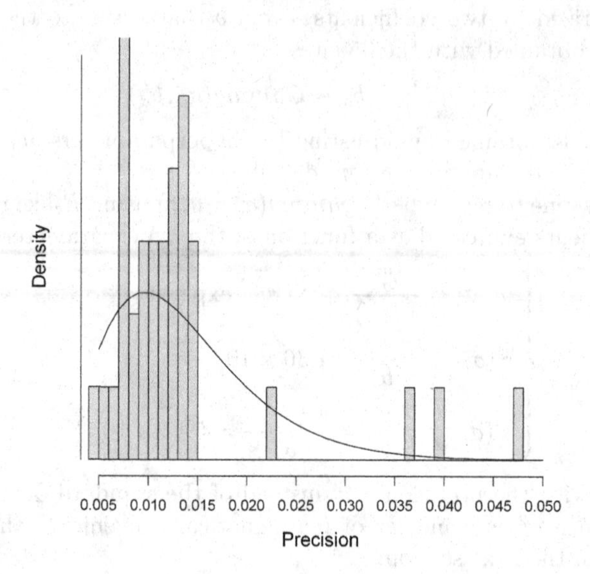


**FIGURE 3.4**: *A priori* knowledge about the precision of the length of Scorff Salmon juveniles. A Gamma with parameters $a_\sigma = 3.4$ and $b_\sigma = 250$ was found to fit reasonably well the empirical distribution (solid line).

When focusing on $(\mu^u, \mu^d)$ in Eq. (3.5), after some algebra, one finds under the exponential, a quadratic form in $(\mu^u, \mu^d)$, which reminds us of a Normal pdf:

$$[\mathbf{y},\theta] \propto \exp\left(-\frac{n^u+\nu_\mu}{2\sigma^2}\left(\mu^u - \frac{\sum_{j=1}^{n^u} y_j^u + \nu_\mu m_\mu}{n^u+\nu_\mu}\right)^2\right)$$
$$\times \exp\left(-\frac{n^d+\nu_\mu}{2\sigma^2}\left(\mu^d - \frac{\sum_{j=1}^{n^d} y_j^d + \nu_\mu m_\mu}{n^d+\nu_\mu}\right)^2\right) \quad (3.6)$$

The remaining terms from Eq. (3.5) do not involve $\theta = (\mu^u, \mu^d)$

neither do the missing terms $\left(\frac{(n^u+\nu_\mu)^{0.5}}{\sigma\sqrt{2\pi}}\right)\left(\frac{(n^d+\nu_\mu)^{0.5}}{\sigma\sqrt{2\pi}}\right)$ for the constant of integration of the Normal pdf. Therefore as $[\mathbf{y},\theta] = [\theta|\mathbf{y}][\mathbf{y}]$, we directly *see* that:

$$\begin{cases} \mu^u|\mathbf{y} \sim N(m'_{\mu^u}, s'^2_{\mu^u}) \\ s'^{-2}_{\mu^u} = s^{-2}_{\mu^u} + n^u\sigma^{-2} \\ s'^{-2}_{\mu^u} \times m'_{\mu^u} = s^{-2}_{\mu^u} \times m_{\mu^u} + n^u\sigma^{-2} \times \bar{y}^u \end{cases}$$

Equivalently, by setting the posterior precision $s'^{-2}_{\mu^u} = \nu'_{\mu^u}\,\sigma^{-2}$

$$\begin{cases} \nu'_{\mu^u} = \nu_{\mu^u} + n^u \\ m'_{\mu^u} = \dfrac{\nu_{\mu^u} \times m_{\mu^u} + n^u \times \bar{y}^u}{\nu_{\mu^u} + n^u} \end{cases}$$

In other words, the posterior precision is the sum of the prior precision and the sample precision while the posterior mean is a linear combination of the prior mean and the sample mean, with weights given by the corresponding precisions. Of course, the Normal Bayesian updating of $\mu^d$ follows the same rules. Numerically:

$$\begin{cases} \bar{y}^u = 105.2, n^u = 12, \nu'_{\mu^u} = 12.735, m'_{\mu^u} = 104.9 \\ \bar{y}^d = 118.1, n^d = 20, \nu'_{\mu^u} = 20.735, m'_{\mu^d} = 117.5 \end{cases}$$

In addition, the couple $(\mu^u, \mu^d)|\mathbf{y}$ is independent (bivariate Normal) since the joint distribution (Eq. (3.6)) factorizes. As a consequence, the difference $\mu^d - \mu^u|\mathbf{y}$ is also normally distributed with mean and variance obtained as follows:

$$\begin{cases} \mathbb{E}(\mu^d - \mu^u|\mathbf{y}) = \mathbb{E}(\mu^d|\mathbf{y}) + \mathbb{E}(-\mu^u|\mathbf{y}) = m'_{\mu^d} - m'_{\mu^u} \\ \mathbb{V}(\mu^d - \mu^u|\mathbf{y}) = \mathbb{V}(\mu^d|\mathbf{y}) + \mathbb{V}(-\mu^u|\mathbf{y}) = s'^2_{\mu^d} + s'^2_{\mu^u} \end{cases}$$

Numerically $m'_{\mu^d} - m'_{\mu^u} = 12.6mm$ and $s'^2_{\mu^d} + s'^2_{\mu^u} = \frac{250}{3.4} \times (\frac{1}{12.735} + \frac{1}{20.735}) = 9.32 = 3.05^2$. A random variable with mean 12.6 and standard deviation 3.05 has practically no chance to be negative (remember that 99% of the probability mass of a Normal pdf lies between $\pm 3$ standard deviations around the mean); therefore, the ecological detective can make the following (posterior) probabilistic judgment with full confidence: *Given the experimental data and the Normal model with known variance, the mean size of a juvenile fish downstream of the aquaculture device is bigger that the corresponding reference measure upstream the fish farm.* Yet $\sigma^2$ is not exactly known and this uncertainty may blur the assessment. Will our previous probabilistic judgment change much? The rest of the chapter is devoted to quantifying the uncertainty of $\sigma^2$ by means of the inverse gamma distribution (taking into account additional data from the third site) and evaluate the influence of this additional uncertainty on the previous statement.

### 3.3.4 Joint prior for $(\mu^u, \mu^f, \mu^d, \sigma^{-2})$

A first additional hypothesis can be made on the priors: they are the same for the three spots and the prior knowledge on $\mu$ does not depend on $\sigma$. Consequently, the prior distribution for $\theta = (\mu^u, \mu^f, \mu^d, \sigma^2)$ is the product of each prior, a mathematical expression that can be written as a function of the hyperparameters that the expertise yielded for the case study:

$$[\mu^u, \mu^f, \mu^d, \sigma^{-2}] = \left(\frac{1}{s_\mu\sqrt{2\pi}}\right)^3 \frac{b_\sigma^{a_\sigma}(\sigma^2)^{-a_\sigma-1}}{\Gamma(a_\sigma)} \times \exp\left(-\frac{\sum\limits_{s=u,f,d}(\mu^s - m_\mu)^2}{2s_\mu{}^2} - \frac{b_\sigma}{\sigma^2}\right) \quad (3.7)$$

with $m_\mu = 100$, $s_\mu = 10$, $a_\sigma = 3.4$ and $b_\sigma = 250$.

Another common but somewhat different hypothesis is to assume that:

- $\sigma^2$ is again inverse-gamma distributed;
- But, given $\sigma^2$, $\mu^u, \mu^f, \mu^d$ are normally independently distributed with a variance $s_\mu^2$ proportional to $\sigma^2$, thus writing as in section 3.3.3 $\sigma^2 = \nu_\mu s_\mu^2$. In other words, the variability of the sizes about the mean, encoded by $\sigma^2$, is proportional to the uncertainty about the value of $\mu$, here written as $s_\mu^2$. To assign a numerical value to the additional hyperparameter $\nu_\mu$, one can notice from the 35 subsamples that the empirical variance for the $\mu$ estimates is $s_\mu^2 = 100$, while the empirical mean of the estimates for $\sigma^2$ is close to $250/3.4 = 73.5$ (*i.e.*, close to the inverse of the mean of the Gamma pdf for the precision). The value of the hyperparameter $\nu_\mu$ can be selected as $73.5/100 = 0.735$, which will be adopted to compute the following results in the chapter.

This leads to a prior distribution for $[\mu^u, \mu^f, \mu^d, \sigma^2]$ (Eq. (3.8)) which differs from Eq. (3.7) since $\mu^u, \mu^f, \mu^d$ and $\sigma^2$ are no longer independent: the bigger $\sigma^2$ tends to be, the more diffuse the prior judgments for

$\mu^u, \mu^f, \mu^d$.

$$
\begin{aligned}
[\mu^u, \mu^f, \mu^d, \sigma^{-2}] &= [\mu^u, \mu^f, \mu^d | \sigma^2][\sigma^{-2}] \\
&= \left(\frac{\nu_\mu^{0.5}}{\sigma\sqrt{2\pi}}\right)^3 \frac{b_\sigma^{a_\sigma} (\sigma^2)^{-a_\sigma - 1}}{\Gamma(a_\sigma)} \\
&\times \exp\left(-\frac{\nu_\mu \sum\limits_{s=u,f,d} (\mu^s - m_\mu)^2}{2\sigma^2} - \frac{b_\sigma}{\sigma^2}\right)
\end{aligned} \tag{3.8}
$$

---

## 3.4 Inference by recourse to conjugate property

### 3.4.1 Bayesian updating in a closed-form

The likelihood (Eq. (3.2)) and the prior (Eq. (3.8)) exhibit common structural features as function of the parameters. Details about conjugate priors in such exponential family models can be found for instance on page 42 of Parent and Bernier ([224]) or in Chapter 8 of Kadane ([153]). Taking the prior in the conjugate family of the likelihood allows for full analytical Bayesian computation. Indeed, with such a conjugate prior, the posterior distribution also belongs to the same Normal-Gamma parametric family, with new parameters which are obtained by updating old (prior) parameters by the data. The joint distribution for unknown and observed quantities is written as:

$$
\begin{aligned}
[\mathbf{y},\theta] = &\left(\frac{1}{\sigma\sqrt{2\pi}}\right)^{n^d+n^f+n^u} \left(\frac{\nu_\mu^{0.5}}{\sigma\sqrt{2\pi}}\right)^3 \left(\frac{b_\sigma^{a_\sigma} (\sigma^2)^{-a_\sigma - 1}}{\Gamma(a_\sigma)}\right) \\
&\times \exp\left(-\frac{b_\sigma}{\sigma^2}\right) \\
&\times \exp\left(-\frac{1}{2\sigma^2}\sum_{j=1}^{n^u}\left(\left(y_j^u - \mu^u\right)^2 + \nu_\mu(\mu^u - m_\mu)^2\right)\right) \\
&\times \exp\left(-\frac{1}{2\sigma^2}\sum_{j=1}^{n^f}\left(\left(y_j^f - \mu^f\right)^2 + \nu_\mu(\mu^f - m_\mu)^2\right)\right) \\
&\times \exp\left(-\frac{1}{2\sigma^2}\sum_{j=1}^{n^d}\left(\left(y_j^d - \mu^d\right)^2 + \nu_\mu(\mu^d - m_\mu)^2\right)\right)
\end{aligned} \tag{3.9}
$$

Rearranging the terms with the sufficient statistics $\bar{y^s} = \frac{1}{n}\sum_{j=1}^{n^s} y_j^s$ for $s = u, f, d$ yields the posterior:

$$
\begin{aligned}
[\theta|\mathbf{y}] &= \frac{[\mathbf{y},\theta]}{[\mathbf{y}]} \propto [\mathbf{y},\theta] \\
&= \frac{{b'_\sigma}^{a'_\sigma}(\sigma^2)^{{a'_\sigma}-1}}{\Gamma(a'_\sigma)} \prod_{s=u,f,d} \left( \frac{\sqrt{\nu'^s_\mu}}{\sigma\sqrt{2\pi}} \right) \\
&\times \exp\left( -\frac{\nu'_\mu \sum_{s=u,f,d} (\mu^s - m'_\mu)^2}{2\sigma^2} - \frac{b'_\sigma}{\sigma^2} \right)
\end{aligned} \tag{3.10}
$$

with updated parameters

$$
\begin{cases}
a'_\sigma = a_\sigma + \dfrac{n^d + n^f + n^u}{2} \\
\nu'^s_\mu = \nu_\mu + n^s \\
m'^s = \dfrac{n^s \bar{y}^s + \nu_\mu m_\mu}{n^s + \nu_\mu}
\end{cases}
$$

and

$$
\begin{aligned}
b'_\sigma &= b_\sigma + \frac{1}{2} \sum_{s=u,f,d} \left( \sum_{j=1}^{n^s} (y_j^s - \bar{y}^s)^2 + n^s(\bar{y}^s - m'^s)^2 + \nu_\mu (m_\mu - m'^s)^2 \right) \\
&= b_\sigma + \frac{1}{2} \sum_{s=u,f,d} \left( \sum_{j=1}^{n^s} (y_j^s - \bar{y}^s)^2 + \frac{n^s \nu_\mu}{n^s + \nu_\mu} (\bar{y}^s - m_\mu)^2 \right)
\end{aligned}
$$

### 3.4.2 Does the fish farm perturb the growth of wild salmon?

The main advantage of deriving the closed-form for the joint posterior (Eq. (3.10)) is that this allows for full analytical Bayesian computation to infer the effect of the fish farm.

From Eq. (3.10), using the same trick as in Section 3.3.3, it appears that given $\sigma$, the three location parameters are independent and normally distributed, respectively $Normal(m'^u, \frac{\sigma^2}{\nu'^u})$, $Normal(m'^f, \frac{\sigma^2}{\nu'^f})$ and $Normal(m'^d, \frac{\sigma^2}{\nu'^d})$. As a consequence, given $\sigma$, the marginal distribution of the difference between any of the two location parameters is also Normal. In particular, given $\sigma$, the marginal posterior pdf of the difference

$\delta = \mu^d - \mu^u$ is Normal:

$$\mu^d - \mu^u | \sigma^2, Y \sim Normal\left(m'^d - m'^u, \sigma^2\left(\frac{1}{\nu'_b} + \frac{1}{\nu'_h}\right)\right) \quad (3.11)$$

Moreover, the marginal posterior distribution of the precision $\sigma^{-2}$ is Gamma with updated parameters $a'_\sigma$ and $b'_\sigma$.

Appendix A details some additional statistical properties of the Normal distribution and its extensions. The most famous one is the Student distribution or $T$ distribution. If, conditionally upon $(\mu, \sigma^2)$, $Z$ is normally distributed with mean $\mu$ and variance $\sigma^2$, and if the precision $\sigma^{-2}$ is distributed as a Gamma$(a, b)$, then the distribution of $\frac{Z-\mu}{\sqrt{b/a}}$ is a standard Student with $\nu = 2a$ degree of freedom.

We can take advantage of this general result to derive the marginal posterior distribution of the difference $\delta = \mu^d - \mu^u$ in a closed-form. Indeed, the conditional posterior pdf of $\delta$ is Normal, and the marginal posterior distribution of the precision $\sigma^{-2}$ is Gamma. Hence, the marginal posterior distribution of the difference $\delta$ is Student. More precisely, the marginal posterior pdf of

$$\sqrt{2\frac{\nu'_h \nu'_b}{\nu'_h + \nu'_b}\frac{a'_\sigma}{b'_\sigma}}(\delta - (m'^b - m'^h))$$

is a standard Student random variable with $2a'_\sigma$ degrees of freedom. Calling it in $R$-like notations, *dStudent*, we can write to make it mathematically short that:

$$dStudent(t, 2a) = \frac{\Gamma(\frac{2a+1}{2})}{\Gamma(a)\sqrt{2a\pi}} \frac{1}{[1 + \frac{t^2}{2a}]^{\frac{2a+1}{2}}}$$

Back to the INRA data of Table 3.1, one gets:

$$a'_\sigma = a_\sigma + \frac{n^d + n^f + n^u}{2} = 3.4 + \frac{20 + 26 + 12}{2} = 32.4$$

and

$$b'_\sigma = b_\sigma + \frac{1}{2} \sum_{s=u,f,d} \left\{ \sum_{j=1}^{n^s} (y_j^s - \bar{y}^s)^2 + \frac{n^s \nu_\mu}{n^s + \nu_\mu} (\bar{y}^s - m_\mu)^2 \right\}$$

with

$$\begin{cases} \bar{y}^d = 118.1, n_d = 20, \nu'_d = 20.735, m'_d = 117.5 \\ \bar{y}^f = 111.6, n_f = 26, \nu'_f = 26.735, m'_f = 111.3 \\ \bar{y}^u = 105.2, n_u = 12, \nu'_u = 12.735, m'_u = 104.9 \\ \dfrac{\sum_{j=1}^{n^u}(y_j^u - \bar{y}^u)^2 + \sum_{j=1}^{n^f}(y_j^f - \bar{y}^f)^2 + \sum_{j=1}^{n^d}(y_j^d - \bar{y}^d)^2}{2} = 4777 \\ \dfrac{\sum_{s=u,f,d} \{n^s(\bar{y}^s - m'^s)^2 + \nu_\mu(m_\mu - m'^s)^2\}}{2} = 173 \end{cases}$$

thus

$$b'_\sigma = 250 + 4950 = 5200$$

With the updated parameters calculated above,

$$\sqrt{2\frac{\nu'_h\nu'_b}{\nu'_h + \nu'_b}\frac{a'_\sigma}{b'_\sigma}}(\delta - (m'^b - m'^h))$$

is a standard Student random variable with 64.8 degrees of freedom. For so many degrees of freedom, a Student distribution is practically indistinguishable from the Normal one. Thus $\delta$ is approximately distributed as a Normal centered on 12.6$mm$ with standard deviation of $\sqrt{\frac{\nu'_h+\nu'_b}{2\nu'_h\nu'_b}\frac{b'_\sigma}{a'_\sigma}} \approx 3.2\ mm$. Compared to Section 3.3.3, here we have to take into account a larger standard deviation (and a larger credible interval), due to the fact that $\sigma^2$ is no longer assumed to be perfectly known. Figure 3.5 shows this bell-shaped pdf.

This distribution is highly concentrated toward positive values of $\delta$, which means that we are almost sure that the mean difference in length between fish downstream and upstream the fish farm is greater than 5 $mm$ (with a most probable value around 12.6 $mm$). In this example, as the probability $p$ that a Normal variable with mean 12.6 and standard deviation 3.2 is larger than 5 is approximately 99%, the posterior odds $\frac{p}{1-p}$ are around 100 against 1 in favor of the preceding judgment: the information conveyed by the dataset is strong enough to come to the conclusion that there is a significant difference in salmon juvenile length between two portions of the river separated by the fish farm.

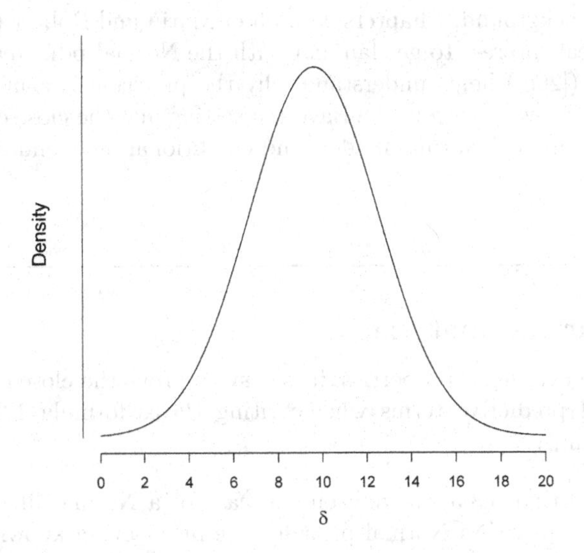


**FIGURE 3.5**: Marginal posterior distribution (approximated by a Normal pdf) of the length difference $\delta$ between Salmon living downstream and upstream of the fish farm.

---

## 3.5 Bibliographical notes

Much has been written about Bayesian aspects of the Normal model. An excellent introduction taking the reader by the hand from elementary notions to very advanced topics is Hoff ([140]). Lee ([172]) is a standard course in applied statistics (hypothesis test, comparison of means, Anova, Linear model, etc.) at the Msc level. Its last edition includes more advanced materials on hierarchical modeling and Bayesian computation. Sivia ([280]) is an intermediate Bayesian tutorial oriented toward engineers and physicists. What can a physicist do when no model seems available? This author also works out in depth Normal distributions and their extensions for the practical reason that they are often met as asymptotic approximations of posterior pdfs. At a more advanced level, Bernardo and Smith ([30]) give theory-oriented explanations on how to understand model construction; reference technical manuals include Raiffa and Schlaifer ([247]) with emphasis on decision theory and experimental planning and Box and Tiao ([36]) for statistical inference based on noninformative priors. Although they need some strong mathematical background, Chapters 2 and 3 of Marin and Robert ([189]) are also of great interest to get familiar with the Normal pdf. Reading Rue and Held ([266]) helps understand why the precision is a much better parameter to work with in the Bayesian setting and the close connection between graphical Normal models and conditional independence.

---

## 3.6 Further material

Due to conjugate properties, it is easy to prove the closed form posterior and predictive terms when writing Bayes formula 1.22 for the Normal family:

- In Section 3.3.3, we rely on the case of a Normal likelihood associated with a Normal prior for the mean but a known variance $(\theta = \mu)$

$$
\begin{aligned}
dnorm(y, \mu, \sigma) \times dnorm(\mu, m, s) = \\
& dnorm(\mu, \frac{\frac{m}{s^2} + \frac{y}{\sigma^2}}{\sigma^{-2} + s^{-2}}, \left(\sigma^{-2} + s^{-2}\right)^{-0.5}) \\
& \times\, dnorm(y, m, \left(\sigma^2 + s^2\right)^{0.5})
\end{aligned}
$$

- Alternatively, one can evaluate the case of a Gamma prior precision with known mean $(\theta = \sigma^{-2})$

$$
\begin{aligned}
dnorm(y, \mu, \sigma) \times dgamma(\sigma^{-2}, a, b) = \\
& dgamma(\sigma^{-2}, a + \frac{1}{2}, b + \frac{1}{2}(y - \mu)^2) \\
& \times \frac{1}{\sqrt{b}} dStudent(\frac{y - \mu}{\sqrt{b}}, 2a)
\end{aligned}
$$

- In Section 3.4.1 and in Appendix A, we developed the case of a Normal likelihood associated with a Normal prior for the mean conditioned on a gamma precision ($\theta = (\mu, \sigma^{-2})$)

$$dnorm(y, \mu, \sigma) \times dnorm(\mu, m, \frac{\sigma}{\sqrt{\lambda}}) \times dgamma(\sigma^{-2}, a, b)$$

$$= dnorm\left(\mu, \frac{\lambda m + y}{\lambda + 1}, \frac{\sigma}{\sqrt{1+\lambda}}\right)$$

$$\times\, dgamma\left(\sigma^{-2}, a + \frac{1}{2}, b + \frac{\lambda}{2(1+\lambda)}(y-m)^2\right)$$

$$\times \sqrt{\frac{\lambda}{(1+\lambda)b}} \times dStudent\left(\frac{y-\mu}{\sqrt{\frac{b(1+\lambda)}{\lambda}}}, 2a\right)$$

These closed forms allow shortcuts when writing conditional pdfs, which is very useful for Bayesian computation.

# Chapter 4

## Working with more than one Beta-Binomial element


### Summary

The quantitative assessment of fish population size is the main issue for both scientific research and resource management. Describing capture-mark-recapture and successive removal techniques, the classical sampling methods for fish surveys in the wild, this chapter illustrates Beta-Binomial constructions and model selection.

1. Beta-Binomial sub-models from Chapter 2 can be assembled to describe capture-mark-recapture and successive removal experiments. Now, we no longer deal with a single parameter: the unknown state of nature becomes bidimensional. We also make a first step toward more elaborate models in this chapter because we recall the Poisson distribution and introduce the Negative binomial pdf.

2. Bayesian analysis offers a coherent deductive framework for model selection and hypothesis testing. In the Bayesian setting, model selection is nothing but making inference about the unknown model index in a supermodel encompassing all possible models. The Bayes Factor is the Bayesian criteria for model comparison that measures the odds ratio of two competing models with regard to the data. In this chapter, we take advantage of the Beta-Binomial conjugate properties to derive a closed-form expression of Bayes Factors and study its sensitivity to the choice of priors.

## 4.1 Bayesian analysis of smolt runs by capture-mark-recapture

### 4.1.1 Motivating example

The Oir is a French index river for Atlantic salmon population dynamics and stock assessment in France and Europe ([17]; [239]).

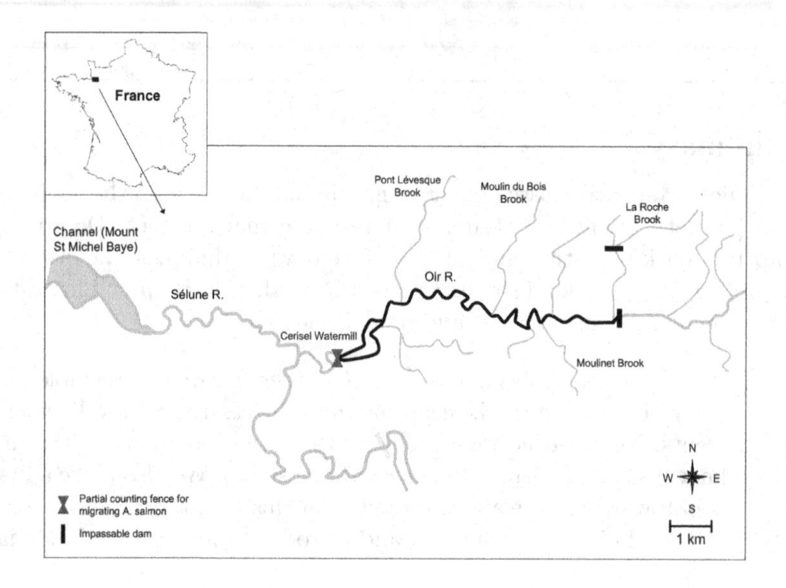


**FIGURE 4.1**: The section of the Oir River considered in this study appears in a solid black line.

It is a spawning tributary of the Sélune River, which flows into the English channel (Fig. 4.1). It is $19.5km$ long with a drainage basin of $85km^2$. The main stream colonized by salmon extends over a $12.3km$ long stretch from a trapping facility (the Cerisel station) at the river mouth to an impassable dam (the Buat watermill). Atlantic salmon juveniles born in the Oir river the previous year turn into smolts around April and undertake a downstream seaward migration to reach the ocean and start their long-distance oceanic migration.

At the Cerisel trapping facility, part of the flow is derived from the river toward a trap (Fig. 4.3) where fishes can be counted (see [17], [255] or [259] for more details). To assess the number of migrating smolts

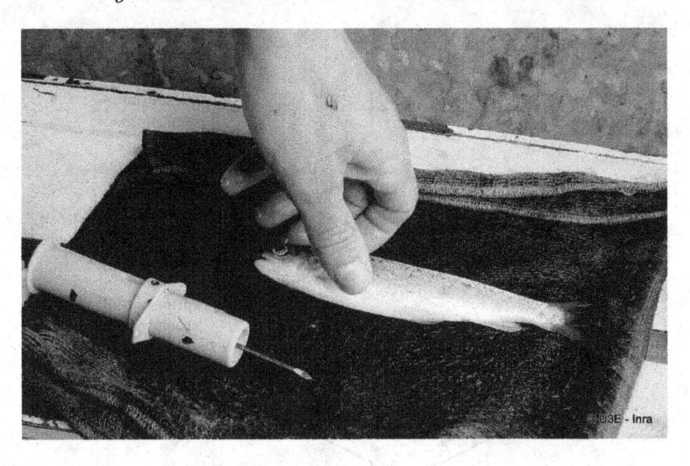


**FIGURE 4.2**: An anesthetized smolt will be marked and released upstream.

$\nu$, records are taken following a capture-mark-recapture scheme at the Cerisel trap (see also Fig. 4.3):

- A total of $\nu$ fish are migrating downstream ($\nu$ is the unknown population size of primary interest).
- A part of these downstream migrating fish gets captured ($y_1$ individuals) at the trapping facility.
- Fishery scientists put tags on $y_2$ individuals taken from the $y_1$ already removed (see Fig. 4.2) and bring them upstream the trapping facility again (the remaining $y_2 - y_1$ are set free downstream so that they can finish swimming to the sea).
- From the $y_2$ tagged and released fish, $y_3$ of them get caught a second time in the same trap.

Based on the observed results from this experiment, fishery scientists would like to make some deductive statements about the number of smolts migrating that year, $\nu$. For instance, the following records (published in [255]) were collected in April 1996, by INRA Rennes, France:

$$\begin{cases} y_1 = 767 \\ y_2 = 76 \\ y_3 = 58 \end{cases}$$

Of course, estimating the number of smolts in the year 1996 is of key

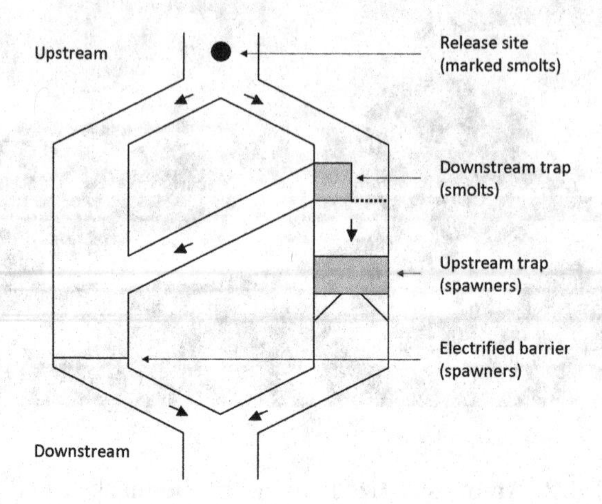


**FIGURE 4.3**: Scheme of the Cerisel trapping facility (Oir River, Lower Normandy, France) working as a double downstream (for smolt runs) and upstream (for spawner runs) partial counting fence. The black circle at the top of the figure indicates the release site of tagged smolts and spawners. The release site and the downstream trap are 1 km away from each other.

importance when trying to understand and quantify the factors that control the population dynamics. This also is a start to develop forecast of the number of adults that will return into the river the following years. To go from the present experimental situation to a formalized mathematical model, one always has to bring answers to the same set of questions:

1. What are the unknowns?
2. What are the observables? (So-called $Y$ in the previous chapter.)
3. Which mechanisms can one imagine that generate the data? What are the links from the unknowns to the observables?
4. What type of hypotheses can be made to model these mechanisms in a probabilistic conditional reasoning framework?

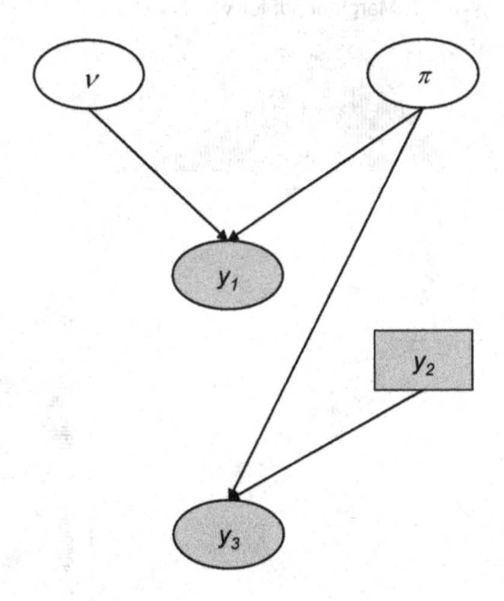


**FIGURE 4.4**: Directed acyclic graph (DAG) for the capture-mark-recapture model for smolts in the Oir River. Priors: $\nu \sim Uniform(0, \nu_{max})$, $\pi \sim Beta(a, b)$. Likelihood: $y_1|\nu, \pi \sim Binomial(\nu, \pi)$ and $y_3|y_2, \pi \sim Binomial(y_2, \pi)$.

### 4.1.2 Sampling distributions and likelihood

Undoubtedly, the data $y_1$ are the realizations of an observable $Y_1$ that one can conceptualize as the possible number of catches at the Cerisel trap for nonmarked smolts (Fig. 4.3). Assuming a constant catchability $\pi$ over time (hypothesis $H_1$) and a homogeneous and independent behavior among individuals (hypothesis $H_2$), we state that $y_1$ is the realization of the random binomial variable $Y_1$ with probability of success $\pi$ and trial number $\nu$:

$$Y_1 \,|\nu, \pi \sim Binomial(\nu, \pi) \tag{4.1}$$

Therefore the unknowns[1] are $(\nu, \pi)$. $\nu$ is the main quantity we want to estimate, (*i.e.*, the target parameter), $\pi$ is here of less interest and was introduced to conveniently design the model. Such a quantity is often called a *nuisance* parameter.

The data $y_2$ have a rather weird status. Given the catches $y_1$, it

[1] We recommend as a good modeling practice to use Greek letters for the unknown parameters.

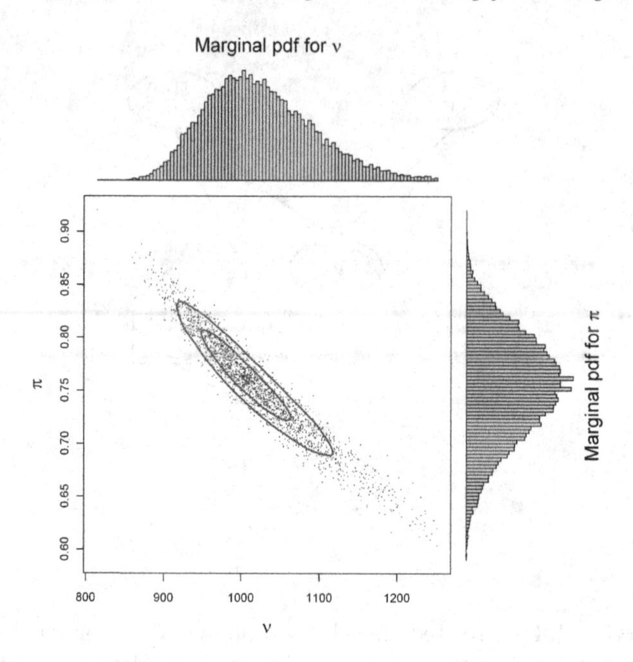


**FIGURE 4.5**: Posterior distributions of $(\pi, \nu)$ from the model in Fig. 4.4. The shape of the joint posterior distribution is shown in the central part (joint MCMC draws and smoothed isodensity contours). The marginal distributions are shown in the top and right panels.

is a covariate, because only part of the catch is marked and released upstream and no stochastic mechanism is involved during this operation: the scientists do not want to put upstream the whole lot but at the same time they will tag enough individuals so that the ratio $\frac{y_3}{y_2}$ will be large enough to provide a good estimate of the trap efficiency $\pi$. More precisely, a second observation equation is introduced to mimic the recapture of marked individuals $y_3$ as a random issue of $Y_3$, a second binomial process with the same trapping efficiency $\pi$:

$$Y_3 \,|y_2, \pi \sim Binomial(y_2, \pi) \qquad (4.2)$$

When writing Eq. (4.2), additional hypotheses are implicitly made. $H_3$: The population is closed during the migration time. There is neither mortality induced by the capture/marking procedure nor natural mortality between the time of marking and recapture for either marked or unmarked smolts; $H_4$: There is no tag shedding and all smolts marked and released will migrate out; $H_5$: All marked and released smolts have the same probability of being recaptured at the downstream trap. We suppose that the capture and marking do not affect the behavior of the smolts in a way that would change their vulnerability to the trap. Thus, the probability of recapture of previously marked and released smolts remains the same as the probability of capture of unmarked smolts exposed to the trap for the first time. This is a crucial assumption for the estimation procedure. Intuitively, Eq. (4.2) will be used to provide information about $\pi$, and Eq. (4.1) will be used to derive $\nu$ conditionally on $\pi$.

From a crude empirical point of view, the main intuition of how the data are used is that the two Eqs. (4.1) and (4.2) specify the two unknowns $\nu$ and $\pi$ (the unknown quantity of interest $\pi$ is thought as the *cause* in the modeling perspective) when observing the consequences, *i.e.*, the number of captured and recaptured fish $(y_1, y_3)$. The *likelihood* $[y_1, y_3|\pi; y_2]$ is obtained by assembling the two Eqs. (4.1) and (4.2) that are *conditionally* independent: once taking out the common explaining factors $\nu, \pi$, and $y_2$, the random events of Eq. (4.1) do not share anything with the ones of Eq. (4.2).

$$\begin{aligned}[y_1, y_3|\nu, \pi; y_2] &= [y_1|\nu, \pi] \times [y_3|y_2, \pi] \\ &= \frac{\Gamma(\nu+1)\Gamma(y_2+1)}{\Gamma(\nu-y_1+1)\Gamma(y_1+1)} \\ &\times \frac{\pi^{y_1+y_3}(1-\pi)^{y_1-y_3+\nu-y_1}}{\Gamma(y_2-y_3+1)\Gamma(y_2+1)} \end{aligned} \tag{4.3}$$

### 4.1.3 Prior distributions

To quantify the uncertainty about the unknowns $(\nu, \pi)$, we need the *a priori* distribution $[\nu, \pi]$ (or *prior*) that encodes the knowledge available to the analyst about the unknowns before observing the data. This prior will be updated by Bayes theorem into a posterior, *i.e.*, knowing the data $(y_1, y_3)$. We will assume for convenience a Beta distribution for $\pi$ and an independent uniform distribution for $\nu$ such that:

$$[\nu, \pi] = [\nu] \times [\pi] \tag{4.4}$$

with prior $[\pi]$ and $[\nu]$

$$\begin{cases} \pi \sim Beta(a, b) \\ \nu \sim Uniform(0, \nu_{max}) \end{cases} \tag{4.5}$$

From an ecological expertise, scientists would only say that with regard to food and living space, there is no room for more than

$\nu_{max}$=10,000 smolts in the Oir River. Not to favor any value of $\pi$ we take the hyperparameters $a = 1$ and $b = 1$ so that the prior $[\pi]$ is the uniform distribution.

Specifying the priors completes the definition of the model. Figure 4.4 gives the directed acyclic graph (DAG) for this model.

### 4.1.4 Getting the posterior distribution

Bayes theorem updates the prior knowledge about $\pi$ and $\nu$ into the posterior distribution $[\nu, \pi | y_1, y_3; y_2]$.

$$
\begin{aligned}
[\nu, \pi | y_1, y_3; y_2] &= \frac{[\nu] \times [\pi] \times [y_1 | \nu, \pi] \times [y_3 | y_2, \pi]}{K} \\
&= \frac{[\nu, \pi] \times [y_1, y_3 | \nu, \pi; y_2]}{K}
\end{aligned}
\tag{4.6}
$$

The denominator $K$ is the constant of integration to be calculated as the double integral:

$$
K = \int_{\pi=0}^{1} \sum_{\nu=0}^{\nu_{max}} [\nu, \pi | y_1, y_3; y_2] d\pi
$$

At first look, its denominator is not straightforwardly computed (on the book's website *hbm-for-ecology.org*, a more detailed study is proposed in a practical section corresponding to this chapter). But it is easy to get draws from the posterior distribution of $\nu$ and $\pi$ relying on Markov chain Monte Carlo simulations. In practice, this can be obtained from a very simple BUGS program:

```
♯Model
model
{
π ~ dbeta(1,1)
nu ~ dunif(0, nu_max)
y_1 ~ dbin(nu, π)
y_3 ~ dbin(y_2, π)
}

♯Data
list(y_1 = 767, y_2 = 76, y_3 = 58, nu_max = 10000)
♯Inits
list(π = 0.5, nu = 5000)
```

Figure 4.5 shows the marginal posterior $[\pi|y_1, y_3; y_2]$ and $[\nu|y_1, y_3; y_2]$ as well as a scatter of $(\pi, \nu)$ draws from the joint distribution $[\nu, \pi|y_1, y_3; y_2]$.

Another practical exercise on the website suggests to check how the results might be sensitive to prior specifications of $\nu$, (*e.g.*, the upper bound $\nu_{max}$).

---

## 4.2 Bayesian analysis of juvenile abundance by successive removals

### 4.2.1 Motivating example

**FIGURE 4.6**: An electrofishing team.

The Nivelle River is a well-studied river in France for Atlantic salmon ([97]). It is a coastal river, which flows from the Pyrennées into the Atlantic Ocean near the Spanish border. Although its drainage basin amounts to nearly 200 km$^2$, the riverine habitat colonized by salmon only extend to 25 km of the Nivelle River and its tributaries due to impassable dams. In 2005, the 0+ juvenile production was surveyed in autumn by successive removal sampling via electrofishing. Successive removal by electrofishing is a commonly used method for deriving estimates of abundance of riverine fish such as salmonids ([33]; [130]). At a single site in shallow waters (< 0.5 m), fish are repeatedly removed from a closed population using constant effort on each pass of an electrode. Removal is performed without replacement, *i.e.*, fish caught during a pass are removed and hold so that they are no longer subject to be caught at the next pass. As an example, three successive passes on riffles with an estimated habitat surface of 564 $m^2$ gave the following series of fish catches: $c^1 = 41$, $c^2 = 28$, and $c^3 = 15$. Within the sphere of influence (the electrical field created by a direct current generator), fish involuntarily swim toward the electrode, gets tetanized and captured. Members of the fishing team (one electrode handler and two dip-netters and the bucket carrier[2]) walk upstream and progressively sweep the whole sector of the river (see Fig. 4.6). The anode operator carries a direct current generator (200 W). The site is swept by progressing upstream to surprise the fish. During the experiment, the sampling sites were not closed with barrier nets. However, we will assume that immigration into or emigration out of the sampling site are negligible since the $2^{nd}$ and $3^{rd}$ passes were performed shortly after the first one.

Based on the observed results of this successive removal experiment ($c^1$, $c^2$, and $c^3$), the scientist will seek to estimate the total number of 0+ juveniles which are present in the site.

### 4.2.2 Sampling distributions

In electrofishing removal experiments, fish caught during a pass are removed and stored so that they cannot be caught at the next pass. In addition to the closeness of the population between each pass, we make the following assumptions as in Carle and Strube ([45]):

- For each removal event, all the fish have the same probability of capture and are independent from each other regarding to the capture process.
- The probability of capture remains the same between successive passes.

Based on these hypotheses, we set a binomial model with a constant probability of capture $\pi$ to mimic the removal experiment at each pass. We refer the reader to Mäntyniemi *et al.* ([187]) for an interesting Bayesian model relaxing this hypothesis. Following Peterson *et al.* ([229]), this author suggested that the probability of capture is likely to decrease in successive removals because:

[2]Carrying the bucket is often the only but essential role awarded to the statistician when he gets out of his office to understand how the data are collected on the field.

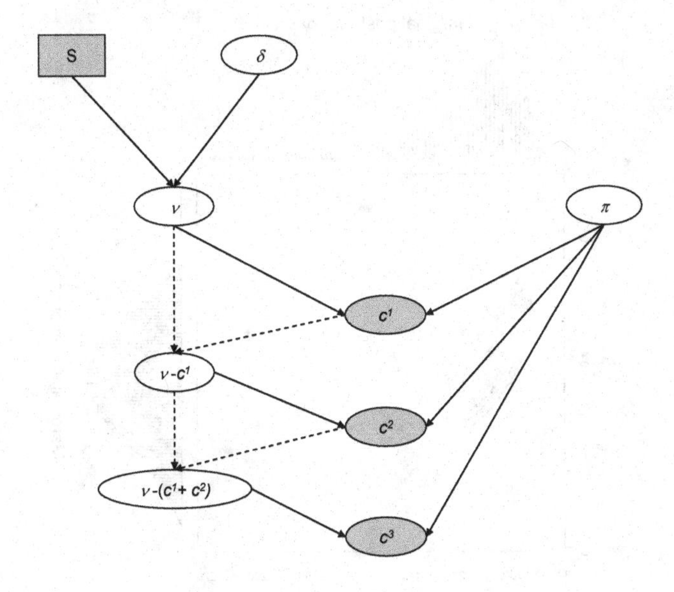


**FIGURE 4.7**: Directed acyclic graph (DAG) for the successive removal model with three passes.

1. fish may exhibit heterogeneous catchability, leading to the most capturable fish being caught first ([93]; [187]);

2. changes in the fish behavior between successive passes and/or a reduced susceptibility to electricity of fish having experienced an electric shock but not been captured ([32]).

Schnute ([271]) proposed a more realistic approach that allows for a decline in the probability of capture between passes. Wang and Loneragan ([309]) developed an over-dispersed model where the probability of capture varies randomly and independently among passes.

Of course, all the violations of a constant $\pi$ hypothesis might be valid but additional sophisticated assumptions and sometimes very tricky *ad-hoc* hypotheses are necessary to develop such nonbinomial models. Making simple assumptions as a first try and checking a posteriori the consistency of the results is our preferred approach. We denote $c_t$ the catch data from the pass $t$ and $\nu$ the unknown population size before the sam-

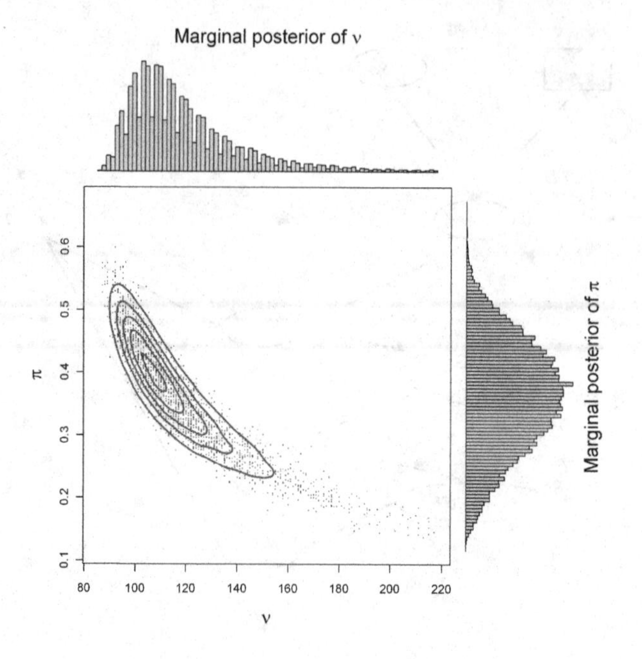


**FIGURE 4.8**: Posterior distributions of $(\pi, \nu)$ from the model depicted in Figure 4.7. The shape of the joint posterior distribution is shown in the central part (joint MCMC draws and smoothed isodensity contours). The marginal distributions are shown in the top and right panels.

pling process began. The sampling distributions of the catch data are:

$$\begin{cases} C^1 & \sim \; Binomial(\nu, \pi) \\ C^2 & \sim \; Binomial(\nu - C^1, \pi) \\ C^3 & \sim \; Binomial(\nu - (C^1 + C^2), \pi) \end{cases} \tag{4.7}$$

### 4.2.3 Prior distributions

We will assume for convenience a Beta distribution for $\pi$ (see Eq. (2.6)) with coefficients $a$ and $b$ related to mean and variance by Eq. (2.7). For instance $a = 6$ and $b = 4$ gives a reasonable prior guess ($\mathbb{E}(\pi) = 0.6$ and $\mathbb{V}(\pi) = 0.15^2$) for the efficiency of common electrofishing experiments in rivers.

We might also assume the population size $\nu$ to be uniformly distributed between 0 and some upper bound $n_{\max}$. As the area $S$ of the sampling site is recorded, we favor an alternative prior model by considering that the population size $\nu$ depends on the expected fish density $\delta$ (fish per m$^2$). We further assume the fish to be randomly distributed in space within an infinitely large surface from which the site with finite surface $S$ is randomly picked. In other words, $\nu$ is hypothesized to be Poisson-distributed with parameter $\delta \times S$:

$$\nu \sim Poisson(\delta \times S) \tag{4.8}$$

The Poisson pdf $dPois(\nu, \lambda)$ of a random variable $\nu$ with parameter $\lambda$ is defined as:

$$dPois(\nu, \lambda) = \frac{(\lambda)^{\nu}}{\Gamma(\nu + 1)} e^{-\lambda} \tag{4.9}$$

We further assign a prior distribution to the fish density $\delta$, via a Gamma distribution with a mean close to 1 fish per 10m$^2$ as currently proposed by biologists and a very large variance to allow for large uncertainty:

$$\delta \sim Gamma(c, d) \tag{4.10}$$

By choosing $d = c \cdot (0.1)^{-1}$, the variance of that Gamma distribution is such that $\mathbb{V}ar(\delta) = \frac{0.01}{c}$ and a sensitivity analysis can be performed by checking various values for $c$. $c = 1$ leads to the exponential distribution, $c$ close to 0 borders up some improper distribution with infinite variance. From a mathematical point of view, if $[\nu|\delta] = dPois(\nu, \delta S)$ and $[\delta|c, d] = dgamma(\delta, c, d)$, $[\nu] = \int\limits_{\delta=0}^{\infty} [\nu|\lambda] \times [\delta|c, d] d\delta$ belongs to the family of negative binomial distributions. One can easily perform the integration to get the explicit expression of this Poisson-gamma convolution (that defines the *negative binomial* pdf):

$$[\nu|S, c, d] = \frac{\Gamma(\nu + 1)}{\Gamma(c)\Gamma(\nu + c)} \left(\frac{d}{d + S}\right)^{c} \left(1 - \frac{d}{d + S}\right)^{\nu} \tag{4.11}$$

with mean and variance

$$\begin{cases} \mathbb{E}(\nu) = \dfrac{c}{d} S \\ \mathbb{V}(\nu) = \dfrac{c(d + S)}{d^2} S \end{cases}$$

For the ease of notations, we will write the pdf given by Eq. (4.11) as $dbinNeg$:

$$[\nu|c, d] = dbinNeg(\nu, S, c, d)$$

Due to conjugate property, the Bayes formula for the Poisson-Gamma model can be written as:

$$dgamma(\delta, c + \nu, d + S) = \frac{dPois(\nu, \lambda S) \times dgamma(\delta, c, d)}{dbinNeg(\nu, S, c, d)}$$

Figure 4.7 gives the corresponding directed acyclic graph (DAG) for the model described by Eqs. (2.6), (4.7), (4.8) and (4.10).

### 4.2.4 Full joint posterior

Updating the prior knowledge about $\pi$ and $\nu$ into a posterior distribution is performed by Bayesian analysis using WinBugs. Figure 4.8 shows the posterior marginals $[\pi|c^1, c^2, c^3]$, and $[\nu|c^1, c^2, c^3]$ as well as a scatter of the joint posterior distribution of $(\pi, \nu|c^1, c^2, c^3)$. A posterior bet about the number of juveniles would be around 110 fish ($\pm 20\%$) but note that the fishing efficiency is not well known; more precise prior knowledge or more data are required. Although they were a priori independent, both quantities become rather correlated when updated through the information conveyed by the data as suggested by the banana-shaped posterior cloud.

---

## 4.3 Testing the efficiency of a new tag for tuna marking

### 4.3.1 Motivating example

Tagging (and recapture) experiments are widely used to study fish biology and to estimate key biological parameters such as migrations, survival and growth. The French Institut de Recherche pour le Développement (IRD) has the scientific responsibility of large scale tagging programs of tropical tuna in the Indian and South Atlantic Oceans. The data used in this section are from Dr. Daniel Gartner, IRD, Sète, France. Conventional "spaghetti"tags used for large tagging campaigns and new tags originally designed for "sport fishing"(Betyp tags) were compared during a tuna tagging program conducted on board Dakar baitboats in 1999. Conventional "spaghetti"tags have a smaller head with only one barb on one side and are generally chosen for tagging large quantities of fish from the boat deck during scientific campaigns. In contrast, Betyp tags have a bigger head with one hook on each side which gives a firmer hold of the tag into the fish. This device is well suited for tagging large sized tunas one at a time directly at sea during sportfishery activities. Both types of tags were placed at the base of the second dorsal fin of the fish in order to firmly attach the barbs of the tag's head into the bones supporting the fin.

The recapture rate of conventionally tagged skipjacks (see Fig. 4.9)

**FIGURE 4.9**: A skipjack tuna.

is well known and has been estimated at $\pi_0 = 0.22$, based on various experiments with more than 4000 tagged individuals. In contrast, only $n = 297$ skipjack tuna were tagged with Betyp tags, and $y = 47$ were recaptured. Can we conclude from that study that Betyp tags have a different (lower?) recapture rate from the conventional spaghetti tags?

## 4.3.2 Translating the question into a Bayesian problem

### 4.3.2.1 Setting a prior pdf for $\pi$

We denote $\pi$ the recapture rate of the new Betyp tags. We seek to compare two alternative prospects ($H_1$ versus $H_2$):

- $H_1$: $\pi = \pi_0$, *i.e.*, the two tags have similar recapture rates.
- $\pi \neq \pi_0$, *i.e.*, the two tags have different recapture rates.

In a Bayesian analysis, these two hypotheses are to be associated with the following prior distributions:

- $H_1$: To formalize the prior assumption $\pi = \pi_0$, the prior distribution for $\pi$ is concentrated around the value $\pi_0 = 0.22$.
- $H_2$: To formalize the prior assumption $\pi \neq \pi_0$, we would assign a prior distribution (for instance a Beta$(a, b)$ distribution) that would reflect our prior beliefs on $\pi$. We postpone the discussion of choosing $(a, b)$ to the end of the section.

As we are indifferent between the two hypotheses $H_1{:}\pi = \pi_0$ and $H_2{:}\pi \neq \pi_0$, we will initially give a weight of $\frac{1}{2}$ to each of them so that the prior assignment of $\pi$ can be written as a mixture of a Dirac and a Beta distribution (using definition (Eq. (2.6)) in proportion 50% – 50%:

$$[\pi] = \frac{1}{2}1_{\pi=\pi_0} + \frac{1}{2}dbeta(\pi, a, b) \tag{4.12}$$

#### 4.3.2.2 Posterior for $\pi$

After observing the result of the binomial recapture experiment for the new Betyp tag ($y = 47$ recaptures (success) out of $n = 297$ marked fish( trials)), the prior for the recapture rate is updated by Bayes rule (see Eq. (2.2)), leading to:

$$[\pi|y, n] = \frac{[\pi] \times [y|\pi; n]}{[y|n]}$$

Let us first focus on the numerator of Eq. (2.2), which is easily obtained through the multiplication of the mixture prior with the Binomial likelihood. Remembering Eq. (2.10) and noting that $dbinom(y, n, \pi_0) = dbinom(y, n, \pi) \times 1_{\pi=\pi_0}$:

$$\begin{aligned}[\pi] \times [y|\pi, n] = & \frac{1}{2}dbinom(y, n, \pi_0) \\ & + \frac{1}{2}dbeta(\pi, a + y, b + n - y) \times dPolya(y, n, a, b)\end{aligned}$$

Let us now focus on the denominator of Eq. (2.2). Since $\int_{\pi=0}^{1} dbeta(\pi, a + y, b + n - y)d\pi = 100\%$:

$$\begin{aligned}[y|n] &= \int_{\pi=0}^{1} [y, \pi|n][\pi]d\pi \\ &= \frac{1}{2}dbinom(y, n, \pi_0) + \frac{1}{2}dPolya(y, n, a, b)\end{aligned}$$

Turning back to the mixture prior to the Betyp recapture rate $\pi$, we obtain as a posterior:

$$\begin{aligned}[\pi|y, n] = & \frac{1}{[y|n]}\{\frac{1}{2}dbinom(y, n, \pi_0) \\ & + \frac{1}{2}dbeta(\pi, a + y, b + n - y) \times dPolya(y, n, a, b)\}\end{aligned} \tag{4.13}$$

Let us denote $\omega(y, n, a, b)$ the weighting factor:

$$\omega(y, n, a, b) = \frac{dbinom(y, n, \pi_0)}{dbinom(y, n, \pi_0) + dPolya(y, n, a, b)}$$

Then the posterior (Eq. (4.13)) is written

$$[\pi|y,n] = \omega(y,n,a,b) \times 1_{\pi=\pi_0} + (1-\omega(y,n,a,b)) \times dbeta(\pi, a+y, b+n-y) \quad (4.14)$$

$\omega(y,n,a,b)$ is the posterior probability that the Betyp tag works the same way as the conventional tagging device. It depends on the values $a, b$ that the fishery scientist would assign to depict the behavior of the new tag, were it different from the old one. A first attempt could be $a=1, b=1$ leading to a uniform distribution that can be considered as poorly informative for which $\omega(y=49, n=297, a=1, b=1) = 0.35$. We therefore are prone to bet in favor of $\pi \neq \pi_0$ in this case.

A more informative guess consists of expecting that the new device works in the vicinity of $\pi_0$, but pessimistically assume that our best guess for the new tag recovery rate $\pi_1$ is 90% of the regular one, *e.g.*, choosing a Beta prior with expected mean $\pi_1 = .9 \times \pi_0 = 0.2$ for the recapture rate of the new device. One degree of freedom remains for setting $(a,b)$. The prior variance $\sigma^2 = \frac{ab}{(a+b)^2(a+b+1)}$ controls the acceptable distance from the old device, and can be set by adjusting $(a+b)$, a quantity that is to be interpreted as a virtual number of prior trials in the context of the Beta-Binomial model. For instance if the prior knowledge is equivalent to $a+b=100$ virtual prior trials, the standard deviation is approximately equal to the difference $\pi_0 - \pi_1$ ($\sigma^2 \approx (.02)^2$, meaning that the prior confidence interval of the expert for $\pi$ does not discard at all $\pi_0$) but the posterior credibility drops from 0.5 to 0.077 in favor of an exactly similar behavior for both tags. Since the choice of the variance seems rather arbitrary, one should wonder about the sensitivity of the result to this tuning parameter.

Figure 4.10 shows the posterior probability that the two tags are identical ($\omega(y,n,a,b)$) as a function of the relative variance for the prior of $\pi_1$, $\sigma/\pi_1$. The U-shape of the curve indicates a strong sensitivity of $\omega(y,n,a,b)$ to the prior variance. When $\sigma$ is small, meaning a strongly peaked prior, a Dirac at $\pi_1$ as a limiting case, the posterior credibility in favor of the hypothesis $\pi = \pi_0$ turns to be $\frac{Binom(n,\pi_0)}{Binom(n,\pi_0)+Binom(n,\pi_1)} = 0.136$ only: although we hypothesize a priori that $\pi = \pi_1 < \pi_0$, the data try to speak partially against this prior assumption by putting weight on $\pi = \pi_0$. If the prior turns to be highly dispersed (as the uniform distribution), not much prior confidence is granted to $\pi$ close to $\pi_1$; the fuzziness around $\pi \neq \pi_0$ for large $\sigma/\pi_1$ blurs the results and prevents the odds in favor of the hypothesis $\pi \neq \pi_0$ from being too strong. Between these two extremes, the posterior credibility in favor of the hypothesis $\pi \neq \pi_0$ passes through a maximum: the prior knowledge equivalent to 100 virtual additional tag trials allows for some uncertainty to get away

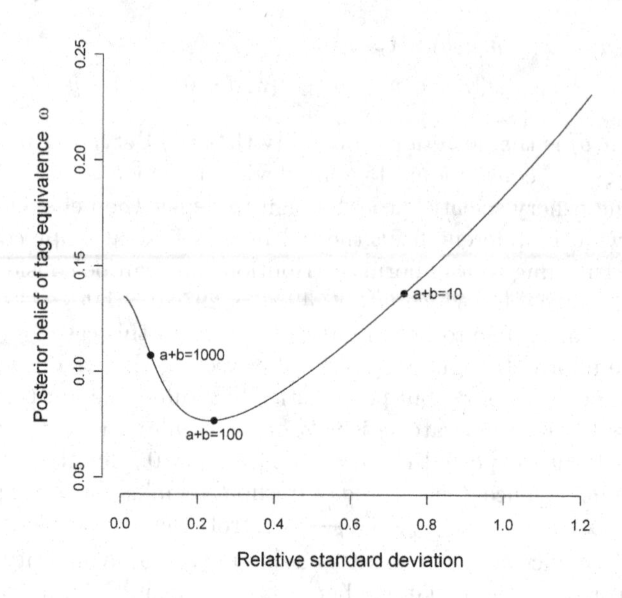


**FIGURE 4.10**: Posterior probability that the two tags are identical depending on the prior coefficient of variation $\sigma/\pi_-$ in the case $n = 297, y = 47$.

from $\pi = \pi_0$ and let the data strongly speak in favor of $\pi \neq \pi_0$ (with a weight of 92%). As a conclusion, the graph shows that even in the range of very pessimistic priors for $\pi$, the data provide strong evidence that the Betyp tag works very differently from the conventional tag.

### 4.3.3 Hypotheses testing via Bayes Factors

Indeed, the recapture rate $\pi_0$ for conventional tags is not exactly known. In this section $\pi_0$ is a random variable. We know that the following recapture rates have been observed during the previous tagging trips with at least ten fish tagged by trip: $0.245, 0.245, 0.157, 0.103, 0.323$ and that $n_0 = 1307$ individuals have been tagged with conventional tags during the comparison experiment, among which $y_0 = 249$ were recaptured. A model for $\pi_0$ would be elaborated first by fitting a beta prior

$(a_0, b_0)$ on the past recaptured rates by the method of moments:

$$\begin{cases} \dfrac{a_0}{a_0 + b_0} = 0.215 \\ \dfrac{a_0 b_0}{(a_0 + b_0)^2 (a_0 + b_0 + 1)} = 0.0073 \end{cases}$$

leading to $a_0 = 4.7$ and $b_0 = 17.2$ and then updating this prior by using the Beta-Binomial conjugate property:

$$[\pi_0] \sim Beta(a_0 + y_0, b_0 + n_0 - y_0)$$

In this setting, we wonder for the Betyp tag experiment on skipjacks with $n = 297$ tagged and $y = 47$ recaptured about which model $M$ is to be considered. Are the data stemming:

- from model $M = M_0$, *i.e.*, a binomial sampling distribution with probability $\pi = \pi_0$, or
- from model $M = M_1$, $\pi = \pi_1 \neq \pi_0$? Note that for model $M_1$, a prior Beta$(a_1, b_1)$ distribution can also be taken for $\pi$.

From a Bayesian point of view, the model itself is to be considered as a parameter to be estimated and such a question can be answered by calculating the Bayes Factor of $M_0$ versus $M_1$, which is nothing more than the odd ratio of the models $M_0$ versus $M_1$.

#### 4.3.3.1 Bayes Factors

More generally, suppose that several competing model structures, $M_i$ with $i = 1, ..., k$, can be proposed with their own set of parameters. Assuming a prior distribution over models structure $[M_i]$ (this prior relates only to the structure of model $M_i$, not to its parameters; $\sum_{m=1}^{m=k} [M_m] = 1$) and a prior distribution $[\pi | M_i]$ on parameters $\pi$ for model $M_i$, Bayes' theorem provides the *a posteriori* distribution for model $M_i$ given the data $\mathbf{y}$:

$$[M_i | \mathbf{y}] = \frac{[\mathbf{y} | M_i] \times [M_i]}{[\mathbf{y}]} \tag{4.15}$$

The first term of the numerator in Eq. (4.15), $[\mathbf{y} | M_i]$, is the marginal likelihood in model $M_i$. It is also called the *predictive* for model $M_i$ as it could be used to predict new data from this model. $[\mathbf{y} | M_i]$ can be rewritten as the integration of the likelihood $[\mathbf{y} | \pi, M_i]$ against the prior $[\pi | M_i]$ :

$$[\mathbf{y} | M_i] = \int_\pi [\mathbf{y}, \pi | M_i] d\pi = \int_\pi [\mathbf{y} | \pi, M_i] \times [\pi | M_i] d\pi$$

The denominator in Eq. (4.15) is obtained by summation over all the competing models:

$$[\mathbf{y}] = \sum_{m=1}^{k} [\mathbf{y}|M_m] \times [M_m]$$

Comparing posterior distributions of the models is useful to find out the most probable model structure among models $i = 1, ..., k$. Comparing two models $M_i$ and $M_j$ is equivalent to computing the Bayes Factor $B_{i,j}$, *i.e.*, ratios of marginal likelihoods between models $M_i$ and $M_j$, corrected by the ratio of priors since:

$$B_{i,j} = \frac{[M_i|\mathbf{y}]}{[M_j|\mathbf{y}]} = \frac{[\mathbf{y}|M_i]}{[\mathbf{y}|M_j]} \times \frac{[M_i]}{[M_j]} \tag{4.16}$$

The Bayes Factor $B_{i,j}$ evaluates the relative increase of evidence (from prior to posterior, given the data) in favor of model $M_i$ over model $M_j$. When the two competing models are given the same a priori weight, *i.e.*, $[M_i] = [M_j]$, then the Bayes Factor $B_{i,j}$ directly gives a comparison between the posterior credibilities of each model.

#### 4.3.3.2 Model selection for the tuna marking experiment

Let us now return to the tuna tagging experiment. The Bayes Factor $B_{0,1}$ can be calculated to compare the two hypotheses $M = M_0$ and $M = M_1$. Let us further assume that both models $M_0$ and $M_1$ are equiprobable a priori, that is $[M = M_0] = [M = M_1] = \frac{1}{2}$. The Bayes Factor $B_{01}$ directly gives a comparison between the posterior credibilities of each model:

$$\frac{[y|M_0]}{[y|M_1]} = \frac{[M_0|y]}{[M_1|y]} \times \frac{[M_1]}{[M_0]}$$

The Beta-Binomial model is a favorable case to compute the Bayes factor as the conjugate property allows for closed-form expressions of the marginal likelihoods $[y|M_0]$ and $[y|M_1]$ :

$$\begin{cases} [y|M_0] = dPolya(y, n, a_0 + y_0, b_0 + n_0 - y_0) \\ [y|M_1] = dPolya(y, n, a_1, b_1) \end{cases}$$

Hence

$$\begin{aligned} B_{01} &= \frac{dPolya(y, n, a_0 + y_0, b_0 + n_0 - y_0)}{dPolya(y, n, a_1, b_1)} \times \frac{\frac{1}{2}}{\frac{1}{2}} \\ &= \frac{\Gamma(a_0 + y_0 + y)\Gamma(b_0 + n_0 - y_0 + n - y)\Gamma(a_0 + y_0 + b_0 + n_0 - y_0)}{\Gamma(a_0 + y_0)\Gamma(b_0 + n_0 - y_0)\Gamma(a_0 + y_0 + b_0 + n_0 - y_0 + n)} \\ &\times \frac{\Gamma(a_1)\Gamma(b_1)\Gamma(a_1 + b_1 + n)}{\Gamma(a_1 + y)\Gamma(b_1 + n - y)\Gamma(a_1 + b_1)} \end{aligned}$$

For a uniform prior ($a_1 = 1, b_1 = 1$), one finds $B_{01} = 7.12$. Thus, $[M_0|y] = \frac{B_{01}}{1+B_{01}} = 0.88$ and $[M_1|y] = 0.12$. Not surprisingly, widening the prior location possibilities for $\pi_0$ unduly reinforces the credibility of the hypothesis $\pi = \pi_0$ as opposed to $\pi \neq \pi_0$ and $\pi$ uniformly distributed. The prior mean $\frac{1}{2}$ is far from what the data show ($\frac{y}{n} = 0.158$) and in this noninformative case we would bet for $\pi = \pi_0$ as a default choice. As previously, one has to study the sensitivity of the posterior probability of $M_0$ (the two tags are identical) to the various possible values for $(a_1, b_1)$.

## 4.4 Further references

Within the Bayesian framework, selecting a model is obtained by a simple extension of the Bayes Theorem to estimate the posterior weight of one additional discrete parameter, the model's label. In the conjugate situation, computing this weight is easy. Indeed, when the likelihood $[y|\pi]$ allows for conjugate prior $[\pi]$ and posterior $[\pi|y]$, the predictive $[y]$ is explicitly obtained in closed form since

$$[y] = \frac{[y|\pi] \times [\pi]}{[\pi|y]}$$

This conjugate property is exemplified in this chapter by the beta-binomial structure, and it is also the case of the Normal-Gamma models of Chapters 3 and 6. For more general structures, getting each competing model's weight is more challenging, because one has to compute the denominator of the Bayes rule, *i.e.*, a possibly intractable multi-dimensionnal integral that we were happy to avoid evaluating when implementing MCMC techniques. Some common importance sampling estimates are described in Appendix B. The interested reader is invited to refer to [46] and [157] for an introduction to Bayes Factors and model selection. Chapter 6 of King *et al.* ([164]) details the most popular methods to compute posterior model probabilities. More technical papers explaining how to evaluate the marginal likelihood for the purpose of Bayesian model comparisons are found in Chib ([55]) and Chib ([57]). Burnham and Anderson ([43]), Johnson and Omland ([148]) or Ward ([310]) give interesting overviews of methods for multimodel inferences in ecology.

# Chapter 5

## Combining various sources of information to estimate the size of salmon spawning run


### Summary

Evaluating the spawners abundance is a major issue to model the dynamics of wild populations. Due to the small number of records or few repetitions as well as many uncontrolled conditions, estimates of population sizes and related demographic parameters are blurred with uncertainty. Conventional statistical analyses can lead to a misrepresentation of this uncertainty while they more often than not ignore additional information readily available to the modeler. This chapter develops a Bayesian analysis to take full advantage of all the information carried by the data collected in the field and by the expertise derived from the fishery manager.

This chapter details a model to estimate the size of A. salmon spawning run in the Scorff River. We point out that:

1. A directed oriented graph is convenient to model the various events that may occur in a salmon population. This graphical representation is based on conditional reasoning and gives a formal representation of the way that three types of quantities interact: Observed variables, latent variables and population parameters. Latent variables represent unobserved or hidden quantities with physical meaning which complement the observed data (systematically measured).

2. Much expertise about the local ecosystem and fishery is brought into the analysis via informative priors.

3. As expected, the dispersions of the unknown quantities are greatly reduced when additional information, such as inter-annual data, is brought into the analysis.

This chapter gives a flavor of the hierarchical models that will be fully developed in Part 2. It is the first attempt to assemble elementary bricks (Beta-binomial submodels describing the successive events during A. salmon spawning run in the Scorff River) in order to design a rather complex structure with many parameters and latent variables. Attention is recentered on the realism of the model to mimic the fish behavior as a probability tree. Bayesian analysis offers a coherent deductive framework to study the various sources of uncertainty. Ecological inferences here are grounded on various kinds of observed data and on local expertise.

## 5.1 Motivating example

Knowing how many adult salmon return to their native river to spawn (see Fig. 5.1) is a major concern for both research programs on population dynamics and stock assessment work for resource management advice ([238]). Both scientists and managers need not only point estimates of the population size (for instance the most likely value) but also a description of their uncertainty. Precision of the population size measurements is crucial to critically assess the reliability of scientific knowledge gained or to propose management decisions based on a precautionary approach.

**FIGURE 5.1**: A spawner swimming back to its native river.

In a natural, (*i.e.*, uncontrolled) environment, only statistical knowledge can be obtained to assess fish stock size. Several statistical techniques are commonly used to estimate salmonids population sizes through well-established *ad hoc* experimental design schemes: capture-mark-recapture methods, successive removals and so on (see [69] or [278] for a review of standard methods and models). Unfortunately many stock size assessment problems do not conform to such standard academic models; founding hypotheses are violated, conditions required to derive estimates do not match the data collected, all the information available is not used. The mark-recapture experiment of the Scorff River presented in the following section illustrates these problems. The hypothesis of a closed population is violated, as unexpected losses occur before marked individuals are recaptured. The conditions to use estimates of variance based on asymptotic properties are not met as only very few marks are recaptured. In addition to the numbers of fish marked and recaptured, information is available through local expertise about the ecosystem and the fishery. In the scientific literature, past years of observations, auxiliary information collected in the field by technicians or fishermen (reporting of dead fish, catch declarations, etc.) are available. Yet, traditional standard estimation procedures often ignore these information sources and treat every yearly estimation as if it were the first one ever done.

### 5.1.1 The dangerous run of spawners

After spending 14 months to 3 years in the Atlantic Ocean, the adult salmon come back to their native river to spawn (see Fig. 5.1). A complete description of the life cycle has been detailed in Chapter 1. On the Scorff River, a management oriented research program is carried out; the population abundance is assessed at several stages in the life cycle. The adult returns are quantified by means of mark-recapture experiments. Marking is done at a trapping facility located at the mouth of the river. Estimation of the returns is conducted separately for the two main age classes in the returns, grilse and spring salmons (which have spent 14 months and 2 or 3 years in the sea, respectively). The case study presented here deals with the grilse run only. Fig. 5.2 sketches the fate of a salmon entering back its originating river when returning after its trip in the Atlantic Ocean. Three main events are likely to occur to the candidate spawner:

1. At their entry into the Scorff River, the salmon may be trapped, marked and released as a first step of a standard stock estimation procedure. Some of them escape from the trap and continue their upstream migration.

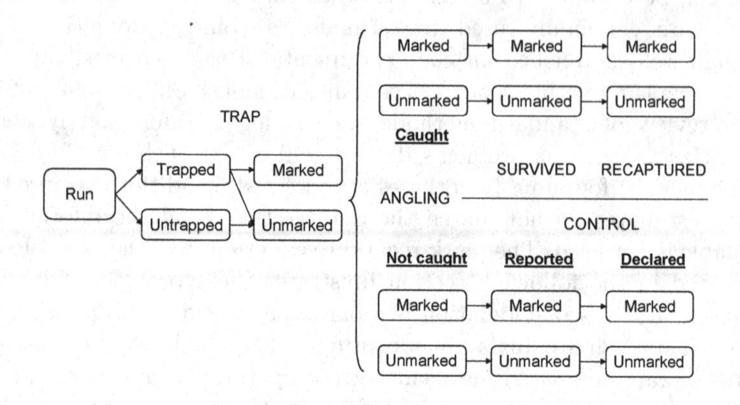


**FIGURE 5.2**: Path diagram showing what the future holds in store for an A. salmon spawner entering the Scorff River.

2. A portion including both marked and unmarked individuals will be removed by the anglers. French law requires that salmon catches be officially declared but this commitment is not always respected. An additional local survey is used to gather further information. Crossing these two sources allow for a first evaluation of the number of salmon "caught for sure." Among the fish known to be caught, some are brought to the research technicians for mark identification.

3. The fish that escaped the angling fishery will have to survive until the spawning season (around December). During spawning time, scientists go to the field and complete the statistical survey by a recapture phase. Some fish (marked or unmarked) are caught during these recapture surveys.

### 5.1.2 Observed variables

The data presented in Table 5.1 consist of six years (in rows) and of six variables (columns). The first year of data (1994) is known to be significantly different from the others. The trap efficiency and the recapture effort during the spawning period were much lower.

The observed variables carry the following information:

- $Y_1$ gives the number of individuals trapped, marked and released;
- $Y_2$ and $Y_3$ are the number of respectively marked and unmarked fish removed by anglers and presented for marking detection;
- $Y_4$ reports the total (marked+unmarked) removed for sure (consequently $Y_4 > Y_2 + Y_3$), as not all fish caught by anglers are presented for mark detection;
- $Y_5$ and $Y_6$ present respectively the marked and unmarked fish that survived and were recaptured during or after spawning.

| Year | $Y_1$ | $Y_2$ | $Y_3$ | $Y_4$ | $Y_5$ | $Y_6$ |
|---|---|---|---|---|---|---|
| 1994 | 156 | 3 | 14 | 42 | 4 | 14 |
| 1995 | 500 | 39 | 10 | 75 | 31 | 28 |
| 1996 | 502 | 25 | 8 | 87 | 45 | 14 |
| 1997 | 320 | 17 | 7 | 33 | 19 | 9 |
| 1998 | 442 | 50 | 5 | 66 | 56 | 13 |
| 1999 | 167 | 16 | 4 | 24 | 16 | 11 |

**TABLE 5.1**: Data available to estimate spawning run in the Scorff River. Only 1SW fish (grilse) are considered. Data are reproduced with permission of Etienne Prevost, INRA, St.-Pée sur Nivelle, France.

## 5.2 Stochastic model for salmon behavior

In this section, we show how parameters, latent variables and observations can be articulated through conditional reasoning to build a Bayesian model for salmon spawning run. The model is first built for one year. Then, in Section 5.2.7 several years are combined within the same model.

### 5.2.1 Bernoulli trials transfer individual behavior into population features

On the basis of Fig. 5.2, a model should mimic the mechanism yielding the data at hand. The observables $Y_1, Y_2, \ldots, Y_6$ are not directly related to individuals; they should be considered as observable quantities concerning the whole population of salmon in the Scorff River.

The observables were modeled as probabilistic issues of Binomial experiments. The underlying hypothesis is that they are probabilistic sums of Bernoulli trials sketching the behavior of each individual, all individuals being independent and sharing the same probability for the Bernoulli event (see Chapter 2 and Eq. (2.4) in this chapter for more details about the Binomial distribution). For instance, hypothesizing that spawners are trapped with a stationary probability $\theta$, $Y_1$ can be considered as a Binomial random variable. More precisely, denoting $\kappa$ the unknown size of the spawning run, $Y_1$ is the sum of $\kappa$ independent Bernoulli trials with coefficient $\theta$.

### 5.2.2 Parameters are assumed stationary

During the modeling task, unknown quantities are to be introduced to describe the stochastic features that govern the behavior of an individual. Such quantities are assumed to remain stationary from one fish to another one. The following technical parameters are unknown but conceptually essential to the methodologist:

- $\kappa$ : the number of spawners to be swimming upstream;
- $\theta$ : probability that a spawner be trapped and marked at the trapping facility;
- $\beta$ : probability that a spawner be removed by anglers;
- $\tau$ : probability that a salmon caught by anglers be reported as a "sure removal";
- $\delta$ : probability that a reported salmon caught by anglers be declared and checked for previous marking by research technicians;
- $\alpha$ : natural salmon survival probability until the reproduction period;
- $\pi$ : probability that a spawner be recaptured during or after the reproduction period.

### 5.2.3 Prior expertise is available for the parameters

Unknown parameters are often "not as unknown" as they seem to be at first glance; even without data, all subsets of possible values do not bear the same prior belief. It is generally not justified to rely on flat priors in stock size assessment, (*i.e.*, noninformative equidistributed bets) since practitioners always have some knowledge about the population features under study (see [135], [198], [241] for arguments about the use of informative vs. noninformative priors). For the Scorff River, prior knowledge (hereafter noted $\underline{K}$) can be summed up as follows:

1. Considering the size of the river, the earlier data on juvenile production ([15]) in the Scorff River and ranges of sea survival ([77]), experts would bet with 9 odds against 1 that the number of spawners entering the Scorff ($\kappa$) stands in the interval $[100, 3000]$ with highly plausible values around 700 individuals.

2. Little is known about the trapping probability $\theta$ at the experimental facility: one would imagine a symmetric repartition with 0.5 as a median and only a 10% chance to be less than 0.1 or greater than 0.9.

3. The first guess for salmon survival rate $\alpha$ in river is above 0.9. An expert would even accept to bet up to 9 against 1 that $\alpha$ is greater than 0.75.

4. The angling exploitation rate $\beta$ is considered to be most likely between 0.1 and 0.3. It is hardly credible (less than 10% chance) that $\beta$ stands above 0.7.

5. The most likely $\tau$, the probability that a salmon catch is known "for sure" is above 0.9 and it seems highly unlikely (5%) that it is below 0.5.

6. Little is known about the probability $\delta$ that a salmon known to be caught is presented for mark control. A symmetric repartition with 0.5 as a median and only a 10% chance to be less than 0.1 or greater than 0.9 would reflect this weak prior knowledge.

7. Considering the number of spawning sites covered and the survey effort during the recapture events, the probability of recapture $\pi$ is most likely below 0.25, unlikely between 0.25 and 0.5 and almost impossible above 0.5. In what follows, "most likely" was interpreted as a bet with 9 odds against 1, "almost impossible" was quantified as less than 1% chance and the remaining probability (around 9%) was used to assess the weight of "unlikely" values.

### 5.2.4 Prior elicitation

Figure 5.3 sketches an acceptable discrete pdf to represent prior belief about $\kappa$ given the expertise $\underline{K}$. This curve has been obtained by a (discrete) Gamma function in Eq. (5.1) with coefficients 2.4 and 1/500, truncated to the interval $[0, 4000]$. The truncation below 4000 is for computational convenience, but a sensitivity analysis shows that it is largely justified. As required when encoding the expertise, this curve exhibits a maximum value around 700 and puts 90% weight on the interval [100, 3000].

$$[\kappa \,|\underline{\mathrm{K}}] = \frac{\kappa^{2.4-1}\exp(-\frac{\kappa}{500})}{\sum\limits_{z=0}^{z=4000} z^{2.4-1}\exp(-\frac{z}{500})} \tag{5.1}$$

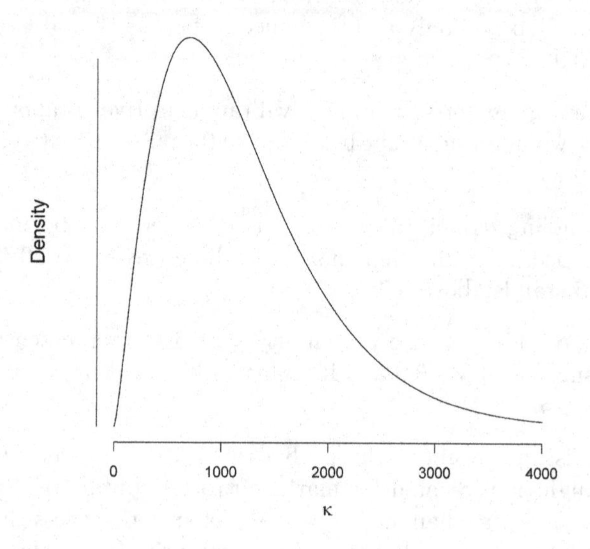


**FIGURE 5.3**: Informative Gamma Prior pdf with parameters 2.4 and 500 (shape and scale parameters, respectively) for the stock size $\kappa$, the mode is around 700 and 90% of the density stands between 100 and 3000.

The six other parameters $\theta, \alpha, \beta, \tau, \delta, \pi$ are probabilities in Bernoulli trials, therefore belonging to the interval [0, 1]. The Beta distribution (see Eq. (2.7) in Section 2) with two fitting coefficients $a_{\underline{\mathrm{K}}}$ and $b_{\underline{\mathrm{K}}}$ already introduced in Chapter 2 can mimic a wide variety of behaviors for a random quantity between 0 and 1. Figure 5.4 and Table 5.2 show the results of the elicitation of Beta($a_{\underline{\mathrm{K}}}$,$b_{\underline{\mathrm{K}}}$) pdf to encode prior expertise for the various technical parameters. The mode of this function occurs for $\frac{a_{\underline{\mathrm{K}}}-1}{a_{\underline{\mathrm{K}}}+b_{\underline{\mathrm{K}}}-2}$, which gives a first linear relation between $a_{\underline{\mathrm{K}}}$ and $b_{\underline{\mathrm{K}}}$ when the expert indicates the parameter most probable value. If a credible interval is given for the prior pdf, the parameters are then adjusted so as to match the probability statement of the expertise by a (unidimensional) procedure of trials and errors.

| Parameter | $\underset{\sim}{K}$={prior expertise} | $a_K$ | $b_K$ |
|---|---|---|---|
| $\theta$ (trapping efficiency) | $p(\theta \mid \underset{\sim}{K})$ sym. | 1.53 | 1.53 |
| | $p(\theta \in [0.1, 0.9] \mid \underset{\sim}{K}) = 0.9$ | | |
| $\alpha$ (survival rate) | $Mode(\alpha \mid \underset{\sim}{K}) \approx 0.9$ | 10 | 1.5 |
| | $p(\alpha > 0.75 \mid \underset{\sim}{K}) = 0.9$ | | |
| $\beta$ (angler efficiency) | $Mode(\beta \mid \underset{\sim}{K}) \approx 0.2$ | 1.3 | 2.2 |
| | $p(\beta > 0.7 \mid \underset{\sim}{K}) \leq 0.1$ | | |
| $\tau$ (prob. of reporting) | $Mode(\tau \mid \underset{\sim}{K}) \approx 0.9$ | 5.5 | 1.5 |
| | $p(\tau < 0.5 \mid \underset{\sim}{K}) = 0.05$ | | |
| $\delta$ (prob. of declaring) | $p(\delta \mid \underset{\sim}{K})$ sym. | 1.53 | 1.53 |
| | $p(\delta \in [0.1, 0.9] \mid \underset{\sim}{K}) = 0.9$ | | |
| $\pi$ (recapture prob.) | $Mode(\pi \mid \underset{\sim}{K}) \approx 0.2$ | 1.6 | 11 |
| | $p(\pi \leq 0.25 \mid \underset{\sim}{K}) = 0.9$ | | |
| | $p(0.25 \leq \pi \leq 0.5 \mid \underset{\sim}{K}) = 0.09$ | | |
| | $p(\pi > 0.5 \mid \underset{\sim}{K}) = 0.01$ | | |

**TABLE 5.2**: Informative Beta distributions for parameters $(\theta, \alpha, \beta, \tau, \delta, \pi)$.

As prior knowledge of each parameter is established independently the joint prior is the product of all univariate priors:

$$[\kappa, \theta, \alpha, \beta, \tau, \delta, \pi \mid \underset{\sim}{K}] = [\kappa \mid \underset{\sim}{K}] \times [\theta \mid \underset{\sim}{K}] \times [\alpha \mid \underset{\sim}{K}] \times [\beta \mid \underset{\sim}{K}] \times [\tau \mid \underset{\sim}{K}] \times [\delta \mid \underset{\sim}{K}] \times [\pi \mid \underset{\sim}{K}] \quad (5.2)$$

### 5.2.5 Introducing latent variables

Unknown parameters and observed variables are not enough to describe the wanderings of salmons. Latent variables, *i.e.*, intermediate quantities related to unobserved variables with a physical meaning, are thus introduced (see also Eq. (1.18) in Chapter 1 for a general introduction of hierarchical modeling using latent variables). They are useful to help understand the intermediate steps of the conditional modeling scheme and will not induce any additional burden for the inference task. Of course, the model should remain well defined: conditional distributions of latent variables given the parameters and the observable variables must be fully specified. The following latent variables are introduced:

- $Z_{uu}$: salmon untrapped, therefore unmarked;

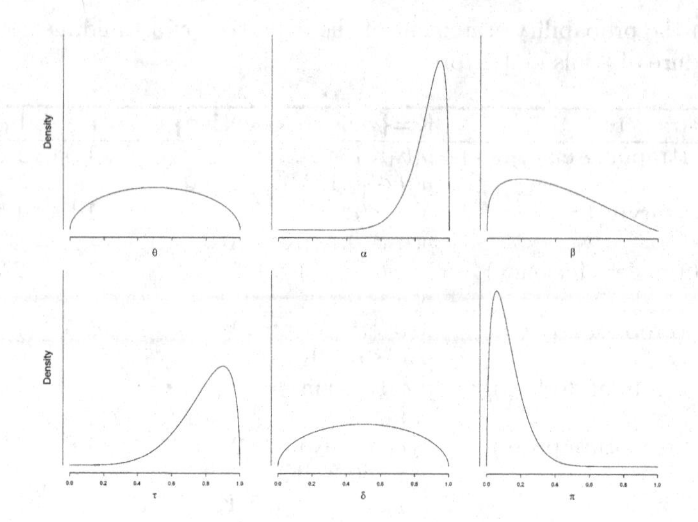


**FIGURE 5.4**: Informative Beta prior pdf for technical parameters $(\theta, \alpha, \beta, \tau, \delta, \pi)$. The characteristics of each prior are given in Table 5.2.

- $Z_{mc}$ and $Z_{uc}$: respectively marked and unmarked individuals caught by anglers;
- $Z_{mf}$ and $Z_{uf}$: respectively marked and unmarked individuals remaining free during the fishing period;
- $Z_{mr}$ and $Z_{ur}$: respectively marked and unmarked salmon being reported as sure removals;
- $Z_{ms}$ and $Z_{us}$: respectively marked and unmarked spawners surviving before reproduction.

Some latent variables combinations are important for scientific reporting. For instance, both scientists and fish managers would like to assess the range of credible values for $Z_{mc} + Z_{uc}$, the total number of salmon caught by anglers. As an other example $Z_{ms} + Z_{us}$, which represents the escapement, appears as a key value to assess the spawning stock status.

### 5.2.6 Stochastic model as a directed acyclic graph

Latent and observed variables and parameters are combined into a stochastic model. The model equations include logical deterministic balance equations and stochastic ones. They read as follows:

$$
\left\{\begin{array}{l}
Z_{uu} = \kappa - Y_1 \\
Z_{mc} \sim Binomial(Y_1, \beta), \ Z_{mf} = Y_1 - Z_{mc} \\
Z_{ms} \sim Binomial(Z_{mf}, \alpha) \\
Z_{us} \sim Binomial(Z_{uf}, \alpha) \\
Z_{uc} \sim Binomial(Z_{uu}, \beta), \ Z_{uf} = Z_{uu} - Z_{uc} \\
Z_{ur} \sim Binomial(Z_{uc}, \tau) \\
Z_{mr} \sim Binomial(Z_{mc}, \tau) \\
\\
Y_1 \sim Binomial(\kappa, \theta) \\
Y_2 \sim Binomial(Z_{mr}, \delta) \\
Y_3 \sim Binomial(Z_{ur}, \delta) \\
Y_4 = Z_{ur} + Z_{mr} \\
Y_5 \sim Binomial(Z_{ms}, \pi) \\
Y_6 \sim Binomial(Z_{us}, \pi)
\end{array}\right. \tag{5.3}
$$

Equations 5.3 allow for the construction of the directed graphical model given in Fig. 5.5. Conditional reasoning with independence assumptions makes the task of getting this posterior pdf easier when using MCMC simulation algorithms such as the Gibbs sampler. As an example, the special structure of the graph points out that $Z_{mf}, Y_1, \alpha, \kappa$ and $\theta$ do not matter when expressing the conditional pdf of $Y_5$:

$$[Y_5 \,|Z_{ms} \,, Z_{mf}, Y_1, \alpha, \pi, \kappa, \theta] = [Y_5 \,|Z_{ms} \,, \pi]$$

### 5.2.7 The interannual model is designed by "piling up slices" of annual models

Figure 5.6 shows that no additional conceptual difficulty arises when designing an interannual model. It must be underlined that a very strong hypothesis of stationarity for basic parameters is made, allowing an interannual coherence (and thus some transfer of information from year to year) by sharing common values for $\theta, \alpha, \beta, \tau, \delta$ and $\pi$. This assumption is especially questionable for the trapping efficiency $\theta$ and the recapture probability $\pi$ that can vary from year to year depending on the river flow and the hydrometeorological conditions.

As the first year of data (1994) is significantly different from the five other ones (with very low trapping efficiency and recapture effort), it was not included in the interannual model.

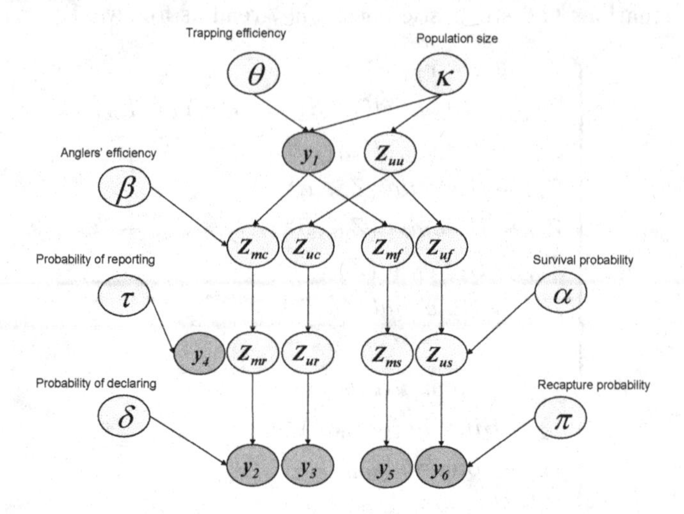


**FIGURE 5.5**: Directed graphical representation for the model describing the life of a spawner in the Scorff River.

## 5.3 Inference with WinBUGS

Inference consists in updating the prior parameter pdf $[\kappa, \theta, \alpha, \beta, \tau, \delta, \pi \,|\underline{\mathrm{K}}]$, (where $bK$ stands for conditioning by the initial knowledge) into the posterior pdf $[\kappa, \theta, \alpha, \beta, \tau, \delta, \pi \,|\underline{\mathrm{K}}, Y_1, Y_2, \ldots, Y_6]$ by taking into account the observations $(Y_1, Y_2, \ldots, Y_6)$. As a multivariate joint conditional pdf, this expression is untractable: even though it can be formally derived from the model (Eqs. 5.3) and the prior pdf $[\kappa, \theta, \alpha, \beta, \tau, \delta, \pi]$ given in Table 5.2, it involves multidimensional normalizing integrals that come from integrating out latent variables and denominators of Bayes formula. Here we rely on WinBUGS programs to perform the Bayesian inference of the model. WinBUGS takes advantage of the model structure in Fig. 5.5 to simplify MCMC sampling via the Gibbs algorithm ([118]; see also [49], for a review). As an example, we simply would like to point out that the Gibbs sampler will be favored for many nodes of Fig. 5.5 that involve only Beta and Binomial distributions. As an example, the prior for $\pi$ is a Beta$(a_{\underline{\mathrm{K}}}, b_{\underline{\mathrm{K}}})$ pdf, with $a_{\underline{\mathrm{K}}} = 1.6$ and $b_{\underline{\mathrm{K}}} = 11$. $Y_5, Y_6$ are Binomial variables with probability $\pi$ and associated number of trials $Z_{ms}$ and $Z_{us}$, respectively. Thus, the likelihood $p(Y_5 = y_5, Y_6 = y_6 \,|\pi, Z_{ms}, Z_{us})$

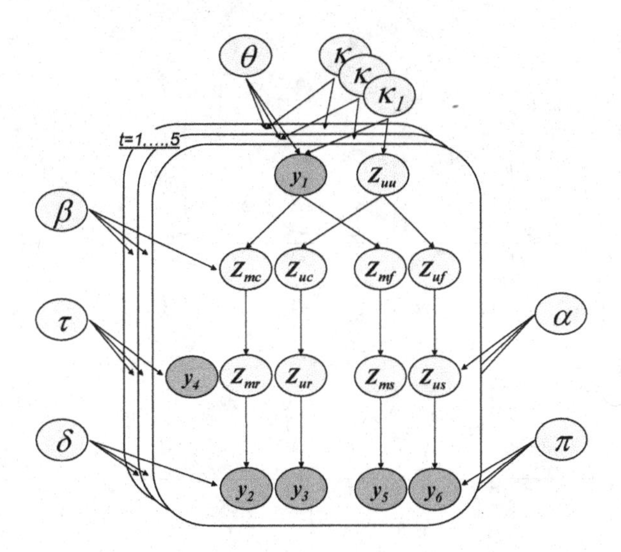


**FIGURE 5.6**: Directed acyclic graphical representation for the inter-annual model describing several years of adult returns in the Scorff River with the same behavior. All parameters except the $\kappa$'s are shared between years.

writes:

$$\begin{aligned} p(Y_5 = y_5, Y_6 = y_6 | \pi, Z_{ms}, Z_{us}) \\ = \Gamma(Z_{ms} + Z_{us} + 1) \times \frac{\pi^{y_5+y_6}(1-\pi)^{Z_{ms}+Z_{us}-y_5-y_6}}{\Gamma(y_5+y_6+1)\Gamma(Z_{ms}+Z_{us}-y_5-y_6+1)} \end{aligned}$$

From Bayes theorem, the full posterior conditional of $\pi$ can be expressed as:

$$\begin{aligned} p(\pi | \underline{K}, Y_5 = y_5, Y_6 = y_6, Z_{ms}, Z_{us}) \\ \propto p(Y_5 = y_5, Y_6 = y_6 | \pi, Z_{ms}, Z_{us}) \times p(\pi | \underline{K}) \\ \propto \pi^{y_5+y_6+a_{\underline{K}}-1} \times (1-\pi)^{Z_{ms}+Z_{us}-y_5-y_6+b_{\underline{K}}-1} \end{aligned}$$

which is a Beta distribution with updated parameters $y_5 + y_6 + a_{\underline{K}}$ and $Z_{ms} + Z_{us} - y_5 - y_6 + b_{\underline{K}}$. When nonexplicit full conditionals are encountered in this problem (consider for instance the node $\kappa$), WinBUGS makes recourse to Metropolis-Hastings techniques.

The Bayesian approach treats latent variables as other parameters. Their full conditional distributions are evaluated as well. Consequently

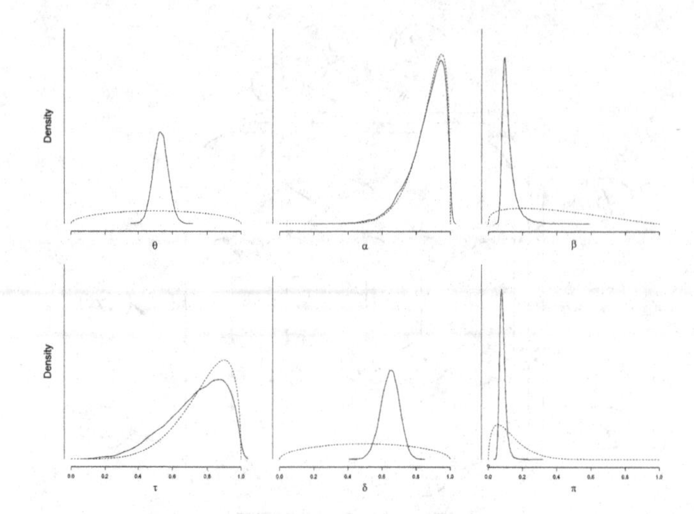


**FIGURE 5.7**: Bayesian inference for basic parameters conditioned upon year 1995 dataset. Dotted line: prior; solid line: posterior.

WinBUGS will provide a sample from the joint distribution:

$$[\kappa, \theta, \alpha, \beta, \tau, \delta, \pi, Z_\star \,|\underline{\mathrm{K}}, y_{1:5}]$$

with

$$Z_\star = (Z_{uu}, Z_{mc}, Z_{uc}, Z_{mf}, Z_{uf}, Z_{mr}, Z_{ur}, Z_{ms}, Z_{us})$$

From this sample, one will simply extract the marginal posterior distribution related to quantities of interest (see Table 5.3).

All results presented below were obtained from a MCMC sample of size 1000 after a 5000 burn-in period. Only one sample over five was kept so as to get rid of the autocorrelation between Gibbs sampler iterations.

---

## 5.4 Results

### 5.4.1 Estimation for the year 1995

Empirical posterior pdf estimates given in Figs. 5.7 and 5.8, and 95% credible intervals in Table 5.3 are directly obtained based on a MCMC sample generated by WinBUGS.

| Parameter | Mean | Sd | 2.5% pct. | 97.5% pct. |
|---|---|---|---|---|
| $\theta$ | 0.53 | 0.04 | 0.45 | 0.62 |
| $\alpha$ | 0.86 | 0.10 | 0.61 | 0.99 |
| $\beta$ | 0.12 | 0.04 | 0.07 | 0.22 |
| $\tau$ | 0.74 | 0.16 | 0.38 | 0.98 |
| $\delta$ | 0.65 | 0.05 | 0.54 | 0.75 |
| $\pi$ | 0.09 | 0.02 | 0.06 | 0.13 |
| $\kappa$ | 941 | 70 | 819 | 1097 |
| $Zmc + Zuc$ | 108 | 33 | 76 | 202 |
| $Zms + Zus$ | 716 | 108 | 487 | 924 |

**TABLE 5.3**: Main statistics of the marginal posterior pdfs of parameters based on year 1995 only. $Zmc + Zuc$ measures angling catches. $Zms + Zus$ measures spawning escapement.

| Correlation (%) | $\kappa$ | $\alpha$ | $\beta$ | $\delta$ | $\tau$ | $\pi$ | $\theta$ |
|---|---|---|---|---|---|---|---|
| $\kappa$ | 100 | 2 | -22 | 0 | 1 | -40 | **-92** |
| $\alpha$ | 2 | 100 | -1 | 0 | 1 | **-65** | -1 |
| $\beta$ | -22 | -1 | 100 | 0 | **-86** | 29 | 20 |
| $\delta$ | 0 | 0 | 0 | 100 | 1 | 0 | 0 |
| $\tau$ | 1 | 1 | **-86** | 1 | 100 | -20 | 0 |
| $\pi$ | -40 | **-65** | 29 | 0 | -20 | 100 | 37 |
| $\theta$ | **-92** | -1 | 20 | 0 | 0 | 37 | 100 |

**TABLE 5.4**: Posterior correlation matrix between parameters based on data collected in 1995.

A simple look at the marginal prior and posterior pdfs for parameters given by Figs. 5.8 and 5.7 shows that for most of the parameters, prior distributions were greatly updated and that prior uncertainty was greatly reduced. Figure 5.8 shows that the size of the grilse spawning run in 1995 is estimated with little uncertainty, with a posterior mode at about 1000 fish. The trap efficiency $\theta$ is certainly greater than 0.5. The trapping device creates a strong flow stream which is preferentially explored by the returning grilse. The ratio removed by anglers is around 10%. Only the survival rate $\alpha$ and the reporting efficiency $\tau$ remain fairly imprecise and their posterior pdfs are essentially similar to their priors. Explanations are given by returning to the influence diagram of Fig. 5.5: no direct data provide information to infer about $\alpha$ and only $Y_4$ is related to the observable effects of $\tau$.

The correlation matrix given in Table 5.4 shows that posterior evaluation of the survival rate $\alpha$ cannot be done independently from the knowledge regarding the recapture efficiency $\pi$. As expected, the anglers

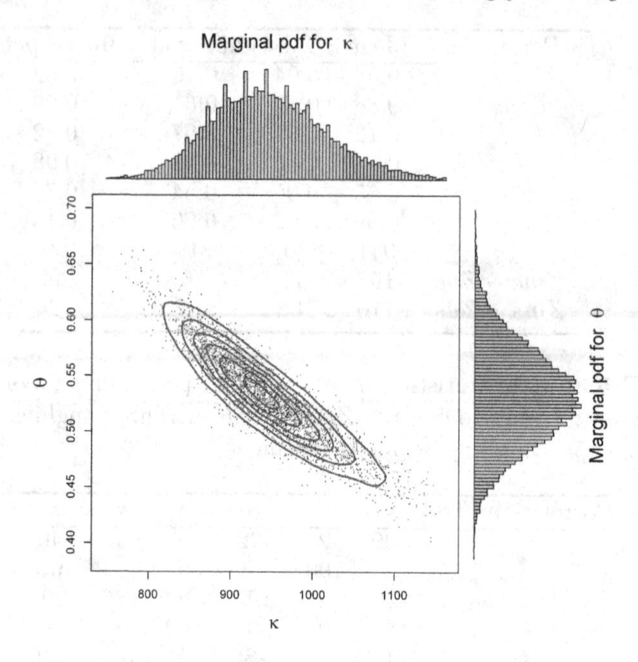


**FIGURE 5.8**: Posterior distributions of $(\theta, \kappa)$ obtained with data of year 1995 only. The shape of the joint posterior distribution is shown in the central part (joint MCMC draws and smoothed isodensity contours). The marginal distributions are shown in the top and right panels.

efficiency $\beta$ and the probability of reporting $\tau$ are partly confounded. The relation between $\theta$ and $\kappa$ stems from the binomial assumption $\mathbb{E}(Y_1 \,|\theta, \kappa) = \kappa\theta$. Figure 5.8 shows how the Gibbs sample for $(\kappa, \theta)$ scatters around the curve: $y_1 = \kappa\theta$.

### 5.4.2 Inference when taking into account 5 years of data

Figures 5.9 and 5.10 report results when incorporating the last 5 years of data (1995-1999) from Table 5.1 according to the interannual model sketched in Fig. 5.6. Posterior distributions of run size in Figure 5.9 points out the between-year variability with posterior means ranging from 256 (year 1999) to 773 (year 1996). A comparison of Table 5.3 and Table 5.5 shows that standard deviations are generally reduced when incorporating more information in the analysis. This is a snowball effect: additional information is transferred from year to year via the common

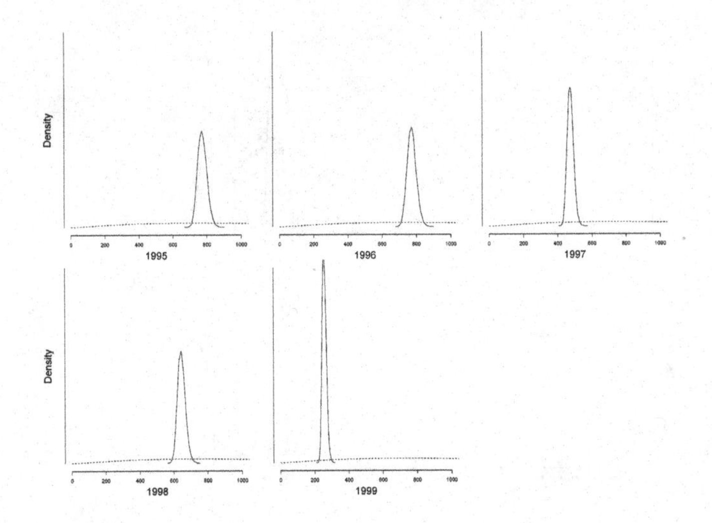


**FIGURE 5.9**: Posterior pdf of the stock size $\kappa$ for years 1995 to 1999. Dotted line: prior; solid line: posterior.

parameters $(\pi, \theta, \alpha, \beta, \tau, \delta)$ to shrink the uncertainty domain attached to the plausible values of the stock sizes.

## 5.5 Discussion and conclusions

This chapter exemplifies some key features of Bayesian modeling of ecological data:

- Setting a probability model is the most subjective element in any statistical analysis, may it be Bayesian or classical. Bayesian studies present the advantage of clearly identifying prior assumptions so they can be criticized and their influence on the results, assessed. From a methodological point of view, this is the main difference between conventional statistics and Bayesian ones. Yet, there is a modest literature on prior elicitation although some advances are presented in [166] and [223]. Further effort should be made to improve the elicitation of prior knowledge integration into the analysis following Berger's advice ([27]). Testing thoroughly how other prior assumptions affect posterior probabilities was not ad-

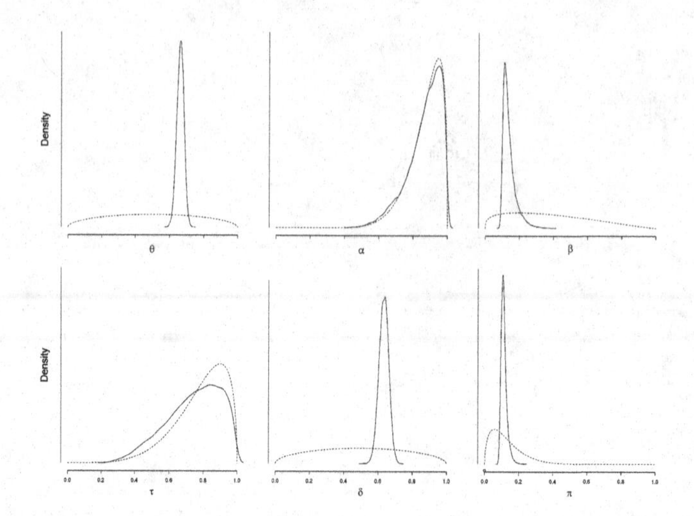


**FIGURE 5.10**: Posterior pdf for probability parameters obtained with the 5-year dataset (1995-1999). Dotted line: prior; solid line: posterior.

dressed in this application to the Scorff data but interested readers can refer to Gelman *et al.* [117].
Prior expertise and data from any source related to the problem, even not mentioned in standard experimental designs, provide valuable information that should be used to reduce model uncertainty. On the Scorff salmon example, the stock size of spawners from year to year and their credible intervals can be evaluated when incorporating such evidence into the study.

- Caution is necessary when handling parameters: there are many of them, at least more than the data points collected during one year of observation. This is a dangerous over-parameterization that would yield some indetermination or confounding effects among parameters in standard frequentist analysis. Luckily enough, in the Bayesian setting, when proper priors are used, the inference task would proceed without computational difficulty and the weakness of the model structure appears only when comparing priors and posteriors. For instance, the model in Eq. (5.3) is over-parameterized since no information except priors is given to assess $\alpha$ and $\pi$ separately from observations. The Bayesian setting can cope with such a linked couple of parameters. More generally the examination of the variance-covariance matrix helps to see which

| Parameter | Mean | Sd | 2.5% pct. | 97.5% pct. |
|---|---|---|---|---|
| $\theta$ | 0.66 | 0.02 | 0.62 | 0.70 |
| $\alpha$ | 0.87 | 0.09 | 0.65 | 0.99 |
| $\beta$ | 0.14 | 0.04 | 0.10 | 0.25 |
| $\tau$ | 0.75 | 0.16 | 0.40 | 0.97 |
| $\delta$ | 0.63 | 0.03 | 0.58 | 0.69 |
| $\pi$ | 0.11 | 0.02 | 0.09 | 0.16 |
| $\kappa_{1995}$ | 772 | 26 | 723 | 827 |
| $\kappa_{1996}$ | 773 | 26 | 725 | 827 |
| $\kappa_{1997}$ | 474 | 18 | 441 | 512 |
| $\kappa_{1998}$ | 644 | 23 | 601 | 692 |
| $\kappa_{1999}$ | 256 | 12 | 235 | 281 |

**TABLE 5.5**: Main statistics of the marginal posterior distributions of parameters obtained with the 5-year dataset (1995-1999).

parameters are confounded, but even severe confounding is not necessary a problem in Bayesian analysis. Ecological modeling aims at an uneasy balance between realistic but often over-parameterized models, and parsimonious models that can be too rough or with coefficients set to arbitrary values found in the literature without possibilities of validation. Bayesian analysis offers a sensible and coherent way of getting away from this dilemma.

- This chapter exemplifies how powerful advantages from conditional structures described by a graphical model can be fully exploited in practice within the Bayesian perspective. The modeling task is simplified once latent variables, model parameters and observed variables have been identified. The use of these three types of constituting elements gives more freedom for designing models that mimic the problem as it stands. Due to the limitations of the observed data, some parameters of such "realistic" models can be confounded. This is not a major impediment for the statistical treatment of the model as far as a Bayesian approach is employed. Diagnostics revealing confounding of parameters are easily obtained. Marginal posterior distributions of quantities of interest account for the uncertainty associated with confounding of parameters in a consistent way. Most often standard statistical approaches solve the problem of confounding of parameters by using fixed values. The Bayesian approach proposes to replace such *ad hoc* solutions by setting prior pdfs on the parameters grounded on the practitioners' expertise.

# Chapter 6

# The Normal linear model

## Summary

This chapter is devoted to the Bayesian analysis of the Normal linear model. The Normal linear model is the backbone of data analysis. It is used to explain how a continuous response variable depends upon one or several explanatory variables. Explanatory variables can be either continuous (regression), categorical (analysis of variance) or both (analysis of covariance).

A large part of the chapter is devoted to the construction of relevant prior distributions (so-called Zellner's priors) and to illustrating how to take advantage of the mathematical properties of conjugacy to derive closed-form expression for model parameter posterior distributions.

As a motivating example, we study the spatio-temporal variability of Thiof abundance in Senegal (Thiof is a grouper fish typically caught on the west coast of Africa).

## 6.1 Motivating example: The decrease of Thiof abundance in Senegal

Thiof (*Epinephelus aenus*; Fig. 6.1) is a grouper species (between 40 and 90 cm long when captured; but see [169] for more details) caught by both artisanal and industrial fisheries along the Senegal coast (Fig. 6.2).

This grouper fish is a cold-water species that can be found migrating forward and backward along the Senegal coast under the influence of marine currents. It is more abundant in periods of upwelling when it migrates from the north. In the warmer season, it moves toward the north. It seems that this fish belongs to a single stock ranging from Cap Blanc in Mauritania to Cap Roxo in the south part of Senegal (Fig. 6.2).

**FIGURE 6.1**: Top. A view of M'bour (Senegal) traditional fishery harbor (picture taken by Etienne Rivot). Bottom. A Thiof fish (*Epinephelus aenus*) caught on the Senegal coast (picture taken by Martial Laurans, Ifremer, France).

Because of its high economic value, the Senegalese stock of Thiof fish has been heavily exploited for years, and the stock is now seriously depleted (see [111] for a complete stock assessment of this species).

We will focus on the data from the artisanal fishery during the period 1974-1999. This artisanal fleet consists of pirogues (Fig. 6.1) using palangra or fishing rods and provides an important part of the harvest since those fish usually swim near the rocky places where industrial trawls are not efficient. The data are from the Oceanographic Research Center of Dakar Thiaroye (CRODT) and have been published in the doctoral thesis of Martial Laurans ([169]).

The data at hand consist in 1870 records of catches collected every year between 1974 and 1999 for the artisanal Thiof fishery along the Senegalese coast (Table 6.1). Each record gives the total catches of Thiof (in kg) for a particular harbor (11 harbors) along the fishing coast and each month of the year, with the associated fishing effort measured as the number of days spent at sea. Instead of taking into account the harbors, a coarser geographical stratification is to consider the four zones along the coast (numbered 1 to 4 from north to south; see Fig. 6.2).

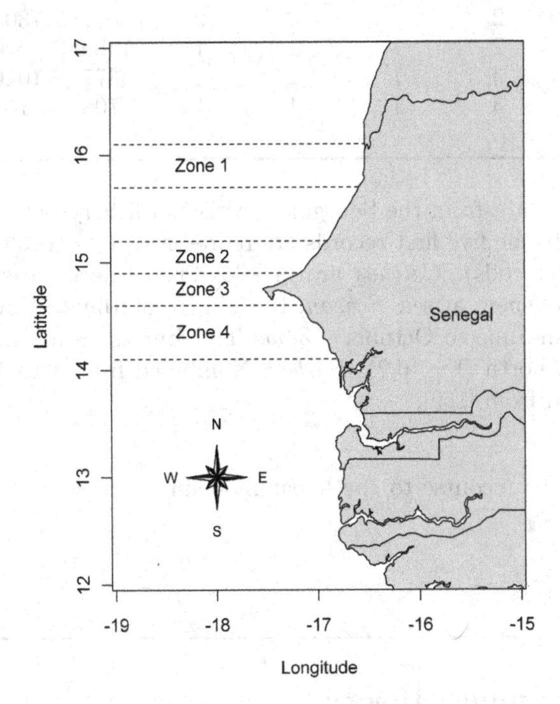


**FIGURE 6.2**: Thiof fishing zones along the coasts of Senegal.

Similarly, instead of considering the 12 months of the year, a coarser time stratification consists of considering two seasons: the cold one (from November to May), and the warm one (from June to October). To keep it simple, the data with zero catches are not modeled in what follows, as they seriously complicate the analysis ([175], [190]).

In fisheries science, the ratio of Catches versus Effort (denoted catch per unit effort or CPUE in the following) is classically interpreted as an index of abundance (see [137]). A quick look at the data in Table 6.1 suggests that Thiof abundance has been seriously depleted over the time series. However, not only the year might influence the abundance, but the other covariates such as the season or the fishing zone could also be influential covariates. What is the influence of regional and seasonal covariates on this apparent decrease of the abundance over time? Is it possible to disentangle the relative influence of these covariates in order to extract the time-trend signal from this dataset? In the following, we

| Year | Month | Season | Harbor | Zone | Effort | Catches |
|---|---|---|---|---|---|---|
| 1974 | 1 | 1 | 2 | 2 | 4089 | 256632.1 |
| 1974 | 2 | 1 | 2 | 2 | 6311 | 73030.0 |
| 1974 | 2 | 1 | 1 | 1 | 1389 | 2366.0 |
| 1974 | 3 | 1 | 2 | 2 | 5377 | 10166.9 |
| 1974 | 3 | 1 | 1 | 1 | 708 | 1676.4 |
| ... | | | | | | |

**TABLE 6.1**: Data from the Senegalese artisanal fishery between 1974 and 1999. Only the five first records are represented (the whole dataset contains 1870 recods). *Catches* are in kilograms. Effort measures the number of days spent at sea. *Season*: 1=Cold (from May to November); 2=Warm (from June to October). *Zone*: The four zones are numbered from south to north (Fig. 6.2). *Harbor*: Numbered from 1 to 11. Data are reproduced from [169].

propose to make recourse to the linear modeling framework to answer these questions.

---

## 6.2 Linear model theory

### 6.2.1 Basics and notations

Let us first recall some basics about the Normal linear model. Many statistical analyses in ecology deal with the linear model: a response variable $y$ (sometimes termed *dependent* variable or *outcome*) is a function of some explanatory variables $\mathbf{x} = (x_1, x_2, \ldots, x_p)$. In our application case, $y$ represent the catches. Depending on the context, the explanatory variables $\mathbf{x}$ could also be called *predictors, covariates, control variables* or *independent variables*; in our application case, $\mathbf{x}$ could be *Effort, Year, Season, Zone* and *Harbor*). When only quantitative variables are used to explain the response, the linear model belongs to the class of *Regression.* When only qualitative variables (factors) are involved, the model belongs to the class of *analysis of variance* (or *ANOVA*). If only one factor is invoked, this is known as a one-way *ANOVA*. When both continuous and qualitative covariates are used together, the model is sometimes referred to as *analysis of covariance* (or *ANCOVA*).

The distribution of $y$ given $\mathbf{x}$ is studied in the context of experimental items $i = 1, ..., n$ (also called statistical units or subjects or observations;

in our case study, these are the 1870 records) on which both the vector of outcomes $\mathbf{y} = (y_1, y_2, \dots, y_n)'$ and matrix of explanatory variables $\mathbf{X} = (\mathbf{x}^1, \mathbf{x}^2, \dots, \mathbf{x}^p)$ are measured. The matrix $\mathbf{X}$ is commonly referred to as the *design matrix*. As justified later in the chapter, we often make the convention to set the first explanatory variable $\mathbf{x}^1$ to a column of $1's$ (what corresponds to adding an intercept in the linear predictor). Not to bother with nonidentifiability, we also assume that $n > p$ and that $\mathbf{X}$ is of full rank (*i.e.*, $\mathbf{X}'\mathbf{X}$ can be inverted), which means that there is no linear redundancy structure among the variables used for explanation.

$$\mathbf{X} = \begin{pmatrix} x_{11} & x_{12} & \dots & x_{1p} \\ x_{21} & x_{22} & \dots & x_{2p} \\ \dots & \dots & \dots & \dots \\ x_{n1} & x_{n2} & \dots & x_{np} \end{pmatrix}$$

The Normal linear model is classically written as:

$$\mathbf{Y}|\mathbf{X}, \theta, \sigma^2 \sim Normal(\mathbf{X}\theta, \sigma^2 I_n) \tag{6.1}$$

It is defined by

1. An expectation of the response $\mathbf{y}$ given the explanatory variables $\mathbf{X}$ under a linear combination of parameters, thus writing in matrix notations
$$\mathbb{E}(\mathbf{Y}|\mathbf{X}, \theta) = \mathbf{X}\theta$$
with $\theta = (\theta_1, \dots, \theta_p)$ the vector of parameters. Here, $\theta_1$ is the intercept associated with the first column of 1's, $\theta_2$ is the second parameter associated with the second column of $\mathbf{X}$, and so on.

2. Independent Normal random noise with the same variance for every observation $i$
$$\mathbb{V}(Y_i|\mathbf{X}, \theta) = \sigma^2$$

The basics of the frequentist treatment of this model by maximum likelihood are recalled in Appendix A; many software routines such as the $lm()$ $R$-procedure are available to launch standard estimation.

### 6.2.2 Bayesian updating of the Normal linear model

Appendix A details the complete Bayesian conjugate treatment of the Normal linear model. To sum up, assuming a typical Gamma as a prior distribution for the precision $\sigma^{-2}$ and a multivariate Normal prior distribution with prior expectation $m_0$ and covariance matrix $\sigma^2 V_0$, for $\theta$ given $\sigma^2$ :

$$\begin{cases} \sigma^{-2} \sim Gamma(\frac{n_0}{2}, \frac{S_0}{2}) \\ \theta \sim Normal_p(m_0, \sigma^2 V_0) \end{cases} \tag{6.2}$$

then the joint posterior distribution of $(\theta, \sigma^{-2})$ obtained with a dataset $\mathbf{y}$ ($\mathbf{y}$ is a vector of size $n$) belongs to the same Gamma-Normal family with updated parameters $n_y$, $S_y$, $\theta_y$ and $V_y$:

$$\begin{cases} \sigma^{-2}|\mathbf{y} \sim Gamma(\frac{n_y}{2}, \frac{S_y}{2}) \\ \theta|(\sigma^{-2}, \mathbf{y}) \sim Normal_p(\theta_y, \sigma^2 V_y) \end{cases} \tag{6.3}$$

with

$$n_y = n_0 + n$$
$$\hat{\theta} = (\mathbf{X}'\mathbf{X})^{-1}\mathbf{X}'\mathbf{y}$$

and

$$\begin{cases} \theta_y = V_y \left( \mathbf{X}'\mathbf{X}\hat{\theta} + V_0^{-1} m_0 \right) \\ V_y^{-1} = \mathbf{X}'\mathbf{X} + V_0^{-1} \\ S_y = S_0 + (\mathbf{y} - \mathbf{X}\hat{\theta})'(\mathbf{y} - \mathbf{X}\hat{\theta}) \\ \quad + (\hat{\theta} - m_0)'(V_0 + (\mathbf{X}'\mathbf{X})^{-1})^{-1}(\hat{\theta} - m_0) \end{cases} \tag{6.4}$$

Further simplifications can be made assuming a simplified structure for the prior variance matrix $V_0$, notably useful in the case of vague prior information (see more details in Appendix A).

### 6.2.3 Defining appropriate Zellner's prior for $\theta$

Some useful benchmarks for defining appropriate priors for parameters $\theta$ in the Normal linear model have been proposed in the literature: Fernandez *et al.* ([104], [105]) give comprehensive examples. These so-called Zellner's priors are especially useful when explanatory variables for the response $y$ include categorical covariates (such as *Season* or *Zone* in the case study). In this section, we explain the seminal ideas of Zellner ([320]) that underline these priors.

#### 6.2.3.1 Defining a system of constraints to ensure statistical identifiability

As an example, let us consider a linear model applied to our case study that would account for the effect of the categorical variable *Zone* to explain the variability of catches. The factor *Zone* takes four categorical values (ranging from 1 to $k = 4$) that would be associated with four coefficients representing the effect, $\alpha_1$, $\alpha_2$, $\alpha_3$ and $\alpha_4$. The usual coding of such effect in the linear model is to introduce dummy variables that will represent the class belonging for each individuals. This could be introduced within the linear model structure as an explanatory subspace of $k$ dimensions generated by the linear combination of indicator variables. The conditionnal expectation of $y$ would then write:

$$\mathbb{E}_y = \alpha_1 \mathbf{1}_{(Zone=1)} + \alpha_2 \mathbf{1}_{(Zone=2)} + \alpha_3 \mathbf{1}_{(Zone=3)} + \alpha_4 \mathbf{1}_{(Zone=4)}$$

Of course, as each statistical unit belongs to a zone and only one, there is an identifiability issue since $\sum_{z=1}^{4} \mathbf{1}_{(Zone=z)} = \mathbf{1}$. Hence, we cannot keep at the same time the first variable $\mathbf{x}^1$ in the design matrix (corresponding to the vector of 1's associated with the intercept) and all the indicator variables. The standard coding imposes one linear constraint on the parameters as a remedy, for instance the omission of one class (*i.e.*, $\alpha_4 = 0$) or a zero mean effect ($\sum_{z=1}^{4} \alpha_z = 0$). Equivalently, this implies considering a linear transformation (known as *contrasts* in the statistical jargon) from the $k-$dimensional space generated by the indicator variables into the $k-1$ dimensional sub-space to be complemented by the constant vector $\mathbf{x}^1$. For instance the $R$ routine *contr.sum*() would create the $k-1$ new dummy variables $\mathbf{x}^2 = \mathbf{1}_{(Zone=1)} - \mathbf{1}_{(Zone=4)}$, $\mathbf{x}^3 = \mathbf{1}_{(Zone=2)} - \mathbf{1}_{(Zone=4)}$ and $\mathbf{x}^4 = \mathbf{1}_{(Zone=3)} - \mathbf{1}_{(Zone=4)}$ to be added to the unchanged first column of $\mathbf{X}$, $\mathbf{x}^1 = \mathbf{1}$ in order to get a set of $k$ vectors that generate the same explanatory space.

Using matrix notations, the linear operator

$$M = \begin{pmatrix} 1 & 1 & 0 & 0 \\ 1 & 0 & 1 & 0 \\ 1 & 0 & 0 & 1 \\ 1 & -1 & -1 & -1 \end{pmatrix}$$

establishes a reversible linear mapping ($M$ is invertible) such that:

$$(\mathbf{x}^1, \mathbf{x}^2, \mathbf{x}^3, \mathbf{x}^4) = (\mathbf{1}_{(Zone=1)}, \mathbf{1}_{(Zone=2)}, \mathbf{1}_{(Zone=k)}, \mathbf{1}_{(Zone=4)})M$$

Consequently, the correspondence between the vectors of initial coefficients $(\alpha_1, \alpha_2, \alpha_3, \alpha_4)'$ and parameters $(\beta_1, \beta_2, \beta_3, \beta_4)'$ of the linear model with an intercept $\beta_1$ can be derived from simple matrix multiplication

$$(\alpha_1, \alpha_2, \alpha_3, \alpha_4)' = M(\beta_1, \beta_2, \beta_3, \beta_4)'$$

leading to the change of parameters

$$\begin{cases} \alpha_1 = \beta_1 + \beta_2 \\ \alpha_2 = \beta_1 + \beta_3 \\ \alpha_3 = \beta_1 + \beta_4 \\ \alpha_4 = \beta_1 - (\beta_2 + \beta_3 + \beta_4) \end{cases}$$

It is straightforward to obtain the reverse relationship

$$(\beta_1, \beta_2, \beta_3, \beta_4))' = M^{-1}(\alpha_1, \alpha_2, \alpha_3, \alpha_4)'$$

with

$$M^{-1} = \frac{1}{4}\begin{pmatrix} 1 & 1 & 1 & 1 \\ 3 & -1 & -1 & -1 \\ -1 & 3 & -1 & -1 \\ -1 & -1 & -1 & -1 \end{pmatrix}$$

Thus one can also write:

$$\begin{cases} \beta_1 = \frac{1}{4}(\alpha_1 + \alpha_2 + \alpha_3 + \alpha_4) \\ \beta_2 = \frac{3}{4}\alpha_1 - \frac{1}{4}(\alpha_2 + \alpha_3 + \alpha_4) \\ \beta_3 = \frac{3}{4}\alpha_2 - \frac{1}{4}(\alpha_1 + \alpha_3 + \alpha_4) \\ \beta_4 = \frac{3}{4}\alpha_3 - \frac{1}{4}(\alpha_1 + \alpha_2 + \alpha_4) \end{cases}$$

There are many ways to introduce a reversible linear mapping $M$ between the set of indicator variables and a set composed by association of the constant vector with new dummy variables. The parameterization $(\alpha_1, \alpha_2, \alpha_3, \alpha_4)$ has four coefficients associated with the four zones, but no intercept added. The (more classical) parameterization $(\beta_1, \beta_2, \beta_3, \beta_4)$, has a constant term $\beta_1$ and three other coefficients that must be interpreted as a whole. This is just a change of base in a vectorial subspace; when qualitative variables are involved in a linear model, the associated indicator variables do not mean anything by themselves, the only important feature is the associated explanatory subspace that must be considered as a whole.

#### 6.2.3.2 Zellner's priors for qualitative effects

Zellner's priors are specifically designed for linear models in order to obtain a posterior expectation of the response which is independent from the system of constraints (or contrast) chosen by the modeler. In a Bayesian setting, suppose we got two possible parametric expressions for the effect of qualitative variables in the linear model, leading to two different expressions for the expected response:

$$\mathbb{E}(\mathbf{Y}|\mathbf{X_0}, \theta_0) = \mathbf{X_0}\theta_0$$

or alternatively

$$\mathbb{E}(\mathbf{Y}|\mathbf{X_1}, \theta_1) = \mathbf{X_1}\theta_1$$

with

$$\begin{cases} \mathbf{X}_0 = \mathbf{X}_1 M \\ \theta_0 = M^{-1}\theta_1 \end{cases} \tag{6.5}$$

For instance, the previous reparameterization of the *Zone* influence would be obtained with

$$\begin{cases} \theta_0 = (\beta_1, \beta_2, \beta_3, \beta_4)' \\ \theta_1 = (\alpha_1, \alpha_2, \alpha_3, \alpha_4)' \end{cases}$$

and

$$\begin{cases} \mathbf{X}_0 = (\mathbf{x}^1, \mathbf{x}^2, \mathbf{x}^3, \mathbf{x}^4) \\ \mathbf{X}_1 = (\mathbf{1}_{(Zone=1)}, \mathbf{1}_{(Zone=2)}, \mathbf{1}_{(Zone=3)}, \mathbf{1}_{(Zone=4)}) \end{cases}$$

Given $\sigma$, the analyst can impose a Normal conjugate structure on $\theta_0$, drawing a priori $\theta_0 \sim N(m_0, V_0)$. But equivalently, another scientist working with parametrization $\theta_1$, might consider a prior $\theta_1 \sim N(m_1, V_1)$. To obtain the same posterior expectation of the response, regardless of the parametrization used, the following coherence constraints on the prior mean and variance must be verified:

$$\begin{cases} m_0 = M^{-1}m_1 \\ \mathbf{X}_0 V_0 \mathbf{X}_0' = \mathbf{X}_1 V_1 \mathbf{X}_1' \end{cases} \tag{6.6}$$

Equation 6.6 is at the heart of the rule used to define Zellner's priors. Adopting Zellner's priors (see Appendix A)

$$\begin{cases} \theta_0 \sim Normal(m_0, \sigma^2 V_0) \\ \theta_1 \sim Normal(m_1, \sigma^2 V_1) \end{cases}$$

with zero mean, $m_0 = m_1 = 0$ and prior variances depending on the design matrix

$$\begin{cases} V_0 = c \times (\mathbf{X}_0'\mathbf{X}_0)^{-1} \\ V_1 = c \times (\mathbf{X}_1'\mathbf{X}_1)^{-1} \end{cases}$$

offers a coherent way to define priors that do not depend on the vectorial base that was chosen for the explanatory subspace generated by a qualitative variable because one can easily check from Eq. (6.5) that:

$$\mathbf{X}_0\left(\mathbf{X}_0'\mathbf{X}_0\right)^{-1}\mathbf{X}_0' = \mathbf{X}_1 M\left\{(\mathbf{X}_1 M)'\mathbf{X}_1 M\right\}^{-1}(\mathbf{X}_1 M)'$$

and

$$\mathbf{X}_1 M M^{-1}\left(\mathbf{X}_1'\mathbf{X}_1\right)^{-1} M'^{-1} M'\mathbf{X}_1' = \mathbf{X}_1\left(\mathbf{X}_1'\mathbf{X}_1\right)^{-1}\mathbf{X}_1'$$

We recall here that the orthogonal projection of $\mathbf{y}$ onto a linear subspace spanned by the columns of the (supposedly full rank) matrix $\mathbf{X}$ is $\mathbf{X}(\mathbf{X}'\mathbf{X})^{-1}\mathbf{X}'\mathbf{y}$. This linear operator is known in the linear model literature as the *hat matrix* and, despite its appearance, this projector does not depend specially on $\mathbf{X}$, the system of coordinates chosen to represent the vectors belonging to that space. Adopting Zelner's priors and following the appendix (Eqs. (A.5) and (A.6)), the posterior pdf given by Eq. (6.4) can be rewritten as:

$$\begin{cases} V_y = \dfrac{c}{1+c}\left(\mathbf{X}'\mathbf{X}\right)^{-1} \\ \theta_y = \dfrac{c\hat{\theta}}{1+c} \end{cases}$$

The constant $c$ expresses the strength of the prior information; the standard frequentist analysis is obtained as a limiting case when $c \to \infty$.

---

## 6.3 Modeling the decrease of Senegal Thiof abundance as a linear model

### 6.3.1 A first look at the data

We want to relate the CPUEs, or catch per unit of effort (the reported catches divided by the number of days spent at sea, interpreted as an indice of abundance), to the year considered and other possible factors of variation such as the geographical locations and the seasonal effects.

A quick first trial reveals that considering the CPUEs as the dependent variable is not a good idea. We performed a (frequentist) linear model under *R* with the *lm*() procedure including all the explanatory variables *Year*, *Season* and *Zone* and the plots of Fig. 6.3 indicate that the estimated residuals behave very badly; although all covariates seem to be interesting explanatory variables for the CPUEs, the variance of the residuals increases with the predicted values of CPUEs, which reveals a clear deviation from the assumption that the variance of the random noise is homoscedastic (does not vary with the levels of the response variable nor explanatory variables). The model also predicts some negative CPUEs, which is nonsense!

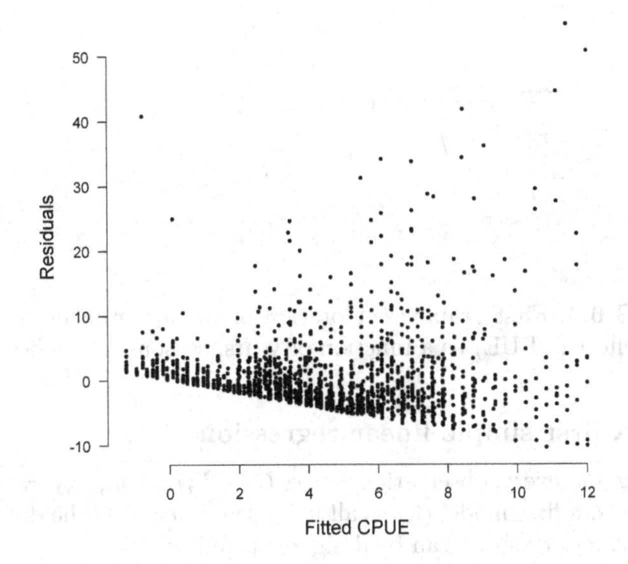


**FIGURE 6.3**: Residuals versus fitted values of CPUEs obtained from a preliminary linear model analysis (using the *lm*() procedure of *R*) without log-transform of the CPUEs.

To get more acceptable residual behaviors with regard to the linear model assumptions, we will therefore work from now on with log-transformed CPUEs that will form the response variable $y$. An exploratory data analysis of log-transformed CPUEs reveals that a declining trend since 1974 is observed on the data (Fig. 6.4). The analysis also reveals that the intensity of the depletion (*i.e.*, the slope) may well depend on the season as shown by the top right panel of Fig. 6.4. Other factors such as the location (*Zones*) seem also to matter when considering the bottom panels of Fig. 6.4.

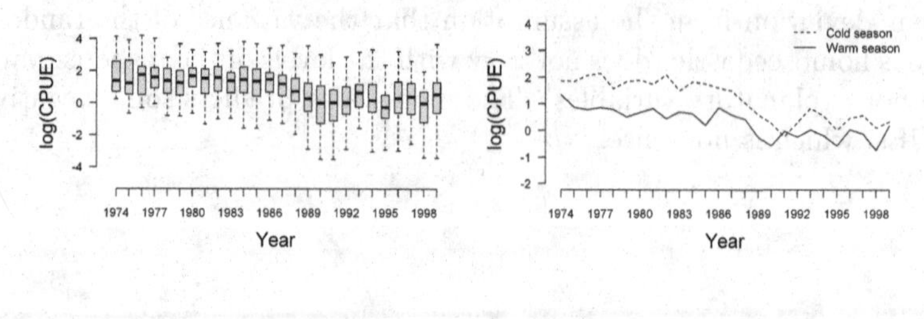


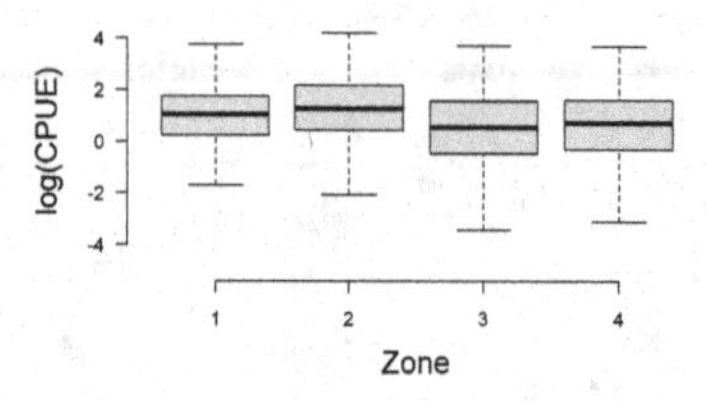


**FIGURE 6.4**: First graphic exploration of the log-transformed catch per unit effort (CPUE) as a function of years, seasons and zones.

### 6.3.2 A first simple linear regression

Setting for every observation $i$, $\log(CPUE_i) = y_i$, $x_{i1} = 1$, and $x_{i2} = Years$, a first model (let's call it $M_0$) with $p = 2$ as the dimension of the explanatory space can be designed as follows:

$$\begin{cases} y_i = \beta_1 + \beta_2 x_{i2} + \varepsilon_i \\ \varepsilon_i \overset{iid}{\sim} Normal(0, 1 \times \sigma^2) \end{cases}$$

This is a standard regression model with $\beta_1$ the intercept and $\beta_2$ the slope. We expect that $\beta_2 < 0$ with regards to the decreasing trend that was graphically identified in Fig. 6.4.

### 6.3.3 Explaining the abundance of Thiof with several factors

This very first model $M_0$ defined by Eq. (6.7) only accounts for $Year$ as a quantitative covariate to explain the variation of the CPUEs. However, Fig. 6.4 reveals that categorical covariates such as *Season* or *Zone* seem also to matter when considering the variation of the CPUE.

#### 6.3.3.1 Selecting relevant covariates

One should avoid considering factors that are too correlated. For temporal within year effects for instance, either a *Month* factor or a *Season* factor can be taken into account, but not both! For spatial stratification, one has to pick either a *Harbor* effect or a *Zone* effect.

The *Year* records range from 1974 to 1999 and can be included in the model as a linear trend (continuous variable) or as a factor with $k = 26$ indicator variables. The choice of coding the effect of the *Year* is an important modeling issue. As we suspect a phenomenological change following the ordering of *Year*, we will consider it as a continuous regressor. This will help to predict future years not in the sample by following the linear trend. Had we chosen a categorical representation, we would be helpless to predict future values as categories not in the explanatory space. This can be addressed using *random effects* but we delay this treatment to the second part of the book where hierarchical models will be explored.

Here, we first proceed with the following linear model with effect of *Season*, *Zone* and *Year* plus an intercept. The model has six parameters $\theta_0 = (\beta_1, \beta_2, \ldots, \beta_6)$. Note that in this model $M_1$, the decreasing trend ($\beta_6$) is considered as being independent upon the season:

$$\begin{cases} y_i = \beta_1 + \underbrace{\beta_2}_{Season} + \underbrace{\sum_{z=3}^{5} \beta_z x_{i,z}}_{Zone} + \underbrace{\beta_6 * Year}_{Year} + \varepsilon_i \\ \varepsilon_i \overset{iid}{\sim} Normal(0, \sigma^2) \end{cases} \tag{6.7}$$

#### 6.3.3.2 Bayesian Inference

The Bayesian inference is performed using Zellner's conjugate priors

$$\begin{cases} \theta_0 \sim Normal(m_0, \sigma^2 V_0) \\ V_0 = c \times (\mathbf{X}_0' \mathbf{X}_0)^{-1} \\ \sigma^{-2} \sim Gamma(1, 1) \end{cases} \tag{6.8}$$

with $m_0 = 0$ and an informative prior information weighting only one data record ($c = n$). The design matrix $\mathbf{X}_0$ is defined using the same contrast as in Section 6.2.3. A Gamma distribution with a large variance is used as a rather vague prior for the precision $\sigma^{-2}$. As a conjugate structure was chosen, the marginal posterior distribution of all parameters are known exactly from Eq. (6.3) and no use of MCMC simulation is needed. Results for model $M_1$ are summarized in Table 6.2.

From this table, one can see that the analyst can bet on $\beta_6 < 0$ with

| Param | Effect | Mean | Sd | 2.5%*pct.* | 97.5%*pct.* |
|---|---|---|---|---|---|
| $\beta_1$ | *Constant* | 156.589 | 8.647 | 173.882 | 139.295 |
| $\beta_2$ | *Season* | 0.412 | 0.028 | 0.468 | 0.357 |
| $\beta_3$ | *Zone* | 0.037 | 0.056 | 0.150 | -0.075 |
| $\beta_4$ | *Zone* | 0.346 | 0.056 | 0.459 | 0.234 |
| $\beta_5$ | *Zone* | -0.328 | 0.042 | -0.245 | -0.412 |
| $\beta_6$ | *Year* | -0.078 | 0.004 | -0.070 | -0.087 |

**TABLE 6.2**: Posterior estimates of the parameters in the linear model 6.7 obtained from the closed-form expression of the posterior distribution.

great confidence. Hence, as expected, there is probably a negative linear trend showing a decrease of the abundance, whatever the *Season* and *Zone*. The *Season* and *Zone* factors are not easy to interpret because the dummy variables created by its indicators must be considered as a whole with the intercept. One can notice a great posterior probability that $\beta_2 > 0$, which corresponds to a greater abundance during season 1 (the cold season).

This model can be put into competition with other ones. For instance, from Fig. 6.4, we see that the slope of the decreasing trend might be considered as different between seasons; this is an example of interaction between the continuous predictor *Year* and the factor *Season*. Thus, one may compare the model $M_1$ with Eq. (6.7) with another one, model $M_2$ that includes a decreasing trend over time depending upon the *Season*:

$$\begin{cases} y_i = \beta_1 + \underbrace{\beta_2}_{Season} + \underbrace{\sum_{z=3}^{5} \beta_z x_{i,z}}_{Zone} + \underbrace{\sum_{s=6}^{7} \beta_s x_{i,s}}_{Season \times Year} + \varepsilon_i \\ \varepsilon_i \overset{iid}{\sim} Normal(0, \sigma^2) \end{cases} \tag{6.9}$$

#### 6.3.3.3 Model selection

Those two competing models ($M_1$ and $M_2$ are respectively defined by Eqs. (6.7) and (6.9)) are compared using Bayes factors. Equation (4.16) needs to evaluate $[\mathbf{y}|M]$, the average of the likelihood with regard to the prior pdf of the unknown, for $M = M_1$ and $M = M_2$. Fortunately enough, because of conjugate properties in the linear regression case, this prior predictive value to be evaluated at the observed data $\mathbf{Y} = \mathbf{y}$, is available in closed form (see Eq. (A.3) of Appendix A). Therefore, model comparison is rendered easy. In addition, as detailed in Chapter 3 of Marin and Robert [189], an improper prior can even be used for

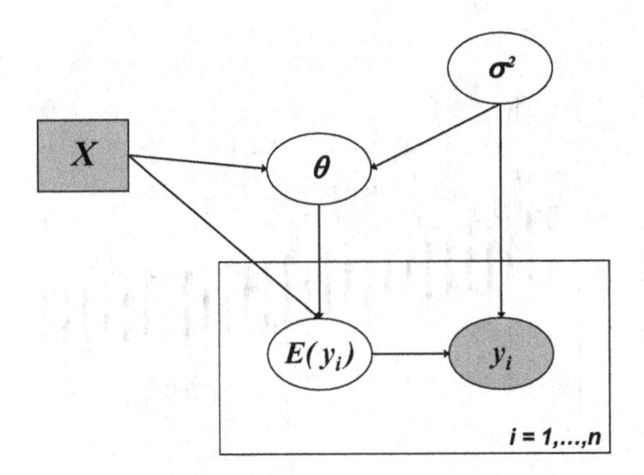


**FIGURE 6.5**: Simplified directed acyclic graph for the linear model defined by Eqs. 6.7 or 6.9. The arrows between $X$ (design matrix), $\sigma^2$ and $\theta$ indicate that Zellner's prior for $\theta$ is a multivariate Normal distribution $Normal_p(m_0, \sigma^2 V_0)$ with $V_0 = c \times (\mathbf{X}'\mathbf{X})^{-1}$.

the weighting coefficient $c$ of the Zellner priors providing the same $c$ is taken for all competing models. As the log-predictive for this new model is only $-7.0 \times 10^{23}$, to be compared with $-3.7 \times 10^{19}$ for model in Eq. (6.7), we clearly do not select this alternative model and prefer the more parsimonious one in which the slope of the decreasing trend is common for both seasons.

For readers who are reluctant to use the algebra in Appendix A, a simple WinBUGS code for the model given by Eq. (6.7) can be straightly written. A simplified directed acyclic graph for this model is presented in Fig. 6.5. Once the design matrix has been built from the dataset (a few lines of $R-$code), the short WinBUGS code takes only a few seconds to be run and to get a reliable MCMC approximation of posterior distributions. It is readily verified that the posterior statistics approximated through MCMC sampling closely match those obtained from closed-form expressions.

We also used WinBUGS to get posterior predictive distributions of the log-CPUE derived from model in Eq. (6.7). Figure 6.6 shows the posterior predictive distribution for the log-CPUE for the two seasons

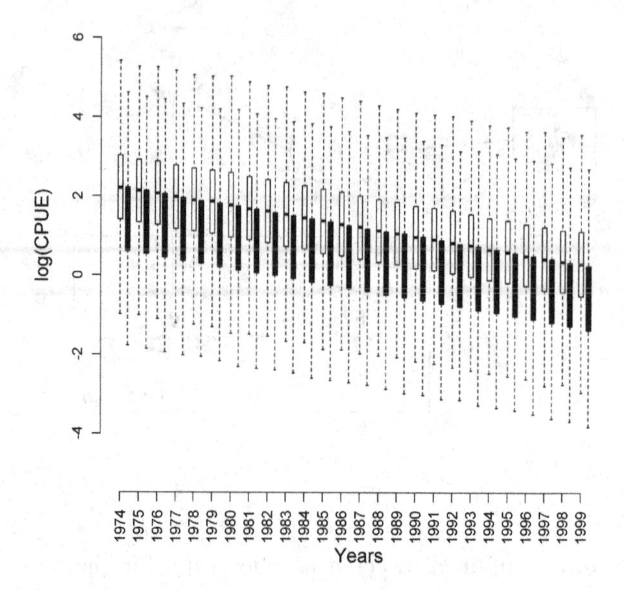


**FIGURE 6.6**: Posterior predictive distributions of the log-CPUE in Zone 1 derived from model in Eq. (6.7). Light gray: Season 1; Dark gray: Season 2.

and the first zone. As assumed in the model, the slope of the decreasing trend is the same for the two seasons.

---

## 6.4 Further reading

### 6.4.1 Linear modeling

There is a huge body of frequentist literature devoted to the linear model; if we were to cite just two authors, essential pieces of the theory can be found in [96] and [270]. The Bayesian perspective on the linear model is also abounding. Raiffa and Schlaifer [247] saw the benefits that could be obtained from Normal-Gamma conjugate properties. Box and Tiao [36] considered noninformative priors in Gaussian modeling; Zellner [319] pointed out workable applications for economists; many aspects of regression techniques are dealt in [117], [140], [172] and [189], etc. It is also worth mentioning a derived branch of rather theoretical investigations, Bayes linear analysis ([123]), that considers partially specified probability models and attempts to work with their linear approximation through their first and second moments.

### 6.4.2 Model selection

Multimodel Bayesian inference is uneasy, but the task deserves efforts. Choosing among models is essential to scientific assessment but it cannot be solved in one snap. A good review can be found in Kadane and Lazar [154]. Textbook treatments of model selection can be found in [43] and [179]. There are an increasing number of applications relying on Bayes factors to select models in applied ecology ([163], [218] and [257]). In this book, we will favor Bayes factors because they are firmly grounded on a coherent probabilistic judgement ([147], [157]). Even though none of the models considered might be actually true, results from Bernardo and Smith [30], Section 6, prove that proceeding that way is still clever. The procedure asymptotically selects the true model if it exists, or the one closest to the truth, using information-based the Kullback-Leibler distance. Unfortunately, Bayes factors are not robust to the specifications of the prior ([177]). Even when this prior is not improper ([230]), caution must be taken to specify good informative priors as a serious part of the modeling task ([155]). Encoding expertise may be very difficult (sometimes, no expert is available by lack of time or interest!), and many authors have searched for remedies to the use of Bayes factors in the case of improper vague priors; their idea is to take into account some slight part of the data to make the prior proper (fractionnal Bayes factors of [222]) or to take expectations on a virtual sample (the minimal training set leading to the intrinsic Bayes factors of [28]). Last but not least, Bayes factors techniques ([113], [129]) and procedures evaluating the posterior probabilities of models ([46],[124]) are generally cumbersome to implement (see also Appendix B), and many authors advocate for other criteria that are easier to compute. Most methods try to find a balance between parsimony (the simpler, the better) and goodness of fit (enough complexity so as to reproduce the main features of the data). The more acceptable from a Bayesian perspective is the BIC (Bayesian Information Criterion), $BIC = -2 \times \log([y|\hat{\theta}]) + \dim(\hat{\theta}) \times \log(\dim(y))$, which can be understood as an asymptotic approximation of the logarithm of the Bayes factor ([112]) around some max-likelihood estimate $\hat{\theta}$. The other popular criterion, because of its systematic evaluation when running WinBUGS, is the *DIC* (see Eq. (B.7) of Appendix B).

### 6.4.3 Variable selection

In a regression context, picking one model is equivalent to selecting a subset of explanatory variables. To give a privileged position to the particular hypothesis that most regressor effects (the $\theta_j$'s) are null, recourse can be made to Lasso techniques in order to penalize estimation by a stringent $L^1$ constraint ([295]). The Bayesian formulation (in terms of a specific mixture of priors) given by Park and Casella [226] and further elaborated by Kyung *et al.* [168] is an interesting matter of mathematical interpretation but remains challenging for practical implementation. Conversely, the Bayesian variable selection method of Mitchell and Beauchamp [208] is straightly understandable in terms of probability that variable $\mathbf{x}^j$ should be included in the model. This stochastic search variable selection ([119]; [213]) tries to switch on and off the effect of each variable $\mathbf{x}^j$ during Bayesian computations. This is made clear by considering a binary vector variable $\gamma$ with spike and slab priors. Its $j^{th}$ component is the index $\gamma^j$ encodes whether $\mathbf{x}^j$ is worth entering the model ($\gamma^j = 1$) or if it can be discarded ($\gamma^j = 0$). A fifty/fifty odd can be set a priori for each explanatory variable to be selected:

$$\gamma^j \sim Bernoulli(0.5)$$

When the $\theta^j$ are a priori independent, one can make recourse to a two-component mixture prior for $\theta^j$

$$\theta \sim (1-\gamma^j) \times \pi_{spike}(\theta^j) + \gamma^j \times \pi_{slab}(\theta^j)$$

Distributions $\pi_{spike}(.)$ and $\pi_{slab}(.)$ are such that the variance of $\pi_{spike}(.)$ is much smaller than the one of $\pi_{slab}(.)$, ensuring that $\theta^j$ belongs to the close vicinity of 0 if $\gamma^j = 0$ , while variable $\mathbf{x}^j$ can be classified as a plausible explanation in the case $\gamma^j = 1$. Ideally, the convenient choice is to adopt a Dirac function (*i.e.*, a mass function at zero) for the spike and some distribution with a large a priori variance for the slab distribution. However, implementing Dirac distributions might cause difficulties in the proposals of the MCMC algorithm to constrain some coefficient to be exactly zero, and in practice recourse can be made to a picky Normal pdf and a very flat one, for instance:

$$\begin{cases} \theta^j \sim Normal(0, V) \\ V = (1-\gamma^j) \times 10^{-8} + \gamma^j \times 10^8 \end{cases}$$

Strictly speaking, classification is only approximate for such a prior with an absolutely continuous spike, because $\gamma^j = 0$ is not exactly equivalent to $\theta^j = 0$, but indicates only that $\theta^j$ is "relatively" close to 0 compared to the other location for $\theta^j$ when $\gamma^j = 1$. Muntshinda *et al.* [213] investigated the demographic variation of moth species in response to global warming with such a Bayesian variable selection technique.

### 6.4.4 Bayesian model averaging

For prediction purposes, the prediction distribution can be obtained by averaging over all models with the weights being the model probabilities, rather than trying to select the *best* one ([139]). This has the effect of combining the predictive capability advantages from all models under consideration. Convincing examples are discussed in Section 9.3 of Hoff [140] and in Section 6.7.3 of Marin and Robert [189]. Attention is drawn to the fact that parameters are nothing but intellectual constructs whose meaning and scale may be highly model dependent. Hence, caution must be taken when averaging over the distributions of *a* parameter even if it carries the same name, say *ratio of efficiency* for instance, for the various competing models.

# Chapter 7

## Nonlinear models for stock-recruitment analysis


### Summary

A considerable effort has been devoted in fisheries science to understand the relationship between the *stock* (*e.g.*, the spawning biomass, the number of spawners, the number of eggs spawned, etc.) and the *recruitment* (*e.g.*, the number of juveniles issuing from the spawning stock). The mathematical modeling of the stock-recruitment process (*SR* in the following) is of fundamental importance in fish population dynamics. *SR* relationships give a synthesis of the population renewal and are critical for setting reference points for sustainable management.

A common pitfall in fisheries science is to consider *SR* models as deterministic relationships explaining how the recruitment varies with stock. We rather consider that *SR* models are conditional probability distributions of the recruitment given the stock. Based on the example of a stock-recruitment model for A. salmon in the Margaree River (Nova Scotia, Canada) we illustrate that:

1. A model is a crude simplification to quantify how unknowns, covariates and observations interact. Probabilistic concepts are helpful to describe the part of randomness. Conditional reasoning provides a helpful interface between the analyst and the ecologist.

2. Choosing on which parameters the model should rely is often overlooked. Encoding prior expertise about these parameters is also a crucial step of the scientific analysis. *SR* models can be parameterized in a way that parameters can be readily interpreted to help decision analysis for stock assessment and management purposes.

3. There is no *best* model. Visual inspection of model fit to observations, ecological knowledge, good sense and quantitative criteria such as the Bayes factor are necessary to help picking a satisfactory model.

## 7.1 Motivating example: Stock-recruitment models to understand the population dynamics and assess the status of A. salmon stocks

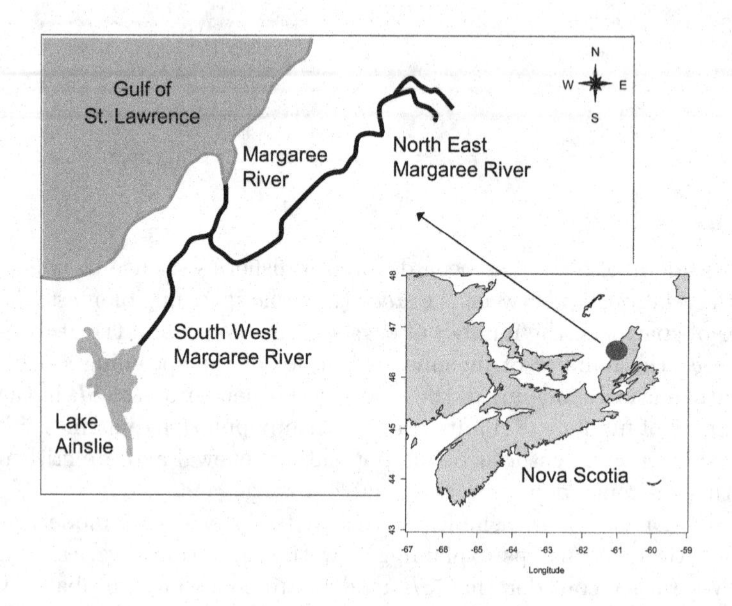


**FIGURE 7.1**: Margaree River area.

The Margaree River mouth is located in the southeastern portion of the Saint Lawrence Gulf, Canada (see Fig. 7.1). It flows across a recreational park, famous for Salmon fishing (fly fishing only). The Salmon fishing season begins June 1 and ends October 31. In 2007, nonresident anglers paid up to \$133 for a seasonal license or \$54 for a 7-day license. Fishing regulations such as bag limits are also enforced: 2 per day and 8 per season (only grilse up to 63 cm may be caught). The Margaree River has been surveyed for a long time. Each year, capture-mark-recapture experiments are made. The number of adult fish homing back to their native river is recorded. These fish will reproduce and die the same year in the Margaree River. Their eggs yield juveniles that will grow in the river and migrate down to the sea as smolts. After their marine journey (mostly one year but sometimes several), fish will ultimately swim back home to their natal river, reproduce and a new cycle will start. In this context, fisheries scientists define the *stock*, denoted $S_t$ as the number of spawners in year $t$, and the *recruitment*, denoted $R_t$ as the number of spawners which are issued from the reproduction of the stock $S_t$. Table 7.1 and Fig. 7.2 give the corresponding measurements of the stock and recruitment in the Margaree River for years 1947 to 1990.

Considerable effort has been devoted in fisheries science to understand the relationship between the stock and the recruitment ([137]; [216]; [244]). *SR* relationships give a synthesis of the population renewal. They are critical for setting biological or management reference points, especially for semelparous species such as salmons ([54]; [234]; [238]; [239]; [273]; [274]; [275]). In particular, the analysis of *SR* relationships help to estimate biological reference points such as the spawning target, a biological reference point for the number of spawners which are necessary to determine an optimal exploitation rate while ensuring the long-term sustainability of the population.

In this chapter, we show how Bayesian analysis of stock-recruitment models can provide some scientific advice to these issues.

| Year<br>Cohort | Stock<br>Spawners | Recruitment<br>Returns |
|---|---|---|
| 1947 | 1685 | 4852 |
| 1948 | 3358 | 7204 |
| 1949 | 1839 | 5716 |
| 1950 | 1744 | 4000 |
| 1951 | 2093 | 2440 |
| ... | | |
| 1985 | 1378 | 5156 |
| 1986 | 3461 | 3484 |
| 1987 | 3899 | 6375 |
| 1988 | 1545 | 3358 |
| 1989 | 2164 | 2900 |

**TABLE 7.1**: Stock and recruitment data for the Margaree River from 1947 to 1990. (Data are reproduced by courtesy of the Department of Fisheries and Oceans from the Canadian Data Report of Fisheries and Aquatic Sciences No. 678.)

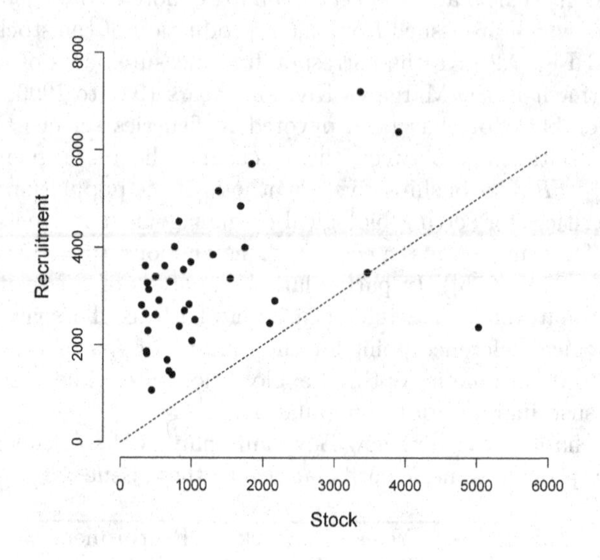


**FIGURE 7.2**: Stock and recruitment data (see Table 7.1) for the Margaree River (Nova Scotia, Canada) from 1947 to 1990. The dotted line is the replacement line (one recruited spawner per spawner).

---

## 7.2 Searching for a *SR* model

A common pitfall in fisheries sciences is to consider *SR* models as deterministic relationships explaining how the recruitment varies with stock. Rather, *SR* models are conditional probability distributions of the recruitment given the stock. From a pure statistical point of view, we are interested in the covariation between the quantities $R$ and $S$. In mathematical terms, we would speak of the joint pdf $[R, S]$. However, most of the time, biologists will be reluctant to adopt this symmetrical point of view. There is some knowledge about the *causal* relationship "*S gives R*": given the number of spawners, biological knowledge may help to predict the corresponding numbers of juveniles $R$. Therefore, in mathematical terms, biologists will favor the conditional decomposition

$$[R, S] = [S] \times [R|S]$$

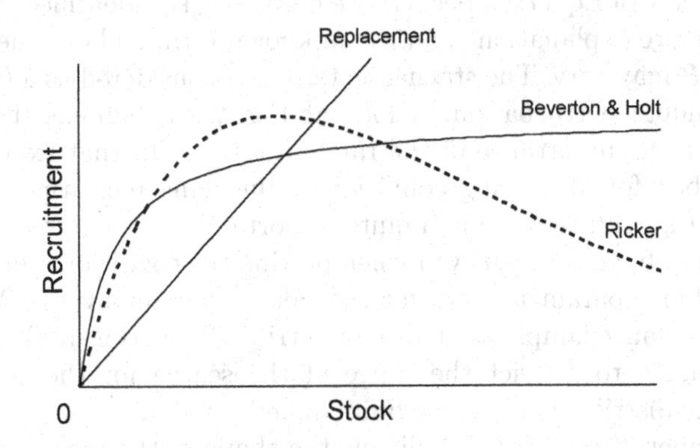


**FIGURE 7.3**: Typical shape for Ricker and Beverton-Holt stock-recruitment relationships. If $S$ and $R$ are expressed in the same unit, then the $SR$ relationship is directly comparable with the replacement line $S = R$.

and, assuming that $S$ is perfectly known (as a covariate), focus onto the modeling of $[R|S]$.

One may imagine $R$ as a smooth function of $S$, say $f(S)$. We would like to have the mathematical relationship $R = f(S)$ idealizing how the phenomenon works, such that the number of recruits would be roughly proportional to the stock, at least when $R$ is small and maybe reaching a plateau or even decreasing when too many spawners, or for that matter recruits, are competing for resources.

But even knowing $S$, $R$ remains uncertain for many reasons. Essentially, the process connecting $R$ to $S$ is so complex that only probability distributions can adequately capture the variation. Let's call $\varepsilon$ the hand of nature, summing up all the unexplained random factors; as a consequence, we can no longer write $R$ as a function of $S$ only. Including the unidentified sources of variation $\varepsilon$ plus the identified explanation variable $S$, one should rather write in a very symmetrical way:

$$R = f(S, \epsilon) \tag{7.1}$$

This functional relationship is formally equivalent to the conditional probability $[R|S]$ since, by a mere change of variable, the random distribution of $\epsilon$ is transferred onto the stochasticity of $R$ given the only available explanatory variable $S$. When proceeding that way, we break the symmetry of Eq. (7.1), because we hope that the identified variable $S$ provides more explanation than the unknown term $\varepsilon$ about the way the response $R$ may vary. The stochastic term $\varepsilon$ is considered as a (wishfully small) random perturbation. A look at Fig. 7.2 enlightens the analyst and shows the importance of the random term $\varepsilon$. In that example, $\varepsilon$ is a brown box for all forcing conditions influencing recruitment and, as shown in Fig. 7.2, they remain quite important.

One step further is achieved when picking the conditional probability $[R|S]$ within a parametric pdf for the sake of parsimony (see Munch *et al.* [211] for an example of non-parametric *SR* relationship). It means that we agree to restrict the range of the search for the conditional probability distribution inside a finite dimensional family of functions. A small number of parameters will rule the shape of the pdf. For instance if $[R|S]$ is chosen to be a Normal pdf, two coefficients (the so-called mean $\mu$ and standard deviation $\sigma$) will be enough to depict the whole Normal family of probability density functions. Of course, both $\mu$ and $\sigma$ will themselves be functions (to be defined) of the conditioning term $S$. Relying on the Normal pdf given on page 68, one might write:

$$[R|S] = dnorm(R, f(S), \sigma(S))$$

Although specifying $f(S, \epsilon)$ or $[R|S]$ are formally equivalent, we may keep on working for a while with Eq. (7.1). For mathematical convenience, it is often assumed that the effects of the explanatory variable $S$ and those of the uncontrolled factors can be disentangled, under either additive form

$$f(S, \epsilon) = f_1(S) + f_2(\varepsilon) \tag{7.2}$$

with $f_2(\varepsilon)$ being a Normal random variable centered at zero or sensibly modeled as a multiplicative effect:

$$f(S, \epsilon) = f_1(S) \times f_2(\varepsilon) \tag{7.3}$$

with $\log(f_2(\varepsilon))$ being a Normal random variable centered at zero[1]. It is not difficult to show that this latter LogNormal multiplicative model is equivalent to defining $[\log(R)|S]$ as a Normal random variable with parameters such that $\mu(S) = \log(f_1(S))$ and $\sigma(S) = \sigma = \sqrt{Var(\log(f_2(\varepsilon)))}$.

Different parametric functions can be proposed to model the systematic effect of the stock $S$ on the recruitment $R$, $f_1(S)$. Competing

[1]There is a tricky technical point for readers unfamiliar with the change of random variables: $\mathbb{E}(log(f_2(\varepsilon)) = 0$ does not imply that $\mathbb{E}(f_2(\varepsilon)) = 1$...

ecological interpretations (*e.g.*, linear relationship between $R$ and $S$, linear relationship for small values of $S$ only, and then a saturation effect) lead to competing formulations (see [137], [216] or [244] for a review). Here again, a step further is made when picking $\mu(S)$ into a parametric family, so as to restrict the number of degrees of freedom of the unknown $\mu(S)$ to a small dimensional space. For the sake of clarity, we only consider here the most widely used models:

- Beverton-Holt form with two parameters $(\alpha, \beta)$ taking into account a saturation effect. With this so-called Beverton-Holt model, the replacement rate $\frac{R}{S}$ is slowly decreasing as the stock increases: $R = \alpha S/(1 + \beta S)$ (see Fig. 7.3);

- Ricker form considering that when too many adults are spawning in the river, the replacement rate $\frac{R}{S}$ is an exponentially decreasing function of the stock: $R = \alpha S e^{-\beta S}$ (see Fig. 7.3). Appendix C provides some bio-mathematical justification for such a choice on page 348.

---

## 7.3 Which parameters?

### 7.3.1 Natural parameters

In this section we adopt the Ricker model with a multiplicative random term:

$$\begin{cases} R_t = f(\alpha, \beta, S_t)e^{\epsilon_t} \\ f(\alpha, \beta, S_t) = \alpha \cdot S_t e^{-\beta \cdot S_t} \\ \epsilon_t \overset{iid}{\sim} Normal(0, \sigma^2) \end{cases} \tag{7.4}$$

Figure 7.4 shows the Directed Acyclic Graph of such a simple *SR* model with natural parameters $\alpha$, $\beta$ and $\sigma$. From a biological perspective, $\sigma^2$ is the variance of the log-ratio of the replacement, *i.e.*, $\sigma$ is the relative standard error of replacement ratio. $\alpha$ is the slope close to the origin $S = 0$, when the relation between $S$ and $R$ is nearly linear. The inverse of the second parameter $\beta^{-1}$ is the stock that produces the maximum recruitment and could be interpreted as (an indicator of) the carrying capacity $S_{max}$. The maximum recruitment is $f(S_{max}) = R_{max} = \frac{\alpha}{\beta}e^{-1}$.

A first suggestion is to use flat noninformative priors for $log(\alpha)$,$\beta$,

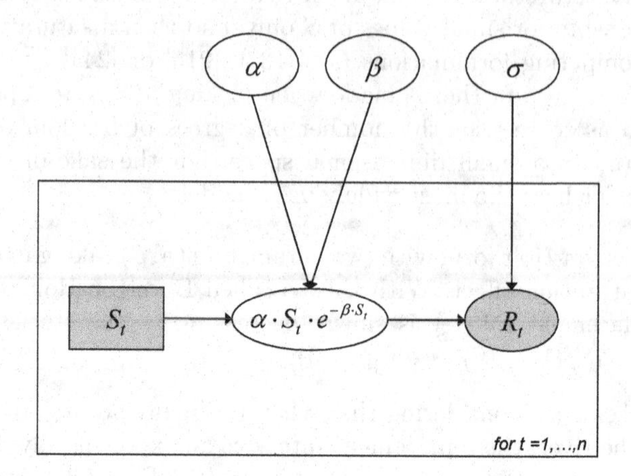


**FIGURE 7.4**: Directed Acyclic Graph for the Ricker model with natural parameters $\theta = (\alpha, \beta, \sigma)$ in Eq. (7.4).

and the precision $\frac{1}{\sigma^2}$ :

$$\begin{cases} log(\alpha) \sim Uniform(-10, 10) \\ \beta \sim Uniform(-10, 10) \\ \sigma^{-2} \sim Gamma(1, 0.25) \end{cases} \tag{7.5}$$

Bayesian analysis was simply run through a WinBUGS code[2] translating the DAG given in Fig. 7.4. Three different chains were launched from different initial seeds. Visual inspection of the three chains indicates that convergence was acceptable after 5000 replicates. These runs were discarded as a burn-in phase and a new chain was started with 50,000 replicates that were thinned 1 over 10 to get a 5000 posterior sample of parameters. Results are summed by the main marginal posterior features given in Table 7.2.

These results show that:

- The Margaree Salmon population has a good replacement ratio; in the vicinity of the origin, 1 adult will be replaced by 6 juveniles

[2]Although the inference of such a model could benefit from an explicit solution following the approach detailed in Chapter 7 (see Eq. (7.15)).

| Parameters | Mean | Sd | 2.5% pct | 97.5% pct |
|---|---|---|---|---|
| $\alpha$ | 6.07 | 0.64 | 4.91 | 7.42 |
| $\beta$ | 0.00049 | 0.00007 | 0.00036 | 0.00062 |
| $R_{max}$ | 4600 | 433 | 3874 | 5562 |
| $S_{max}$ | 2084 | 298 | 1613 | 2777 |
| $\sigma$ | 0.43 | 0.05 | 0.34 | 0.54 |

**TABLE 7.2**: Main features of marginal posterior distributions obtained with a Ricker-type recruitment function, noninformative prior on natural parameters (see Eq. (7.5)) and logNormal random variations. $S_{max}$ is the value of stock producing the maximum recruitment $R_{max}$.

and we will bet that this ratio stands between 5 and 8 with high confidence;

- The knowledge of the carrying capacity $S_{\max}$ remains rather uncertain (a 95% posterior credible interval spans between 1628 and 2760) with a posterior mean around 2100 adults;

- The environmental stochasticity is quite high. Recalling that the coefficient of variation of a logNormal distribution is $\sqrt{(e^{\sigma^2}-1)} \approx \sigma$ when $\sigma^2$ is small, a standard deviation of $\sigma = 0.4$ for the log-value is roughly equivalent to a 40% standard deviation for the relative error between the quantity of interest $R$ and its phenomenological prediction!

### 7.3.2 Working with management parameters

Fisheries scientists often prefer working with reformulations of *SR* relationships given by Eq. (7.4) involving parameters directly related to management. For instance, Schnute and Kronlund [274] (also used by [258] and [275]) suggested that the Ricker and the Beverton-Holt functions could advantageously be rewritten in terms of management-related parameters such as the stock, $S^*$, producing the maximum sustainable yield, $C^*$. Let us consider a sustainable population (*i.e.*, $\alpha > 1$), with a *SR* relationship that is stable over time (constant parameters), and submitted each year to a constant exploitation of recruitment equal to $C$ that produces an equilibrium state $R - C = S$, where $S$ and $R$ are expressed in the same unit (adults, eggs, ...) (see Fig. 7.5). There is a single equilibrium state for which captures are maximum, denoted $C^*$, obtained for a stock $S^*$, verifying $R^* - C^* = S^*$. The transformation is one-to-one: for each parameter vector $(C^*, S^*)$ in $]0, +\infty[\times]0, +\infty[$, a

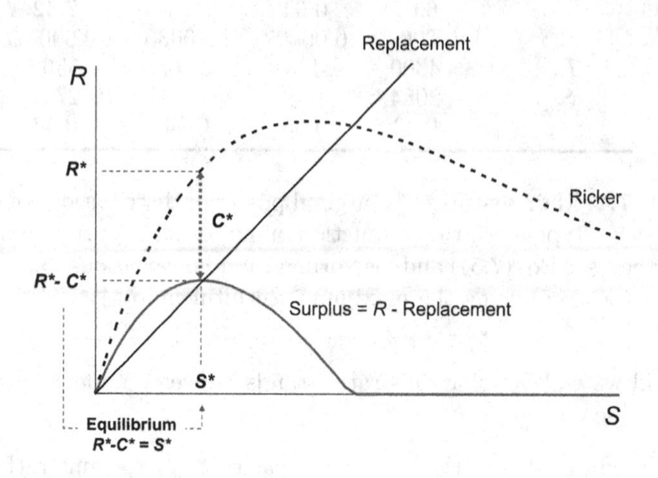


**FIGURE 7.5**: Management-related parameters for the Ricker stock-recruitment relationship ($S$ and $R$ expressed in the same unit).

unique pair $(\alpha, \beta)$ can be deduced through a closed transformation. Alternatively, one might prefer working with the couple $(S^*, h^*)$ instead of $(S^*, C^*)$, where $h^* = \frac{C^*}{C^*+S^*}$ is the harvest rate at equilibrium. $(S^*, C^*)$ or $(S^*, h^*)$ are reference points for stock assessment and management. Under a fixed escapement strategy, $S^*$ is the spawning escapement that should be reached to maximize the average long-term catch (see Appendix C for a sketch of the proof). Thereby, reformulating $SR$ relationships as a function of $(S^*, C^*)$ or alternatively of $(S^*, h^*)$ leads to straightforward inferences on reference points helpful to fisheries managers.

For the Ricker model, the natural parameters $(\alpha, \beta)$ are linked to the management parameters $(S^*, C^*)$ and $(S^*, h^*)$ by the following relations:

$$\begin{cases} \alpha = \dfrac{(S^* + C^*)}{S^*} \cdot e^{\frac{C^*}{S^*+C^*}} \\ \beta = \dfrac{C^*}{S^*(S^* + C^*)} \end{cases} \tag{7.6}$$

or equivalently

$$\begin{cases} \alpha = \dfrac{1}{1-h^*} \cdot e^{h^*} \\ \beta = \dfrac{h^*}{S^*} \end{cases} \tag{7.7}$$

Equations (7.6) and (7.7) are obtained by searching $((S^*, C^*))$ as solutions of the system:

$$\begin{cases} Equilibrium\ conditions: \ R - C = S \\ Maximization\ of\ catches: \ \dfrac{\partial(R-S)}{\partial S}|_{S=S^*} = 0 \end{cases} \tag{7.8}$$

Working with management parameters makes it easier to define an informative prior. For instance, the biologist would *a priori* (*i.e.*, without seeing the data) say:

- that his best guess for the Margaree River stock at maximum sustainable yield is $S^* = 1000$ individuals,
- that $S^*$ lies between 700 and 1300 as a 70% credible set,
- and that the optimal sustainable exploitation rate $h^*$ is likely to lie around 0.75.

We may tentatively model these prior knowledge by the following prior structure:

$$\begin{cases} h^* \sim Beta(3,1) \\ S^* \sim Normal(1000, 300^2) \end{cases} \tag{7.9}$$

The full prior structure must be completely specified by adding a prior on the variance $\sigma^2$. The Gamma distribution $Gamma(p,q)$ is commonly chosen for the precision $\sigma^{-2}$ . Improper priors are obtained by letting $p$ and $q$ go toward zero. However, we will assume in the following sections $p = 1; q = 0.25$, *i.e.*, an exponentially decreasing precision but still some possibly large environmental noise (with an essential relative possible variation not far from 50% of the signal as a prior bet).

Relying on these informative priors, Bayesian inference of the same Ricker model with logNormal error is obtained through WinBUGS with the same MCMC configuration than the one used in Section 7.3.1. Figure 7.6 shows the shape of the posterior distribution for the main parameters $(S^*, h^*, \sigma)$, and the main statistics of marginal posterior pdf are summed up in Table 7.4.

When comparing the prior and posterior distributions for the stock $S^*$ at sustainable yield and for the sustainable harvest ratio $h^*$, one can see that the data are compatible with the expert prior but they make us

| Parameters | Prior |
|---|---|
| $S^*$ | $\sim Normal(\mu = 1000, \sigma = 300)$ |
| $h^*$ | $\sim Beta(3,1)$ |
| $\frac{1}{\sigma^2}$ | $\sim Gamma(p,q),\ p = 1, q = 0.25$ |
| $\alpha$ | $= \frac{1}{1-h^*} \cdot e^{h^*}$ |
| $\beta$ | $= \frac{h^*}{S^*}$ |

**TABLE 7.3**: Prior distribution on management related parameters applied to the Margaree stock-recruitment data.

| Parameters | Mean | Sd | 2.5% pct | 97.5% pct |
|---|---|---|---|---|
| $C^* = R^* - S^*$ | 2862 | 288 | 2330 | 3459 |
| $R^*$ | 4209 | 351 | 3571 | 4948 |
| $\alpha$ | 6.2 | 0.62 | 5.08 | 7.50 |
| $S_{max}$ | 1989 | 229 | 1596 | 2488 |
| $S^*$ | 1347 | 124 | 1132 | 1621 |
| $h^*$ | 0.68 | 0.02 | 0.63 | 0.72 |
| $\sigma$ | 0.43 | 0.05 | 0.34 | 0.54 |

**TABLE 7.4**: Main features of marginal posterior distributions obtained with a Ricker-type recruitment function, informative priors on management parameters (see Table 7.3) and logNormal random variations.

more confident when considering the range of uncertainties. Compared with Table 7.2, $S_{\max}$ tends to be smaller and the environmental noise level $\sigma$ remains in the same range.

Figure 7.7 visualizes the main results from the Bayesian inference of the Ricker model with logNormal noise. Uncertainty about the mean Ricker model $f(S) = \alpha S e^{-\beta S}$ stems from the posterior pdf of parameters $\alpha$ and $\beta$ and a light gray band is formed by the 90% credible interval for each value of $S$. The upper 95% and lower 5% quantiles of the posterior predictive distribution of the Ricker model with lognormal noise have also been drawn in Figure 7.7 and theses two lines encompass 90% of the possible values for the data (indicated by crosses). These lines are not as smooth as they theoretically should be since they have been derived empirically from the MCMC sample from $[\alpha, \beta, \sigma | R, S]$. A cloud of 5000 simulated values of $(S^*, R^*)$ locates the management parameters on the graph with their bivariate posterior distribution. Figure 7.7 points out that the multiplicative logNormal model is a rather poor (over-dispersed) choice for the environmental noise; there is an unexplained hump of $S$ in the neighborhood of $S^*$ and no data at all is laying out of the 95%

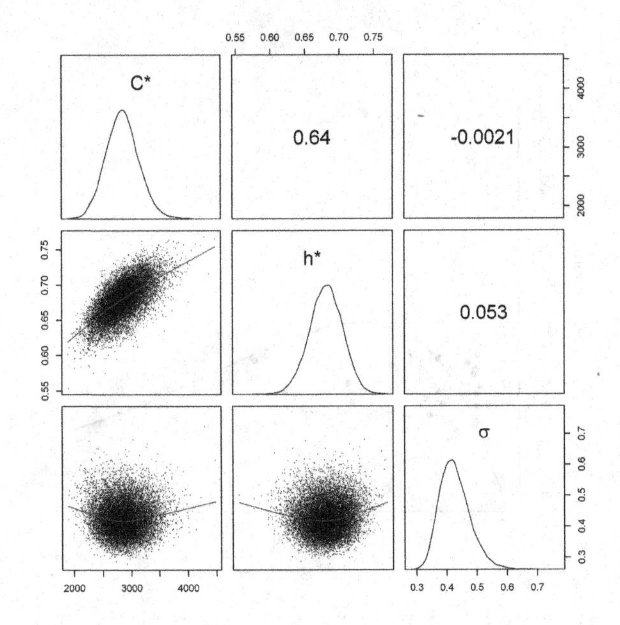


**FIGURE 7.6**: Posterior distributions of key parameters $(C^*, h^*, \sigma)$ obtained using the Ricker-type SR model with logNormal random variations. Marginal distributions are shown in the diagonal. Joint MCMC draws are shown in the lower part. The upper part shows linear correlations between the MCMC draws.

predictive range (although it should concern approximately 5% of the sample).

## 7.4 Changing the error term from logNormal to Gamma

Under the standard logNormal formulation (Eq. (7.4)), the recruitment variance is proportional to the square of the Ricker function $f(S) = \alpha S e^{-\beta S}$. Consequently, for $S > \frac{1}{\beta}$, *i.e.*, for large values of $S$, the recruitment variance behaves like $f(S)$ does, that is, it decreases with $S$. We are willing to explore an alternative hypothesis to render the recruit-

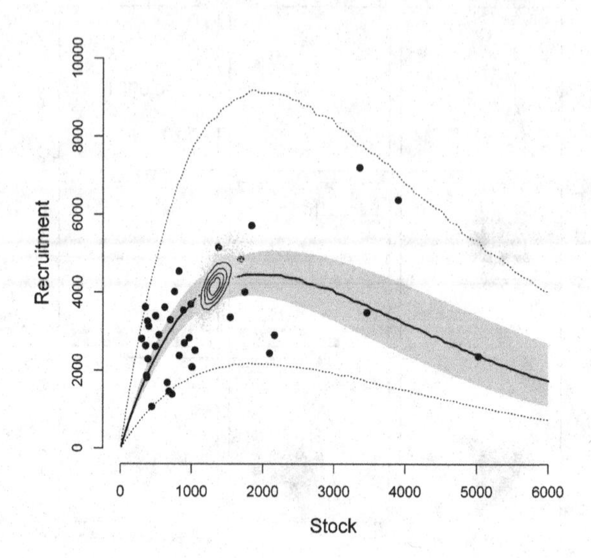


**FIGURE 7.7**: Ricker model fitted with logNormal noise. The gray zone shows the 90% credible interval for the model (uncertainty around parameters $(S^*, h^*)$ only), the upper and lower lines (dotted) gives a 90% posterior predictive interval for the data (including the environmental noise $\sigma$.) The joint posterior distribution of management parameters appears as a cloud showing the uncertainty about $(S^*, R^*)$.

ment variance proportional to $S$ for the whole range of $S$, something like $\sigma^2(R) = \delta S$ with $\delta$ an unknown constant. Implementation is easy when assuming a Gamma distribution for the recruitment given the stock. The gamma pdf also belongs to a two-parameter family of distributions, with parameters usually denoted by $a$ (shape) and $b$ (inverse scale). They are uneasy to interpret directly, but they are related to the expectation $\mu$ and the variance $\sigma^2$ by the following relationships:

$$\begin{cases} \mu = \dfrac{a}{b} \\ \sigma^2 = \dfrac{a}{b^2} = \dfrac{\mu}{b} \end{cases} \tag{7.10}$$

Therefore, if we want the mean recruitment to be the Ricker function and the recruitment variance proportional to the stock, we simply have to take:

$$\begin{cases} \mu(R) = \dfrac{a(S)}{b(S)} = f(S) = \alpha \cdot S \cdot e^{-\beta \cdot S} \\ \sigma^2(R) = \dfrac{\mu(R)^2}{a(S)} = \delta \cdot S \end{cases} \quad (7.11)$$

so that:

$$\begin{cases} R \sim Gamma(a(S), b(S)) \\ with \\ a(S) = \dfrac{\alpha^2 \cdot S \cdot e^{-2\beta \cdot S}}{\delta} \\ b(S) = \dfrac{\alpha \cdot e^{-\beta \cdot S}}{\delta} \end{cases} \quad (7.12)$$

Table 7.5 contains the Bayesian inferences obtained with the formulation (7.12). The same priors as previously were used. We chose $\frac{\delta}{S*} \sim Gamma(1, 4)$ to keep with the prior belief that the relative error ratio $\frac{\sigma(S^*)}{S*}$ stands around 50% since $\frac{\sigma^2(S^*)}{S^{*2}} = \frac{\delta S^*}{S^{*2}}$ will be expected to lie around the mean value of a $Gamma(1, 4)$ distribution, *i.e.*, $\frac{1}{4} = (0.5)^2$.

| Parameters | Mean | Sd | 2.5% pct | 97.5% pct |
|---|---|---|---|---|
| $C^* = R^* - S^*$ | 2972 | 301 | 2426 | 3623 |
| $R^*$ | 4295 | 411 | 3570 | 5151 |
| $\alpha$ | 6.54 | 0.66 | 5.33 | 7.93 |
| $S_{max}$ | 1916 | 285 | 1408 | 2525 |
| $S^*$ | 1323 | 164 | 1018 | 1660 |
| $h^*$ | 0.69 | 0.02 | 0.64 | 0.74 |
| $\sigma$ | 0.86 | 0.12 | 0.64 | 1.09 |

**TABLE 7.5**: Main features of marginal posterior distributions obtained with a Ricker-type recruitment function, informative priors on management parameters (see Table 7.3) and a Gamma random variations assuming the variance is proportional to the stock.

Once again, prior and posterior distributions for the stock $S^*$ at sustainable yield and for the sustainable harvest ratio $h^*$ do not point out major discrepancies between expert judgment and information conveyed by the data (see Table 7.5 which is quite comparable to Table 7.4). But a much better fit of environmental noise is shown in Fig. 7.8. The data look much more in agreement with their predictive distribution when working with the gamma error term than with the logNormal noise. The hump around $S^*$ has vanished and the environmental noise increases with the number of spawners, not with the number of recruits; unfortunately, there few data with a large number of spawners to strongly validate this

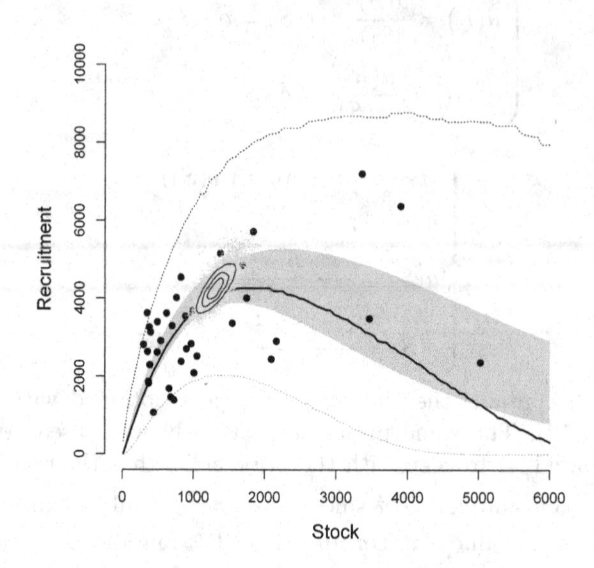


**FIGURE 7.8**: Ricker model fitted with Gamma random variations. The gray zone shows the 90% credible interval for the model (uncertainty around parameters $(S^*, h^*)$ only), the upper and lower lines (dotted) give a 90% posterior predictive interval for the data (including the environmental noise $\sigma$.) The joint posterior distribution of management parameters appears as a cloud showing the uncertainty about $(S^*, R^*)$.

hypothesis. Additionally, for large values of the stock, the distribution is highly skewed: the curve of the posterior mean for recruitment (solid line) lies out of the 90% credible interval for the model.

## 7.5 Changing the explanatory structure from Ricker to Beverton-Holt

In this section, we turn back to Eq. (7.1) and its multiplicative simplification Eq. (7.3). We seek to incorporate a saturation effect $R = \alpha S/(1 + \beta S)$ as a guideline (Beverton-Holt) for the explanatory structure $f_1(S)$. Consider a logNormal model for the unexplained part of the phenomena $f_2(\varepsilon)$ :

$$\begin{cases} R = \dfrac{\alpha}{1+\beta \cdot S} \cdot e^{\varepsilon} \\ \varepsilon \sim Normal(0, \sigma^2) \end{cases} \tag{7.13}$$

The management parameters can be derived from the following relationship:

$$\begin{cases} \alpha = \dfrac{R^{*2}}{S^{*2}} \\ \beta = \dfrac{R^* - S^*}{S^{*2}} \end{cases} \tag{7.14}$$

| Parameters | Mean | Sd | 2.5% pct | 97.5% pct |
|---|---|---|---|---|
| $C^* = R^* - S^*$ | 2294 | 206 | 1918 | 2719 |
| $R^*$ | 3191 | 319 | 2622 | 3883 |
| $\alpha$ | 13.75 | 5.02 | 8.11 | 26 |
| $R_{max}$ | 4453 | 596 | 3416 | 5772 |
| $S^*$ | 897 | 185 | 559 | 1292 |
| $h^*$ | 0.72 | 0.04 | 0.65 | 0.80 |
| $\sigma$ | 0.39 | .05 | 0.31 | 0.49 |

**TABLE 7.6**: Main features of marginal posterior distributions obtained with a Beverton-Holt recruitment function, informative priors on management parameters (see Table 7.3) and logNormal random variations.

Table 7.6 sums up the Bayesian inference for the Beverton-Holt model with a logNormal noise. For the management parameters $S^*$ and $h^*$ and the environmental noise $\sigma$, the same informative priors as previously have been used. The sustainable stock $S^*$ for this model seems to be much lower than the other models with a more intensive harvest ratio $h^*$ (see Tables 7.4 and 7.5). But compared to the Ricker model with logNormal noise, a lower posterior estimate of $\sigma$ seems to indicate a better fit of the environmental noise. Shapes of the posterior distribution for the main parameters ($C^*$,$h^*$,$\sigma$) are shown in Fig. 7.9.

Figure 7.10 shows how the model behaves: the general shape differs much from the Ricker structure since there is an asymptotic number of recruits around 4000 individuals (with a rather large posterior uncertainty), the predictive confidence interval increases with $S$ (and $R$ which, on average, increases monotonically as $S$), but the 95% predictive quantile might be far from the data and rather overpessimistic.

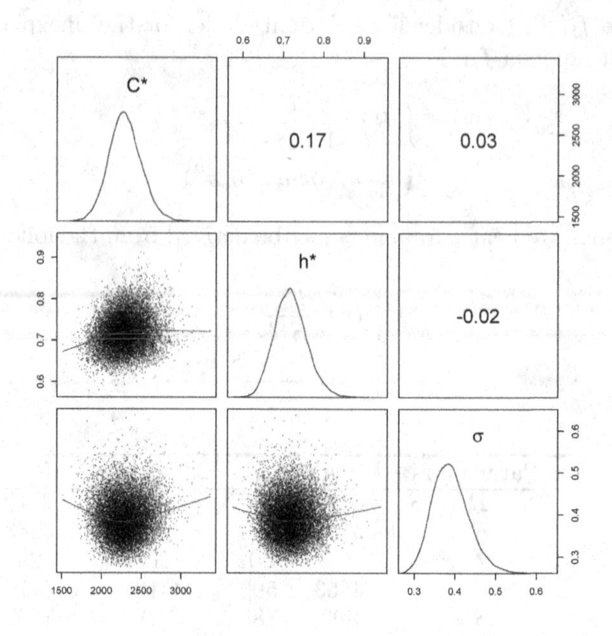


**FIGURE 7.9**: Posterior distributions of key parameters $(C^*, h^*, \sigma)$ obtained using the Beverton-Holt *SR* model with logNormal random variations. Marginal distributions are shown in the diagonal. Joint MCMC draws are shown in the lower part. The upper part shows the linear correlation between the MCMC draws.

---

## 7.6 Model choice with informative prior

One may wish to consider a formal test to choose between the three proposed models: Beverton-Holt-type with logNormal random variations $(M_1)$, Ricker-type and logNormal random variations $(M_2)$, Ricker-type and Gamma random variations $(M_3)$. Such comparisons can be made through Bayes Factors. Bayes Factors were already introduced in Chapter 4 (Section 4.3.3.1). The Bayes Factor $B_{i,j}$ given in Eq. (4.16) is the tool of choice to evaluate the relative increase of evidence (from prior to posterior, given the data) in favor of model $M_i$ over model $M_j$. Indeed, given the data $\mathbf{y}$, $B_{i,j}$ is calculated as the ratio of marginal likelihoods between models $M_i$ and $M_j$, corrected by the ratio of priors. The parameters $\theta = (S^*, h^*, \sigma^{-2})$ have the same meaning for the three com-

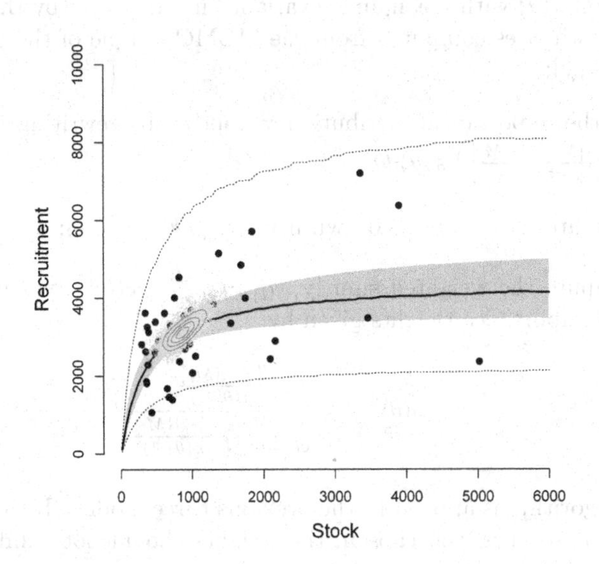


**FIGURE 7.10**: Beverton-Holt model fitted with logNormal random variations. The gray zone shows the 90% credible interval for the model, the upper and lower lines gives a 90% posterior predictive interval for the data. The posterior of management parameters appears as a cloud showing the uncertainty about $(S^*, R^*)$.

peting stock-recruitment models; they are management reference points and relative environmental noise level. Informative prior knowledge is encoded via proper priors for $\theta$ given in Table 7.3; therefore, the predictive $[\mathbf{y}|M_i]$ (for models $i = 1, 2, 3$) themselves will also be proper. For Beta-Binomial or Normal models presented respectively in Chapter 4 or 6, a closed-form of the marginal likelihood was available thanks to the conjugate properties (recall the mathematical miracle happening in Section 4.3.3.2 for the Beta-Binomial model and Section 6.3.3.3 for the Normal one). For stock-recruitment models, no closed-form expression for the marginal likelihoods is available. An estimation of the marginal likelihood must be derived through numerical integration of the likelihood via a Monte Carlo sampling methods ([157]). As $[\theta|M_i][\mathbf{y}|\theta, M_i]$ is proportional to $[\theta|M_i, \mathbf{y}]$, a good numerical approximation can be obtained by importance sampling techniques as follows:

1. For each model $i$, approximate $[\theta|M_i, \mathbf{y}]$ by the multi-Normal distribution $\pi(\theta)$ with mean and covariance matrix given by the empirical estimates computed from the MCMC sample of the posterior $[\theta|M_i, \mathbf{y}]$;

2. Let the importance distribution $\pi$ appears by rewriting $[\mathbf{y}|M_i] = \int_\theta \left(\frac{[\theta|M_i][\mathbf{y}|\theta,M_i]}{\pi(\theta)}\right)\pi(\theta)d\theta$;

3. Generate a $G-$ sample drawn from $\pi$, $(\theta^{(g)})_{g=1,\ldots,G}$;

4. Compute the weighted sum $\widehat{[\mathbf{y}|M_i]} = \sum_{g=1}^{G} \omega(\theta^{(g)})[\mathbf{y}|\theta^{(g)}, M_i]d\theta$, with importance weights given by:

$$\omega(\theta^{(g)}) = \frac{\frac{[\theta^{(g)}|M_i]}{\pi(\theta^{(g)})}}{\frac{1}{G}\sum_{g=1}^{G}\frac{[\theta^{(g)}|M_i]}{\pi(\theta^{(g)})}}$$

This algorithm is applied to the previous three models. If the importance distribution is well chosen, the weights should not exhibit large variance, allowing for a stable computation of $\widehat{[\mathbf{y}|M_i]}$. As an example, Figure 7.11 shows that the 5000 weights obtained for the logNormal Ricker model are relatively balanced since, when plotting their cumulative distribution, one can see that it is not too far away from the Uniform distribution (first diagonal). Similarly good computational behavior is observed for the other models (not shown here). Finally, Table 7.7 points out that the Beverton-Holt model with LogNormal environmental noise appears to be the best choice among the three competing proposals.

| Model | Rank | $Log(\widehat{[\mathbf{y}|M_i]})$ | $BF_{1vsi}$ |
|---|---|---|---|
| Beverton-Holt + LogNormal noise | 1 | −335.75 | 1 |
| Ricker + LogNormal noise | 2 | −340.23 | 88 |
| Ricker + Gamma noise | 3 | −340.56 | 123 |

**TABLE 7.7**: Marginal likelihood $\widehat{[\mathbf{y}|M_i]}$ for the three competing models with informative priors. Bayes Factors of model $M_1$ (Beverton-Holt-type with logNormal random variations) *versus* all other models were calculated.

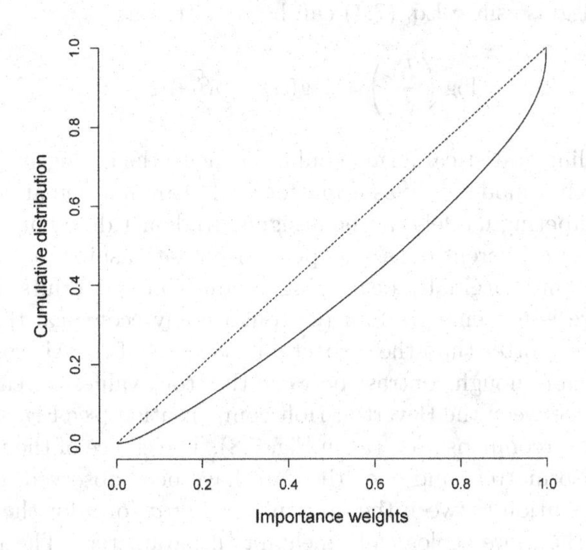


**FIGURE 7.11**: Checking the IS computation for the Ricker model with logNormal noise. Cumulative distribution of the importance weights (solid line) compared with cumulative distribution of a Uniform distribution (dotted line).

---

## 7.7 Conclusions and perspectives

Ecological models have to account for the environmental stochasticity. When investigating the Margaree River case study, the part due to the noise in the data can amount up to 40% of the signal! In the present stock-recruitment model, noise is a convenient all-in-one-bag concept for unexplained variations; it encompasses both:

1. Stochasticity in survival of eggs that may vary due to environmental factors such as water temperature or high and low riverflows;
2. Experimental or observation errors when counting the recruits, such as the ones resulting from successive removal techniques and extrapolation of abundance estimates from a few sampling sites to the whole river.

The *SR* Ricker model with logNormal noise is a linear regression model in disguise since Eq. (7.4) can be rewritten as

$$\log\left(\frac{R_t}{S_t}\right) = \log(\alpha) - \beta S_t + \varepsilon_t \tag{7.15}$$

This appealing statistical form should not incite the ecologist to loosely embrace such a model only as a matter of (linear) mathematical convenience. Competing models can be designed to depict different ecological behavior. Here different density dependence relationships were encoded in the different marginally decreasing production of recruits. Unfortunately, there is not enough data (although many ecological time series data sets are shorter than the respectable 38 years of the Margaree data collection!) nor enough contrast between the stock values to really make a difference between the Beverton-Holt compensation (see Fig. 7.10) and the Ricker overcompensation (as in Fig. 7.8). The choice of the noise distribution also matters and once the data have been observed, there appears a correlation between the noise parameter $\sigma$ (or $\delta$ for the Gamma noise) and the more ecologically meaningful parameters. The posterior predictive distribution allows to explain the discrepancy between a tentative model and the data. We feel more satisfied with Fig. 7.8 than Fig. 7.7, while a formal Bayes factor helps to make a choice between a set of competing models.

In this chapter, we focused on the estimation of stock-recruitment parameters but we deliberately forgot two important sources of bias:

- *The errors in variable problem.* In the DAG of Fig. 7.4, we put the stock values within a shaded box, which means that we hypothesize that these explanatory variables are known without error. Indeed, these stock data are not covariates, they are observations of stock (that may be far from the "*true* " value stock itself), *i.e.*, stemming from an experiment. In a more realistic model, such observed stock values should themselves be put within shaded ellipses and linked to unknown "*true*" stock that would influence the reproduction phenomenon. In other words, we forgot to take into account a source of randomness which might lead to a complete misconception of the whole structure of the stock recruitment relationship. In turn, the influence of these neglected observation errors can be damaging when performing the estimation of the unknown $(\alpha, \beta, \sigma^2)$.

- *The time series bias.* We cut the ecological dynamics into supposedly independent pieces. Indeed, there is a temporal link between the recruits of one year and the adults of the same cohort returning to their home river to spawn (via a survival rate). These overlooked dynamic aspects might cause dangerous overestimation of the sustainable harvesting rate.

We postpone exploring these important issues until Chapter 11, which we devote to a discussion of dynamic structures and state-space models. We conclude the present chapter by urging the reader to tackle exercises which are available at our website *hbm-for-ecology.org*. This will help the reader to further understand the consequences of model mispecification, a common problem encountered by ecological detectives.

# Chapter 8

## Getting beyond regression models


### Summary

This chapter exemplifies how to go beyond the limitations imposed by Normal linear models (see Chapter 6); the Normal distribution cannot handle binary or count data, for instance. The trick is to use so-called *link functions* to transform a linear combination of the explanatory covariates into the location parameter of the appropriate response distribution.

In previous chapters, we learned how to handle Bernoulli and Binomial pdfs for categorical variables, Poisson distributions for discrete quantities, and Normal, Gamma and Beta distributions for continuous ones. Let's play again, this time by linking a linear regression-like explanation term and

- Bernoulli observables (the so-called logistic and Probit regressions);
- Poisson observables (Poisson regression);
- Ordered categorical responses (ordinal Probit regression).

The remarkable feature is that such hybrid structures do not require much additional effort for inference, especially in the Probit case which can be interpreted as working with a hidden Gaussian layer in the model.


## 8.1 Logistic and Probit regressions

### 8.1.1 Motivating example: Changes in salmon age at smoltification

Atlantic Salmon populations are characterized by an intra-population variation in the reproductive life span. In particular, juveniles spend one or several years in their home river before running to the sea as smolts, and adults may spend one or several winters at sea before returning to their home river as spawners (see the salmon life cycle in Fig. 1.9, page 26 and Table 8.1).

We only care about the four main life histories encountered in the Northwest of France (Table 8.1). Almost all juveniles smoltify after 1 or 2 years spent in freshwater ([15]), they are denoted as 1+ and 2+ smolts, respectively. Fish from the two smolt age classes can return as spawners after one or two winters spent at sea ($1SW$ or $2SW$).

| Sea age | 1 year ($1SW$) | 2 years ($2SW$) |
|---|---|---|
| River age | | |
| $1 + Smolts$ | $n_{11,t}$ | $n_{12,t}$ |
| $2 + Smolts$ | $n_{21,t}$ | $n_{22,t}$ |
| | $n_{1,t}$ | $n_{2,t}$ |

**TABLE 8.1**: 1+ and 2+ smolts spend either one or (at least) two years at sea; see the complete life cycle in Fig. 1.9 of Chapter 1. $n_{ij}$ denotes the number of adults in the river-age class $i$ and sea-age class $j$ that return for spawning.

Inter-annual variation of environmental conditions is known to influence Atlantic salmon demography by changing the balance point between these various anadromous reproductive strategies. In particular, early growth conditions of juveniles are known to influence the probability of smoltification ([15]; [220]; [294]; [302]). Studying the time variations of the age at smoltification helps to understand how Salmon populations react to potential changes in juvenile growth performance, for instance in response to warming (climate change) or in response to local nutrient enrichment and stream productivity due to eutrophication.

For each cohort born on year $t$, the proportions of returning spawners that have migrated as 1+ smolts relative to the total returning spawners (that have migrated as 1+ and 2+ smolts) are considered as indicators of the mean age of smoltification ($MAS$). These proportions were considered for both sea-age classes, and are then estimated as $\frac{n_{11}}{n_{11}+n_{21}}$ and $\frac{n_{12}}{n_{12}+n_{22}}$ for $1SW$ and $2SW$ fish, respectively (see Table 8.1). Data used to compute these proportions were collected as part of the survey of the A. Salmon rod-and-line fishery in Brittany rivers. The original dataset is composed of more than 24,500 reported catches of salmon (and associated archived scales) caught during their spawning migration in rivers of the Armorican Massif (Northwest France; denoted AM in the following) from 1972 to 2005 (see Fig. 8.1). The proportions of 1+ smolts were directly computed from the number of reported catches.

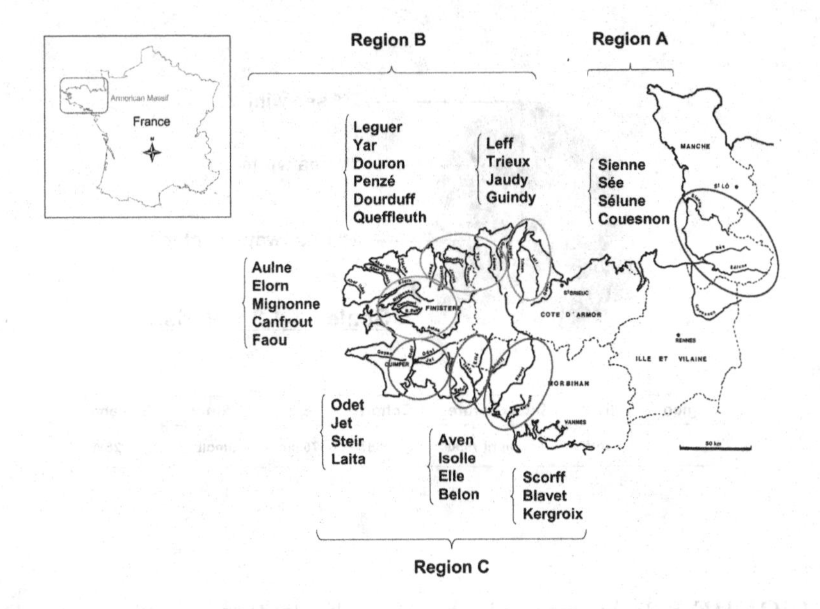


**FIGURE 8.1**: The three geographical units for the rivers in the Armorican Massif, Brittany, France.

Detailed information is available on each captured fish. The date and the place (river) of capture are recorded. The size and weight of each fish are also known, and scale reading ([14]) provides key information on the life history of each fish. The river age ($1 + Smolt$, $2 + Smolt$), sea age ($1SW$ or $2SW$) and other biological information from which the birth date (cohort) of each captured fish can be easily inferred (Fig. 8.2).

We assume that the survey of the rod-and-line fishery of migrating spawners provides a representative random sample for the proportion of 1+ smolts in each sea-age class (this hypothesis will be discussed later in the chapter).

Figure 8.3 shows an increasing trend of the proportion of adults that have migrated as 1+ smolts over the period. It also shows some synchronous fluctuations between the two sea-age classes. By the end of the period, the proportion of 1+ smolts in the $2SW$ component seems to be greater than in the $1SW$ component.

As shown in Fig. 8.1, the data have been grouped according to three geographical units of the Armorican Massif in France. These rivers are located at the southern edge of the species distribution in Europe, which offers a good opportunity to study life history variants in a changing en-

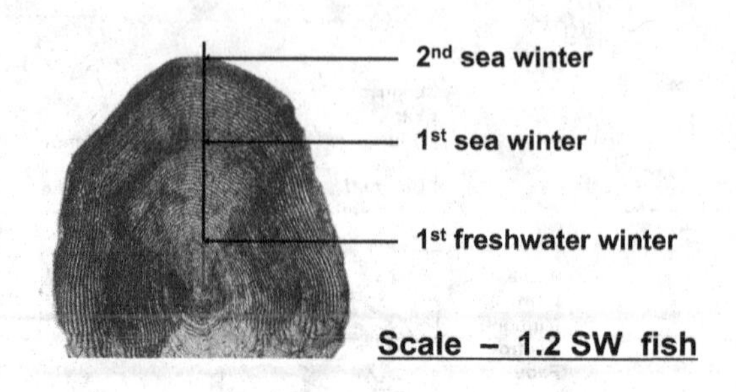


| Region | River | Date capture | Cohort | Size | Smolt | Sea age |
|---|---|---|---|---|---|---|
| C | Scorff | april 1990 | 1987 | 75 cm | Smolt 1 | 2SW |

**FIGURE 8.2**: An archived scale from the database. (Picture from J-Luc Bagliniere, INRA, Rennes.)

vironment (local and global) where unfavorable conditions may occur rapidly. Figure 8.4 shows that the proportion of 1+ smolts is different between regions, with region $A$ (lower Normandy) exhibiting larger proportions of 1+ smolts than regions $B$ (northern and western Brittany) and C (southern Brittany).

The catches were declared by fishermen on a voluntary basis prior to 1986, and they have become mandatory since that date. Figure 8.5 highlights how the sample sizes $n(r, a, t)$ subsequently increased over time.

### 8.1.2 Formalization: The Binomial-logit model

In this section, we propose a Binomial-logit model for the variability of the proportion of 1+ smolts as a function of time (year of cohort birth, between 1973 and 2002, sea-age classes and regions).

Let's call $\pi(r, a, t)$ the probability that a Salmon, born in year $t$, with a sea age $a$ ($1SW$ or $2SW$), from region $r$ (A,B or C), has left its native river as a 1+ smolt. Note that in the sequel, $t$ is the year of the cohort birth: $t = 1$ stands for year 1973 and $t = 30$ stands for year 2002.

The observables are the numbers of fish caught corresponding to the ones in Table 8.1 decomposed by regions. $n_{r,a,t}$ denotes the total number

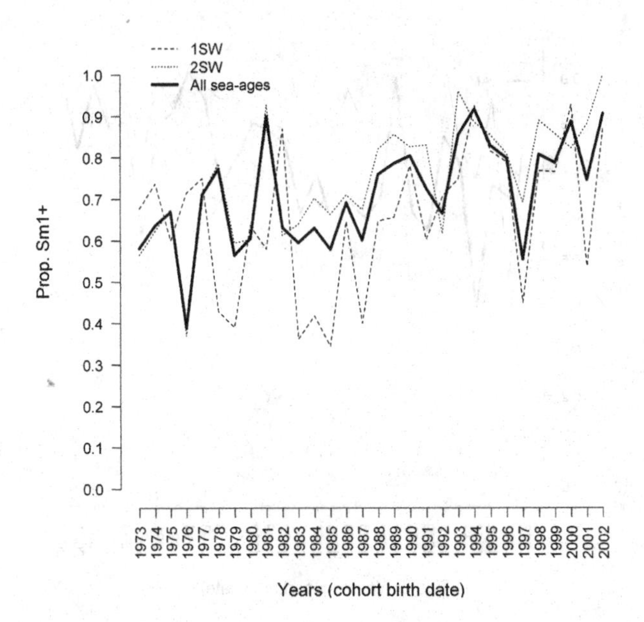


**FIGURE 8.3**: Proportion of fish (empirical values) that have migrated as 1+ smolts for both sea-age classes, for the cohorts born between 1973 and 2002.

of fish born in year $t$ with a sea age $a$, from region $r$, among which $n_{1,r,a,t}$ are 1+ smolts. Under the classical (but questionable) assumption that all fish are independent $n_{1,r,a,t}$ is considered as the result of a Binomial sampling process with $n_{r,a,t}$ *trials* and a probability of success $\pi_{r,a,t}$.

$$n_{1,r,a,t} \sim Binomial(\pi_{1,r,a,t}, n_{r,a,t}) \tag{8.1}$$

Note that in this model, because the sample sizes $n(r,a,t)$ increased over time (Fig. 8.5), a poor precision is to be expected from the data in the very beginning of the time series.

We now propose a *logit* relation to assess how $\pi(r,a,t)$ varies depending on year $t$ (quantitative), and according to the categorical covariates *Region* $r$ and *Sea-age class*. The logit function maps the interval $]0,1[$ into $]-\infty,+\infty[$ such that:

$$p \in ]0,1[ \longmapsto logit(p) = \log\left(\frac{p}{1-p}\right) \tag{8.2}$$

It can be interpreted as the log-odds ratio of the event; it is negative

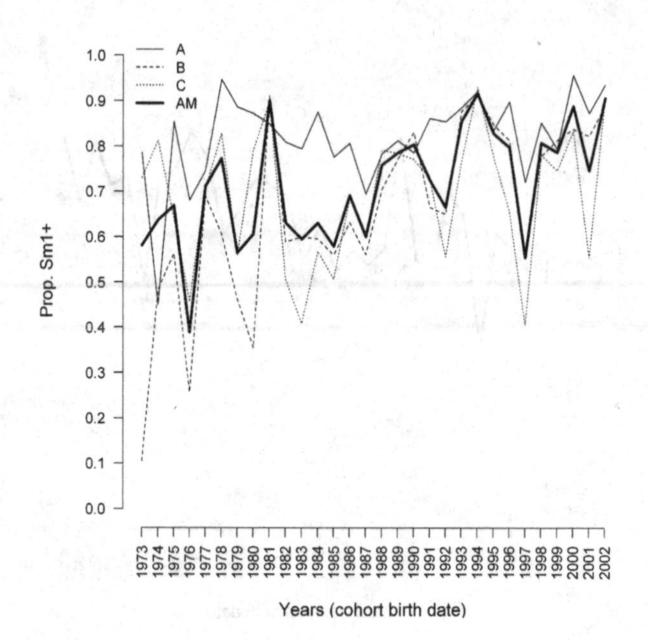


**FIGURE 8.4**: Proportion of fish (empirical values) that have migrated as 1+ smolts for the three regions of the Armorican Massif, Brittany, France, for the cohorts born between 1973 and 2002.

when the odds against the event are bigger than the odds in favor of the event, balanced when $p = 0.5$, and positive otherwise ($p > 0.5$).

As a baseline model $M_0$, the systematic effects of the three covariates are modeled as a linear function in the *logit* scale without any interactions:

$$logit(\pi_{r,a,t}) = \mu + \alpha_t + A_a + B_r \tag{8.3}$$

We assume a linear trend with years:

$$\alpha_t = \alpha \times t$$

The age at sea effect will be centered such that:

$$\begin{cases} A_a = +\beta \text{ if } a = 1SW \\ A_a = -\beta \text{ if } a = 2SW \end{cases} \tag{8.4}$$

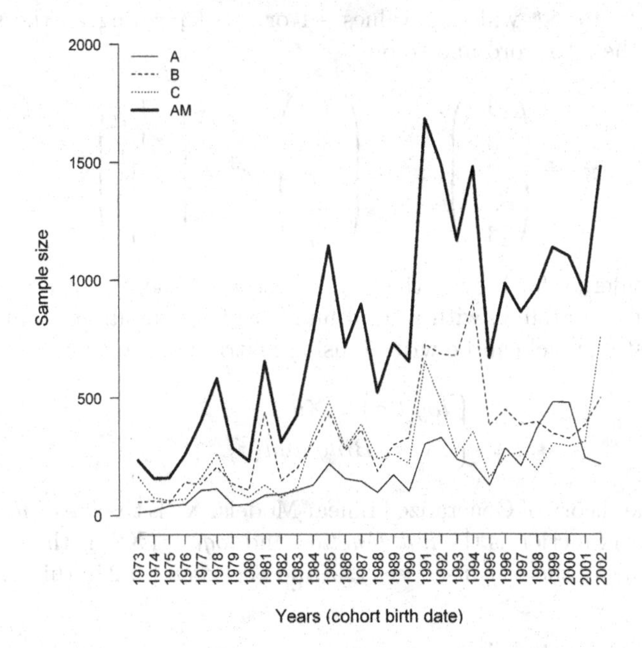


**FIGURE 8.5**: The sample sizes drastically increased over time, driven by the increase in the catch declaration rate over the period.

The regional effect will be modeled as:

$$\begin{cases} B_{r=A} = \gamma \\ B_{r=B} = \delta \\ B_{r=C} = -\gamma - \delta \end{cases} \quad (8.5)$$

It is worth noting that the model $M_0$ described above can be viewed as an extension of the classical linear model such as the one analyzed in Chapter 6. Considering the vector $\theta$ such that $\theta' = (\mu, \alpha, \beta, \gamma, \delta)$, the relationship 8.3 for each record $i$ (a record $i$ is defined by a combination of the three covariates $(r, a, t)$) can be put under the form of a linear regression-like expression in the logit scale:

$$logit(\pi_i) = (X\theta)_i \quad (8.6)$$

in which the design matrix $\mathbf{X}$ is such that each of its five columns $\mathbf{x}^j$, $(j = 1, ..., 5)$ corresponds to an explanatory variable and each row will depict a record. For instance, the column $\mathbf{x}^1$ will correspond to the constant, the $i^{th}$ row of the vector $\mathbf{x}^2$ to the birth year of the $i^{th}$ record, the third column vector $\mathbf{x}^3$ will take values $+1$ or $-1$ depending on the sea-age class of the $i^{th}$ record and so on.

$$\mathbf{x}^1 = \begin{pmatrix} 1 \\ 1 \\ 1 \\ \cdots \\ 1 \end{pmatrix}, \mathbf{x}^2 = \begin{pmatrix} t_1 \\ \cdots \\ t_j \\ \cdots \\ t_n \end{pmatrix}, \mathbf{x}^3 = \begin{pmatrix} 1 \\ 1 \\ -1 \\ \cdots \\ -1 \end{pmatrix}, \ldots$$

Considering for brevity that the $i^{th}$ record is encoded $x_{\mathrm{i,j}}$ for the $j^{th}$ explanatory variable, with a total number of 1+ smolts $n_{1,i}$ among $n_i$, the logistic model can be written using matrix notations:

$$\begin{cases} logit(\pi) = \mathbf{X}\theta \\ n_{1,i} \sim Binomial(\pi_j, n_i) \end{cases}$$

In the theory of Generalized Linear Models, $\mathbf{X}\theta$ is the *linear predictor*, the *logit*() function is the *link function* and $logit^{-1}(\mathbf{X}\theta)$ is the predictor in the scale of the *response* (see [199] and Section 8.1.5.2 in this chapter).

## 8.1.3 Bayesian inference

### 8.1.3.1 Noninformative Zellner priors for parameters

The expression $X\theta$ in the previous equation includes both qualitative factors (belonging to a region, member of a sea-age class) that would be encountered in classical analysis of variance models and quantitative explanatory variables (trend with birth year of the cohort) as in standard regression models. As in the linear model (see Chapter 6, Section 6.2.3), we wish that prior distributions on parameters $\theta$ do not depend upon the system of constraint assigned to the parameters (Eqs. 8.4 and 8.5). We therefore set a Zellner multivariate Student distribution on the parameters $\theta$ (see Appendix D):

The model was run using conjugate Zellner's prior distributions

$$\begin{cases} \theta \sim Normal(m_0, \sigma^2 V_0) \\ V_0 = c \times (\mathbf{X}_0'\mathbf{X}_0)^{-1} \\ \sigma^{-2} \sim Gamma(1,1) \end{cases} \tag{8.7}$$

with $m_0 = 0$ and an informative prior information weighting only one data record ($c = n$). The design matrix $\mathbf{X}$ is defined using the zero-mean system of constraints as defined in Section 6.2.3. The Gamma(1,1) distribution used for the precision $\sigma^{-2}$ is an exponential pdf with mean 1, which scales the prior variance to the range of the data.

#### 8.1.3.2 Posterior distributions

Posterior distributions were obtained through WinBUGS (see the DAG of the model in Fig. 8.6).

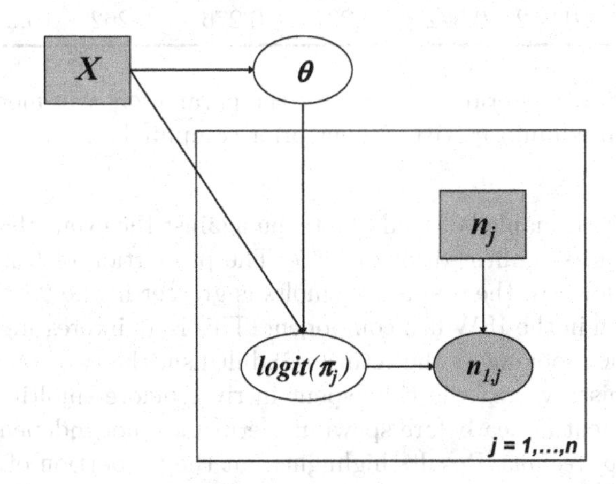


**FIGURE 8.6**: Directed acyclic graph for the model defined in Eq. (8.7) with Zellner prior distributions (Eq. (8.7)). The arrows between $X$ (design matrix) and $\theta$ indicate that Zellner's prior for $\theta$ is a multivariate Normal distribution $Normal_p(m_0, \sigma^2 V_0)$ with $V_0 = c \times (\mathbf{X}'\mathbf{X})^{-1}$.

Table 8.2 presents the marginal posterior statistical features of the components of $\theta$. Figures 8.7 and 8.8 also show the posterior predictive of the proportion of 1+ smolts over time, for the two sea-age classes and the three regions. The posterior predictives are plotted along with the empirical proportions of 1+ smolts. Unsurprisingly, these empirical curves exhibit a larger variability than the corresponding predictions for the proportion of 1+ smolts. This variability depicts the natural dispersion attached to Binomial experiments.

*Time trend.* The posterior variance of the effect of the parameter $\alpha$ is very low since nearly 40 years are employed to fit the linear trend. This indicates that this parameter is well known a posteriori. The effect of the year is positive for sure, indicating an increasing linear trend of the proportion of A. salmon smolts that smoltify after one year only spent in the rivers of Brittany.

*Effect of sea-age classes.* The effect of sea age is also markedly as-

| Param. | Mean | Sd | 5% | 25% | 50% | 75% | 95% |
|---|---|---|---|---|---|---|---|
| $\mu$ | −0.367 | 0.051 | −0.452 | −0.401 | −0.367 | −0.332 | −0.282 |
| $\alpha$ | 0.056 | 0.001 | 0.053 | 0.055 | 0.056 | 0.058 | 0.059 |
| $\beta$ | −0.261 | 0.016 | −0.288 | −0.271 | −0.261 | −0.249 | −0.234 |
| $\gamma$ | 0.457 | 0.026 | 0.416 | 0.439 | 0.457 | 0.475 | 0.500 |
| $\delta$ | −0.196 | 0.020 | −0.229 | −0.210 | −0.196 | −0.182 | −0.162 |
| $-\delta-\gamma$ | −0.262 | 0.022 | −0.297 | −0.276 | −0.262 | −0.248 | −0.225 |

**TABLE 8.2**: Posterior statistics for the parameters $\theta$ of model $M_0$ obtained with noninformative Zellner prior computed with $c = n$.

sessed: no reasonable bet could be taken against the event that $\beta$ is negative, which is confirmed by Fig. 8.7. The proportion of fish that have migrated down to the sea as 1+ smolts is greater in the 2SW fish component than in the 1SW fish component. This is an interesting ecological result. When looking at the return of adult fish, the two components of their life history, *i.e.*, the time spent in river before smoltification and the time spent at sea before spawning return are not independent.

*Effect of regions.* Results highlight that the proportion of smolts 1+ is always greater for the region A (lower Normandy). Credible intervals made from Table 8.2 also show that the values of $\gamma$ and $-\delta-\gamma$ may overlap,suggesting a common effect for the last two regions (B: northern Brittany and C: southern Brittany) as opposed to the first one, the lower Normandy. This is confirmed by Fig. 8.8 that presents the prediction of the proportion of 1+ smolts over time by region.

### 8.1.4 Model comparison

In the baseline model $M_0$ described in the previous section, the systematic effect of the three covariates was modeled as a linear function on $logit(\pi(r,a,t))$ without any interactions (Eq. (8.3)). A strong *a priori* linear behavior underlines this model structure, and one could suspect the model $M_0$ not to be the right one. Would a simpler model collapsing the effects of the first two regions within a single one work better? Would a more complicated model allowing for different time trends for the three regions work better? As seen in the previous chapters, a proper comparison of models requires Bayes factors. In this section, we compare the following models:

- Model $M_1$ is like model $M_0$ but without a regional effect;
- Model $M_2$ is like model $M_0$ but without a sea-age effect;

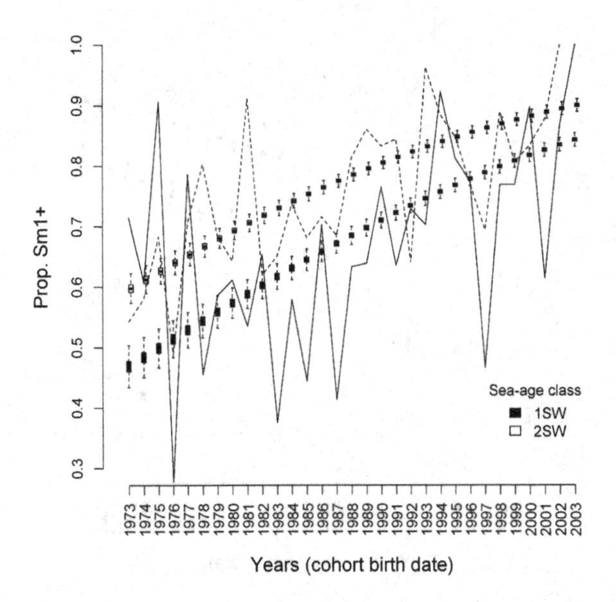


**FIGURE 8.7**: Posterior predictive and empirical values of the proportion of 1+ smolts over time, for the two sea-age classes $1SW$ and $2SW$.

- Model $M_3$ includes a common mean effect, a linear trend with year for each of the sea age and a shared region influence;
- Model $M_4$ is like model $M_3$, but allows for different mean effects corresponding to each sea-age component.

The importance sampling technique described in Section 7.6 of Chapter 7 (see also Appendix B) works well to compute the predictive $[y|M]$ for $M = M_1, ..M_4$. As an example, Fig. 8.9 checks the empirical cumulative distribution of the weights used to compute the predictive distribution of model $M_4$: as expected, the weights cumulate along the first diagonal confirming that they are rather uniformly distributed.

As Bayes Factors are known to be sensitive to prior distributions, therefore, we computed the Bayes Factors for two contrasted Zellner priors. The first Zellner prior structure ($c = n$) gives much weight to the data. In the second one ($c = 1/n$), the prior is much more informative, weighting less than the data. Table 8.3 shows the values of the log-predictive distributions for all competing models for the two alternative Zellner priors. Results highlight that whatever the sensitivity to the prior

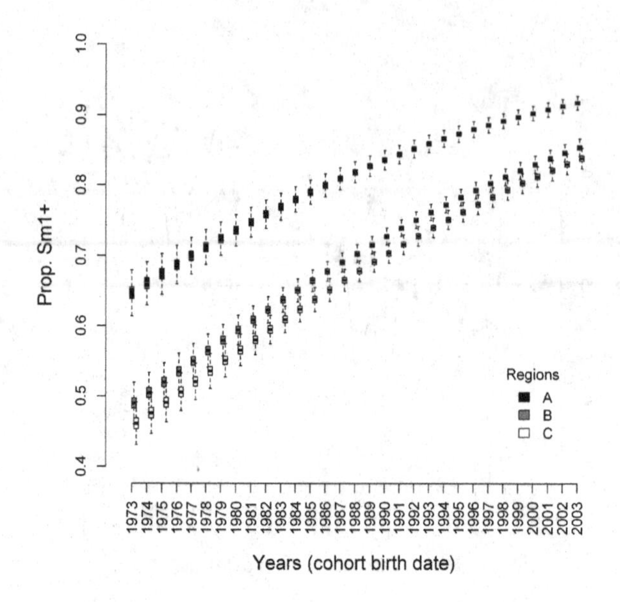


**FIGURE 8.8**: Posterior predictive of the proportion of 1+ smolts over time, for the three regions $A$, $B$ and $C$. Empirical values are not drawn for the sake of clarity.

choice, the model $M_4$ is the best choice among the competing models (see Table 8.4).

### 8.1.5 Discussion

#### 8.1.5.1 Ecological implications of the results

In the analysis, we assumed that the survey of the rod-and-line fishery of migrating spawners provides a representative random sample for the proportion of 1+ smolts in each sea-age class. The fishing process is indisputably selective for the sea-age character. However, the hypothesis we made is likely to be verified, since for a given sea-age class, one can hardly think about a reason why fishing could be considered as a selective sampling process for the river-age characteristics.

However, the proportion of 1+ smolts in returning spawners is not necessarily a good indicator of the proportion of 1+ smolt measured at the smolt stage. Indeed, 1+ and 2+ smolts have different survival rates

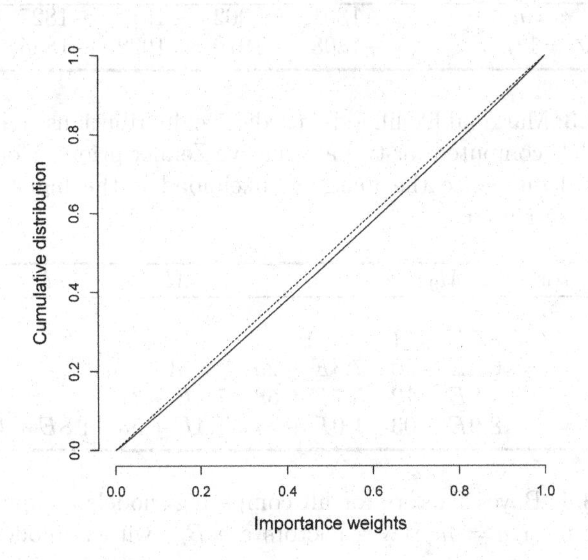


**FIGURE 8.9**: Model $M_4$. Cumulative distribution of the importance weights (solid line) compared with cumulative distribution of a Uniform distribution (dotted line).

during their sea sojourn. The survival rate of 2+ post-smolts during the first months at sea is likely to be higher than for 1+ post-smolts. As a consequence, the proportion of 1+ smolts in returning adults is likely to be a negatively biased estimate of the proportion of 1+ smolts measured at the smolts stage. Moreover, the survival rate of 1+ post-smolts during the first months at sea is likely to be more variable than the survival rate of 2+ post-smolts. Hence, the short- to medium-term fluctuations of the proportion of 1+ smolts observed at the spawners stage could rather result from fluctuations of the post-smolts survival rate at sea than from real fluctuations in the demographic composition at the smolt stage.

Beyond these remarks, results highlight that an important change has occured in the A. salmon populations of the Armorican mountain range over the last 30 years. In the Armorican mountain range, the time trend of the proportion of 1+ smolts per cohort over the period 1973-2002 (year of cohort birth) combines a long-term increasing trend with medium-term fluctuations that are synchronous across the three regions and across sea-age classes (not analyzed here but clearly visible in the series). The increasing trend observed in the proportion of 1+ smolts is

| Log marginal likelihood | $M_0$ | $M_1$ | $M_2$ | $M_3$ | $M_4$ |
|---|---|---|---|---|---|
| $c = n$ | −1801 | −1962 | −1914 | −1827 | −1792 |
| $c = 1/n$ | −1808 | −1970 | −1922 | −1835 | −1799 |

**TABLE 8.3**: Marginal likelihoods (predictive distributions) for all competing models computed for two alternative Zellner priors. Model $M_4$ is the preferred one since the marginal likelihood is the higher one, no matter the prior used.

| Bayes Factors | $M_0$ | $M_1$ | $M_2$ | $M_3$ | $M_4$ |
|---|---|---|---|---|---|
| $M_0$ | 1 | | | | |
| $M_1$ | $8.5E-71$ | 1 | | | |
| $M_2$ | $6.2E-50$ | $7.3E+20$ | 1 | | |
| $M_3$ | $4.8E-12$ | $5.7E+58$ | $7.8E+37$ | 1 | |
| $M_4$ | $8.9E+03$ | $1.0E+74$ | $1.4E+53$ | $1.8E+15$ | 1 |

**TABLE 8.4**: Bayes Factors for all competing models computed with the Zellner prior $c = n$. Bayes Factors are $B_{i,j}$ with $i$=model in line, $j$=model in column.

similar for both $1SW$ and $2SW$ sea-age class components, and cannot be explained by the increasing proportion of the $1SW$ component in the samples. Indeed, the proportion of 1+ smolts is greater in the $2SW$ component than in the $1SW$ component, and the decreasing trend of the $2SW$ component in the sample would rather have caused the proportion of 1+ smolts to decrease. The synchrony of the medium term signal between the three regions suggests a response to an environmental forcing variable acting at the regional scale, such as climate. Moreover, the synchrony between the two sea-age classes suggests the response to an environmental forcing acting during a period when 0+ juveniles or 1+ post-smolts, respectively experience the same environmental conditions at the same time, *i.e.*, the first year of the fresh-water phase (0+ juveniles) or during the first year at sea for 1+ post-smolts. Following these results, further analysis and model development could be initiated toward the research of a biological or ecological mechanism explaining the response of A. salmon populations to fluctuations of the environmental variables.

#### 8.1.5.2 About GLM

This section tries to integrate the model previously developed in a more general perspective and may be skipped at first reading. The previous models $M_0, ..., M_4$ belong to the so-called family of *GLM*, *i.e.*,

Generalized Linear Models ([199]). The Bayesian approach to handle such models is presented in Dey *et al.* [89]. More generally a *GLM* is specified by two functions:

1. $\mu(\mathbf{X}) = \mathbb{E}(y|\beta, \mathbf{X})$ depends on a linear function of the covariates $\mathbf{X}$ and the parameters $\beta$ through a so-called *link* function $g$:

$$g(\mu) = \mathbf{X}\beta$$

For an identifiability reason, $g$ is a one-to-one mapping and equivalently, one can write $\mu$ as a function of the *linear predictor* $\mathbf{X}\beta$

$$\mu = g^{-1}(\mathbf{X}\beta)$$

2. The random component specifying the distribution of the observed variable $Y$ with mean $\mathbb{E}(y|\beta, \mathbf{X}) = \mu(\mathbf{X})$ and possibly another parameter to tune dispersion. Standard models include Poisson for counts, Binomial for binary (or sums of binary) data, Normal as a special case for continuous data. Only in the latter case, a second parameter is required for the variance. The previous pdfs belong to the *exponential family.* This family encompasses most of the pdfs presented in introductory probability courses: notably the Normal, exponential, Binomial, geometric and Poisson distributions. The practical section of this chapter (see our website *hbm-for-ecology.org*) proposes to study other *GLMs*, in particular the Poisson model with a logarithm link function. The general representation of a pdf belonging to the *exponential family* with parameter $\theta$ is

$$[y|\theta] = \exp(a(y) + b(\theta) + c(y)d(\theta))$$

$c(y)$ is the sufficient statistics, $b(\theta)$ is defined by the constraint that $\int [y|\theta]dz = 100\%$. Note[1] that attention is not necessarily focused onto the mean and there are many ways of parameterizing such an expression. When $c(y) = y$, an interesting reparametrization

[1]One can demonstrate with a bit of algebra that

$$\begin{cases} \mathbb{E}(c(Y)) = -\dfrac{\frac{\partial b(\theta)}{\partial \theta}}{\frac{\partial d(\theta)}{\partial \theta}} \\ \mathbb{V}ar(c(Y)) = \dfrac{-\frac{\partial^2 b(\theta)}{\partial \theta^2}\frac{\partial d(\theta)}{\partial \theta} + \frac{\partial^2 d(\theta)}{\partial \theta^2}\frac{\partial b(\theta)}{\partial \theta}}{\left(\frac{\partial d(\theta)}{\partial \theta}\right)^3} \end{cases}$$

known as the natural parameter is to make the transformation $\phi = d(\theta)$. Of course, one may also choose to adopt the mean $\mu$ as the model parameter. In such a case, the *canonical* link is the function $d(\mu)$ of the mean parameter that then appears in the exponential form of the probability density. Following tradition, a logistic link function has been used in models $M_0, ..., M_4$ with a Binomial likelihood (the canonical link!). There is nothing especially compelling about using the canonical link, only remarkable for its mathematical aesthetics. Other link functions could have been chosen; any member of the general class of the reciprocal of continuous cumulative distribution functions mapping the interval $]0,1[$ into $]-\infty,+\infty[$ could have been used as a link function. Taking the $N(0,1)$ cumulative function $\Phi$ leads to the so-called *Probit* model that will be developed in the next section.

---

## 8.2 Ordered Probit model

### 8.2.1 Motivating example: Which skate species can be seen at various depths?

We consider here skate fish caught during 5 years of September scientific surveys (1980-1985) in the Gulf of St. Lawrence (courtesy of Hugues Benoit, Fisheries and Ocean Canada, Moncton). Three types of skates inhabit the Gulf: thorny, winter and smooth stakes (see Fig. 8.10). Marine scientists know that these categories are naturally *ordered* according to water depth: winter skates (*Leucoraja ocellata*) tend to occur in shallow waters, thorny skates (*Amblyraja radiata*) can be encountered at intermediate depths while smooth skates (*Malacoraja senta*) seem to feel at ease in deeper waters. In the following, we seek to build an ordered Probit model to quantify these habitat preferences according to water depth.

The available dataset consists of a bottom trawl survey of 310 sites $t = 1, ..., T$, $T = 310$. For each site $t$, we will consider the logarithm of the water depth as the explanatory variable $X_t$. Here only one covariate is considered for possible explanation, *i.e.*, the water depth, but one can think of many other covariates (prey abundance, sediment composition, year effect, etc.) favoring the presence or the absence of a given species. In Fig. 8.11, the water depth has been grouped into bins corresponding to 21 classes of logarithmic depth. For each site, the response variable $Y_t$ consists of the type of skates that was observed. $Y_t$ is a categorical

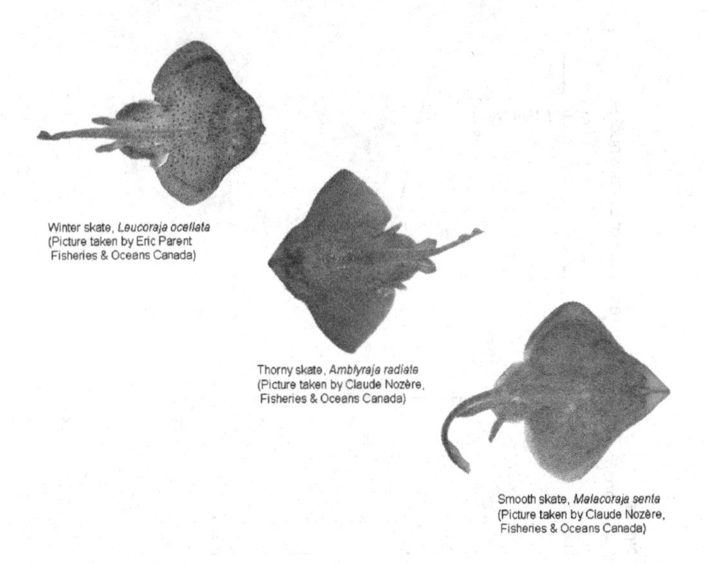


**FIGURE 8.10**: Winter, thorny and smooth stakes. (Pictures have been taken by Éric Parent and Claude Nozère. Courtesy of Fisheries and Ocean Canada.)

variable which can take on values $j = 1, 2..., J$, with J=3 because only 3 species of skates can be encountered. We aim at linking the category of skates $Y_t$ with the covariate $X_t$.

It is worth noting that only the sites where at least one skate was seen are retained in the analysis. All sites where no skates were seen were discarded. In other words, the model will not seek to explain the presence/absence of skates, but rather to explain the probability that an observed skate belongs to species $j$, as a function of depth. Moreover, for each site, the response variable does not account for the quantity of skates that was seen. For instance, a response $y_t = (0, 1, 0)$ at site $t$ indicates that skates of type 2 were seen, but not the observed quantity. In other terms, the model is designed to explain the variability of the various skate species presence according to depth, but not the skate abundance.

Figure 8.11 displays empirical presence frequencies of these 310 observations grouped to bins corresponding to 21 classes of logarithmic depth. In addition, the bottom panel of Fig. 8.11 gives the number of observations in each bin, which indicates that the shape of the curves of presence is not that reliable in shallow or in deep waters. We therefore need a model to depict the presence of each species.

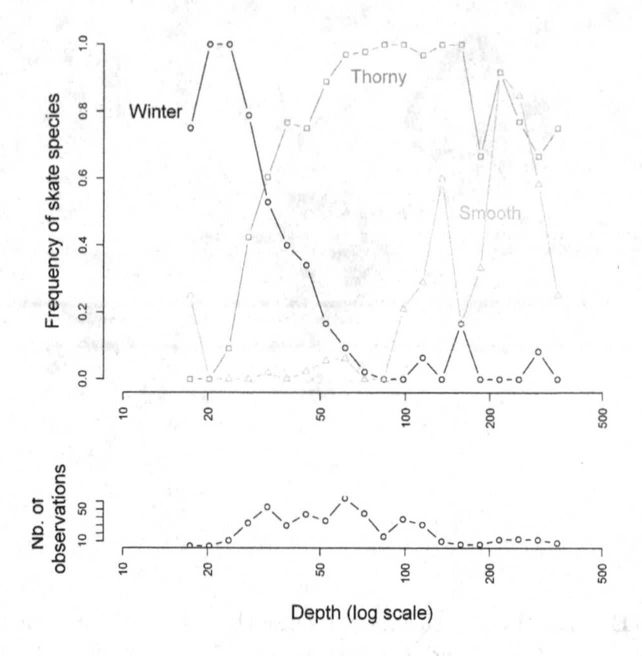


**FIGURE 8.11**: Empirical frequencies of the three skate species according to 21 classes of logarithmic depth. Square: winter; triangle: smooth; diamond: thorny.

### 8.2.2 Formalization

The data augmentation approach presented in Albert and Chib [3] provides a general framework for analyzing ordered multinomial response models. The basic idea for the model is to imagine that for every site $t$, there is an explanation summed up in the quantity $\mu_t$, that depends upon the covariates $X_t$ and which characterizes the skates' strength of presence and takes continuous values on the subsequent intervals defined by parameters $\gamma_0 = -\infty < \gamma_1 < ... < \gamma_{J-1} < \gamma_J = +\infty$. As a first try, we may suggest a linear two-parameter relationship with the logarithmic depth $X$:

$$\mu_t = \beta_0 + \beta_1 X_t$$

but there are many other possible competing structures. We are searching for a probabilistic allocation mechanism of $Y_t$ to a category $j$ such that the larger $\mu_t$, the more likely $Y_t$ will take on a large categorical value. Consider $\pi_{tj}$, the probability that observation $t$ fall into category $j$ obtained by inverting a cumulative probability function. Take for in-

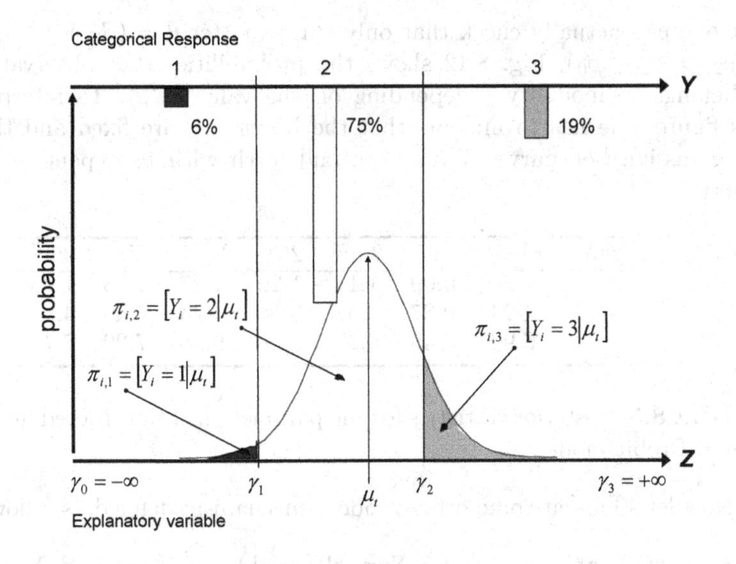


**FIGURE 8.12**: Random mechanism for an ordered categorical response with three levels depending on the explanatory covariate $\mu_t$.

stance a Normal distribution $\mathbf{N}(0,1)$ with cumulative density function $\Phi$ such that

$$[a < Z < b | Z \sim \mathbf{N}(0,1)] = \Phi(b) - \Phi(a)$$

and consider the multinomial trial

$$[Y_t = j \,| \mu_t] = \pi_{tj} \tag{8.8}$$

with

$$\pi_{tj} = \Phi\left(\gamma_j - \mu_t\right) - \Phi\left(\gamma_{j-1} - \mu_t\right) \quad (j = 1, \ldots, J)$$

For the skate application, we take of course $J = 3, \gamma_0 = -\infty; \gamma_3 = +\infty$. There appears to be four parameters $(\beta_0, \beta_1, \gamma_1, \gamma_2)$ in the model and the likelihood reads:

$$[\mathbf{Y} \,| \mu] = \prod_{t=1}^{T} \left( \Phi\left(\gamma_{y(t)} - \mu_t\right) - \Phi\left(\gamma_{y(t)-1} - \mu_t\right) \right)$$

with

$$\mu_t = \beta_0 + \beta_1 X_t$$

But one can actually check that only three matter $\theta = (\beta_1, \delta_1 = \gamma_1 - \beta_0, \delta_2 = \gamma_2 - \beta_0)$. Fig. 8.12 shows the probabilities that observation $y_t$ belongs to modality $j$ depending on the value of $\mu_t$. To interpret this figure, one has to imagine that the bounds $\gamma_i$ are fixed and that the Gaussian bell curve moves back and forth with $\mu_t$ depending on experiment $t$.

| Param. | Mean | Sd | 2.5% | 25% | Med. | 75% | 95% |
|---|---|---|---|---|---|---|---|
| $\beta_1$ | 1.27 | 0.09 | 1.11 | 1.20 | 1.27 | 1.33 | 1.43 |
| $\gamma_1$ | 4.24 | 0.37 | 3.62 | 3.98 | 4.23 | 4.49 | 4.86 |
| $\gamma_2$ | 6.99 | 0.45 | 6.25 | 6.68 | 6.98 | 7.29 | 7.75 |

**TABLE 8.5**: Posterior statistics for the parameters of the ordered multinomial Probit model.

Now let's look at some other random mechanism defined as follows:

- Draw $Z_t$ at random as a Normal variable, with mean $\beta_1 X_t$ and variance 1,
- See where $z_t$ will occur (on which of the intervals bounded by the $\delta_j' s$, $\delta_0 = -\infty < \delta_1 < ... < \delta_{J-1} < \delta_J = +\infty$),
- Take $Y_t = j$ if $\delta_{j-1} < z_t \leq \delta_j$.

From a statistical point of view, this latter event and the model defined by Eq. (8.8) are identical. Indeed

$$[Y_t = j \,|\delta, \beta_1 X_t] = [\delta_{j-1} < z_t < \delta_j \,|\delta, \beta_1 X_t] = \int_{z=\delta_{j-1}}^{z=\delta_j} [z \,|\beta_1 X_t] \, dz$$

and we find $[Y_t = j \,|\delta, \beta_1 X_t] = \pi_{tj}$ since

$$\int_{z=\delta_{j-1}}^{z=\delta_j} [z \,|\beta_1 X_t] \, dz = \Phi(\delta_j - \beta_1 X_t) - \Phi(\delta_{j-1} - \beta_1 X_t) = \pi_{tj}$$

To sum up, we have introduced a latent Normal variable $Z_t$ centered on $\beta_1 X_t$ that drives the categorial phenomenon (described by Eq. (8.8)).

$$\begin{cases} Z_t = \beta_1 X_t + u_t \\ u_t \sim_{iid} N(0,1) \end{cases}$$

One can rewrite the likelihood as:

$$[\mathbf{Y}\,|\beta_1 X_t, \delta] = \prod_{t=1}^{T} \int_{z_t} [y_t, z_t\,|\beta_1 X_t, \delta]\, dz_t$$

$$= \prod_{t=1}^{T} \left( \int_{z_t} dnorm(z_t, \beta_1 X_t, 1) \times 1_{\delta_{y(t)-1} < z_t < \delta_{y(t)}} dz_t \right)$$

We are just adding a layer of categorical data generation to a linear model. Conversely to the categorical $Y_t$, the latent $Z_t$ is not observable. The so-called *complete* likelihood is defined as the joint probability of latent and observable variables:

$$[\mathbf{Y}, \mathbf{Z}\,|\delta, \beta_1, \mathbf{X}] = \prod_{t=1}^{T} \left(dnorm(z_t, \beta_1 x_t, 1) \times 1_{\delta_{y(t)-1} < z_t < \delta_{y(t)}}\right) \tag{8.9}$$

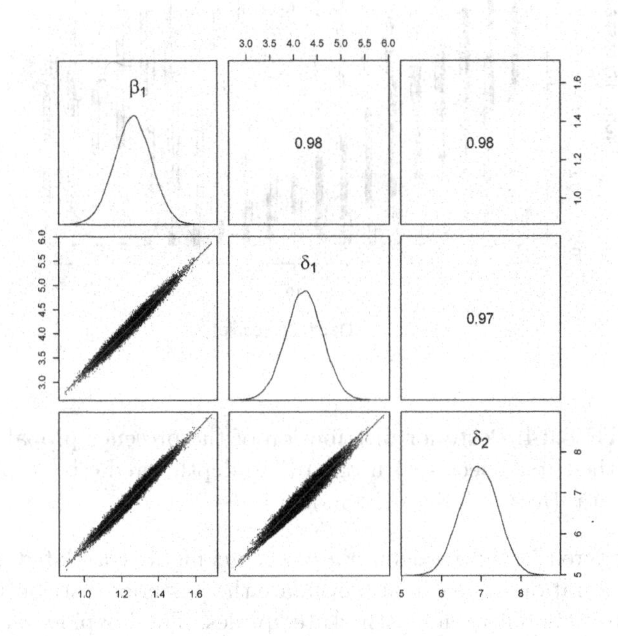


**FIGURE 8.13**: Joint *a posteriori* distribution for $\beta_1, \delta_1, \delta_2$. Marginal pdfs are shown in the diagonal. Joint MCMC scatters are shown in the lower part. The upper part points out strong correlations among the components of the posterior parameter pdf.

### 8.2.3 Bayesian inference of the ordered multinomial Probit model

Priors can be set as Normal distributions for $\beta_1$, $\delta_1$ and $\delta_2$ so as to benefit from conjugate properties. Appendix A shows that a Gibbs sampler can be easily implemented in that case. A WinBUGS program has been used here to perform the posterior inference with flat priors for $\delta_1, \delta_2$ and $\beta_1$. The model parameters (see Eq. (8.8)) of Table 8.5 have been estimated with the 100,000 last MCMC iterations after a 50,000 burn-in period. Figure 8.13 shows that the three parameters $(\beta_1, \delta_1, \delta_2)$

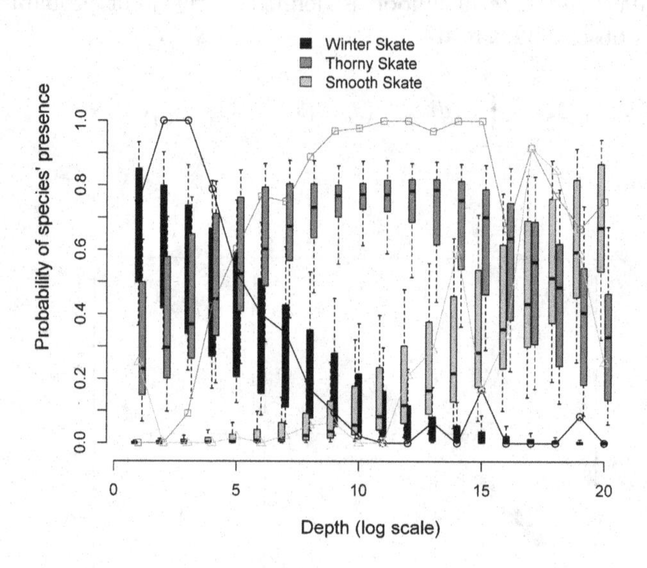


**FIGURE 8.14**: Posterior distribution of the presence probabilities for each of the three species with regards to depth; predictive probabilities versus data: How well does the model fit?

of the ordered Probit model are *a posteriori* highly correlated. From the MCMC iterations, one can also evaluate the posterior distribution of the presence probabilities for each skate species. The boxplots of Fig. 8.14 show their statistical features with large ranges of uncertainty corresponding to depths levels with scarce data.

## 8.3 Discussion

Many questions have not been treated in this chapter, in particular:

- How can we test that a model is significant? More specifically Fig. 8.14 also includes data (frequencies by depth levels) versus the corresponding predictive probabilities. Some observations stand out of their confidence bounds, shall we conclude that the model is to be rejected?
- How robust to *prior* specifications are the results?
- How can one use such results to develop sampling plans? Can the next scientific campaign be optimized?

All these questions are noteworthy, but in this chapter we rather focus on the model making issue from real case studies. For many aspects, it is like a game of *LEGO* and it is not difficult! One has to think conditionally; complex models are built by adding latent structures that bring into the analysis conceptual key variables. In the following chapters of part 2, we keep on assembling *LEGO* blocks of elementary model components, but the game gets more sophisticated with temporal, hierarchical and spatial aspects to be taken into account.

# Part II

# Setting up more sophisticated hierarchical structures

# Chapter 9

## Hierarchical Bayesian Modeling I: Borrowing strength from similar units


### Summary

In this chapter, two different examples, both issued from our research experience in salmon ecology, are developed to introduce Hierarchical Bayesian Models (HBM), that make up the backbone of today's Bayesian modeling. Hierarchical (also called *multilevel* or *random effect*) models assume that the dataset being analyzed consists of a hierarchy of different groups within which records look more alike than between groups. Random effects or latent variables are probabilistic objects which are introduced to capture the variability between those groups. They are considered to be *a priori* drawn from a probability distribution with parameters (typically mean and variance) that will adjust to the data. A small variance will express a strong resemblance between groups, a large one will mean that the groups do not look like one another. This probability structure ties together the various layers of such a multilevel construction.

In the first example, we revisit the Binomial analysis of capture-mark-recapture data introduced in Chapter 4. Here, a model to estimate the number of salmon spawners returning each year in the Oir River is developed within the hierarchical framework.

The second example extends the salmon stock-recruitment analysis developed in Chapter 7 so as to design hierarchical analysis of stock-recruitment models and study a group of 13 rivers in Europe with datasets varying from data-rich to sparse situations. The hierarchical structure between rivers is designed conditionally on some available covariates, namely the latitude and the riverine wetted area accessible to salmon. In both examples, we show how the hierarchical structure organizes the transfer of information between different units. It successfully accommodates large but sparse datasets containing poorly informative data for some units, and its conditional structure enables to borrow strength from data-poor to data-rich units.

---

## 9.1 Introduction

HBM have been paid considerable attention in statistical ecology ([75]; [183]; [312]). For instance, in forestry or fisheries sciences, HBM should reveal fruitful in the hands of a skilled ecological detective to distinguish between population (*e.g.*, forests, fish populations) and individuals (*e.g.*, trees, fish) with eventually some grouped data effects that will not be nested into the previous ones (such as tree species or sub-populations of fish). The key idea is to express also dispersion between units by conditional probabilities (it is worth noting that up to this chapter, pdfs were rather mimicking variations within a statistical unit). Such models are for instance useful to depict:

- *Cohort effects* in correlated or familial survival data as opposed to individual behaviors within a group;
- *Site effects* in meta-analyzes or in *spatially* structured phenomena.

Due to its historical importance in the hierarchical Bayesian modeling approach, we detail in Appendix E the mathematical treatment of the famous baseball example from [98], pointing out Stein's paradox ([286]). We do encourage the reader to take some time to work out this example even though it has nothing to do with ecology! Stein's article brought an irreversible change of mind among the statistical community: computing the average of past events had long been considered by many statisticians as the best guess about the future but this example was highlighting random effect circumstances in which Bayesian thinking helped devise much better estimators! Winning bets exploited the so-called *shrinkage effect* (that pulls back empirical estimates toward the grand mean and "dampers" the sample size effects).

Simple hierarchical (or multilevel) models as sketched in Fig. 1.13 assume that the dataset being analyzed consists of different groups $k = 1, ..., n$ with differences in some characteristics modeled within a hierarchical structure. Random effects $Z_{1:n} = (Z_1, ..., Z_n)$ are latent variables used to capture both the variability (or speaking positively the *similarity*) between the $n$ groups. The $Z_{1:n}$ are independently drawn from the same distribution (let's call it the urn of resemblance) with parameter $\theta$ tuning that resemblance between groups:

$$[Z_{1:n}|\theta] = \prod_{k=1}^{n}[Z_k|\theta] \tag{9.1}$$

Each latent variable $Z_k$ is associated with a data subset $y_k$ through a (partial) likelihood term $[y_k|\theta, Z_k]$, so that the likelihood can be obtained as:

$$[y_{1:n}|\theta] = \int_{Z_{1:n}} \prod_{k=1}^{n}[Z_k|\theta] \times \prod_{k=1}^{n}[y_k|Z_k, \theta] : dZ_{1:n} \tag{9.2}$$

Knowing $Z_{1:n}$, the $y_k$'s are independent because

$$[y_{1:n}|\theta, Z_{1:n}] = \prod_{k=1}^{n}[y_k|Z_k, \theta].$$

When $Z_{1:n}$ is unknown, the $y_k$'s are *exchangeable*: as shown by Eq. (9.2), any permutation of the group indices $1:n$ will leave the joint distribution $[y_{1:n}|\theta]$ unchanged. This is why this simple hierarchical structure is named an exchangeable hierarchical model (see also Appendix E). The joint posterior distribution of all unknowns (parameter $\theta$ and latent variables $Z_{1:n}$) writes:

$$[Z_{1:n}, \theta|y_{1:n}] \propto [\theta] \times \prod_{k=1}^{n}[Z_k|\theta] \times \prod_{k=1}^{n}[y_k|Z_k, \theta] \tag{9.3}$$

As emphasized in Figures 1.12 and 1.13 in Chapter 1 , random variables $Z_{1:n}$ constitute a latent layer that ties together the parameters $\theta$, common to all groups, and the data specific to each group. The hierarchical structure captures the variability between groups by the latent variable $Z_{1:n}$ but also the similarity by organizing the transfer of information between groups thanks to the vector of parameters $\theta$. As developed in Rivot and Prévost [255], the data of all groups $k = 1, ..., n$ are included to estimate the latent variable for any particular group $k$. As shown in Eq. (9.4), the marginal posterior $[Z_k|y_{1:n}]$ can be written as an average (or an integral) of the conditional distribution $[Z_k|\theta]$ over the posterior distribution of the parameters $\theta$ conditioned on observed data for all units $k = 1, ..., n$:

$$\begin{aligned}[Z_k|y_{1:n}] &= \int_{\theta} [Z_k, \theta|y_{1:n}] : d\theta \\ &= \int_{\theta} [Z_k|\theta] \times [\theta|y_{1:n}] : d\theta \end{aligned} \tag{9.4}$$

Hierarchical models also propose a consistent probabilistic rationale for prediction. Inferences about the random effect for a new group, $Z^{new}$, can be derived through the predictive distribution conditional on the observed data ([117]). It is an average of the conditional population distribution $[Z|\theta]$ over the posterior distribution of the parameters $\theta$ conditioned on all observed data:

$$\begin{aligned}[Z^{new}|y_{1:n}] &= \int_{\theta} [Z^{new}, \theta | y_{1:n}]\, d\theta \\ &= \int_{\theta} [Z^{new}|\theta] \times [\theta | y_{1:n}]\, d\theta \end{aligned} \quad (9.5)$$

The difference between the marginal prior predictive $[Z] = \int_{\theta} [Z|\theta] \times [\theta]\, d\theta$ and the posterior predictive$[Z|y_{1:n}] = \int_{\theta} [Z|\theta] \times [\theta|y_{1:n}]\, d\theta$ in Eq. (9.5) reflects the amount of information brought by the data $y_{1:n}$ of all groups $k = 1..n$ to update the prior distribution common to all groups. Additionally in the examples, writing $y_k = (y_k^1, ... y_k^{p_k})$, the number $p_k$ of elementary data may differ between groups.

## 9.2 Hierarchical exchangeable Binomial model for capture-mark-recapture data

This section is a natural followup to Chapter 4. The main idea is to link the capture-mark-recapture (CMR) models for yearly observations together by a hierarchical structure as the one represented by the DAG of Fig. 1.13. In this case study inspired by the article of Rivot and Prévost ([255]), the years are the statistical units that look alike and that hypothesized resemblance allows for transferring information from a given year to the other years. The only differences with the baseball players' model from Appendix E are that the data structure is a little more sophisticated and that the latent variables are two component vectors instead of simple real numbers.

### 9.2.1 Data

Relatively long but sparse (small sample size) series of data are quite common when dealing with CMR surveys aimed at estimating the abundance of wild populations over a series of years. For instance on the Oir

**FIGURE 9.1**: Marking a spawner entering the Scorff River.

River, already presented for the smolt runs in Chapter 4, the rangers from the French National Research Institute for Agronomy (INRA) and from the National Office of Water Management (ONEMA) have collected CMR data about adult salmon that swam back to spawn in the Oir River for the years 1984–2000.

Data are shown in Table 9.1. For each year $t$ from 1984 to 2000, $y_{1,t}$ denotes the number of fish trapped at the Cerisel station (close to the mouth of the river, see Fig. 4.3 of Chapter 4 for more details about the trapping device). $y_{2,t} + y_{3,t}$ individuals from the captured ones are not replaced upstream, either because they died during manipulation or because they are removed for experimental use or for hatchery production. Let $y_{4,t} = y_{1,t} - (y_{2,t} + y_{3,t})$ the number of (tagged) fish released. These spawners are individually marked before they keep on swimming upstream (Fig. 9.1). The recapture sample is gathered during and after spawning (see more details on recapture conditions hereafter). Let us denote as $y_{5,t}$ and $y_{6,t}$ the number of marked and unmarked fish among recaptured fish, respectively.

### 9.2.2 Observation submodels for the first phase (Cerisel trapping place)

Let denote $\nu_t$ the unknown of interest, *i.e.*, the population size of spawners at year $t$ and $\pi_t^1$ the unknown trapping efficiency. Assuming all

| Years | $y_1$ | $y_2$ | $y_3$ | $y_4$ | $y_5$ | $y_6$ |
|---|---|---|---|---|---|---|
| 1984 | 167 | 10 | 3 | 154 | 12 | 10 |
| 1985 | 264 | 37 | 11 | 216 | 21 | 4 |
| 1986 | 130 | 28 | 9 | 93 | 5 | 4 |
| 1987 | 16 | 3 | 1 | 12 | 2 | 22 |
| 1988 | 226 | 35 | 8 | 183 | 12 | 0 |
| 1989 | 235 | 31 | 5 | 199 | 56 | 0 |
| 1990 | 15 | 4 | 4 | 7 | 2 | 15 |
| 1991 | 44 | 0 | 0 | 44 | 23 | 1 |
| 1992 | 31 | 10 | 1 | 20 | 4 | 5 |
| 1993 | 100 | 17 | 2 | 81 | 4 | 3 |
| 1994 | 32 | 12 | 2 | 18 | 1 | 4 |
| 1995 | 109 | 6 | 1 | 102 | 39 | 7 |
| 1996 | 70 | 13 | 2 | 55 | 25 | 57 |
| 1997 | 56 | 19 | 3 | 34 | 12 | 3 |
| 1998 | 34 | 3 | 1 | 30 | 6 | 30 |
| 1999 | 154 | 5 | 1 | 148 | 13 | 22 |
| 2000 | 53 | 0 | 0 | 53 | 4 | 33 |

**TABLE 9.1**: Capture-mark-recapture data for spawners by spawning migration year in the Oir River. $y_1$: Number of fish trapped at the counting fence during the upstream migration time; $y_2$, $y_3$: One sea-winter (resp. two sea-winter) fish removed from the population; $y_4$: Tagged and released fish; $y_5$, $y_6$: Number of marked (resp. unmarked) recaptured fish.

of the $\nu_t$ spawners are independently and equally catchable in the trap, with a probability $\pi_t^1$ considered constant over the migration season, the migration of the $\nu_t$ spawners are independent Bernoulli experiments with probability of "success" $\pi_t^1$. Accordingly, $y_{1,t}$ is the observed result of a Binomial experiment as given by Eq. (9.6):

$$y_{1,t} \sim Binomial(\nu_t, \pi_t^1) \tag{9.6}$$

### 9.2.3 Observation submodels for the second phase (recollection during and after spawning)

The fate of a salmon swimming upstream to spawn is described in great detail in Chapter 5. The recapture sample is obtained by three methods: electrofishing on the spawning grounds, collection of dead fish after spawning, and trapping of spent fish at the downstream trap of the Cerisel facility. Due to the available data here, we adopt a simplified

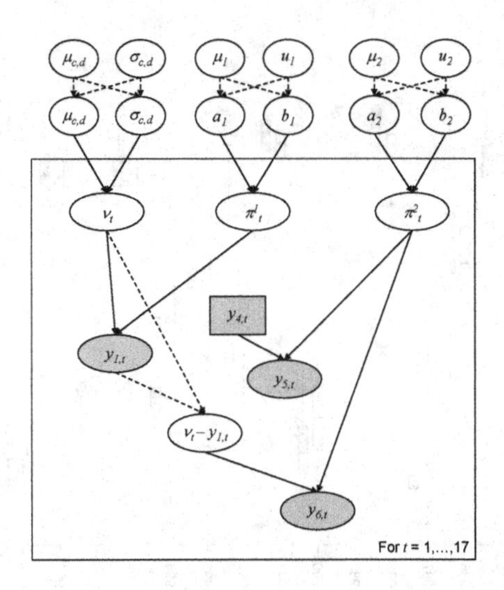


**FIGURE 9.2**: Directed Acyclic Graph representation of the hierarchical structure for the joint modeling of capture-mark-recapture experiments for the 17 years.

version of the model. As a first approximation to this rather complicated scheme, recapture Binomial experiments with efficiency $\pi_t^2$ for marked fish and untagged ones are acceptable, providing one is willing to assume the three following hypotheses: No spawner runs downstream after getting over the trap ($H1$); there is no tag shedding ($H2$); the recapture probability $\pi_t^2$ is the same for all the fish whether or not marked ($H3$):

$$\begin{cases} y_{5,t} & \sim Binomial(y_{4,t}, \pi_t^2) \\ y_{6,t} & \sim Binomial(\nu_t - y_{1,t}, \pi_t^2) \end{cases} \tag{9.7}$$

### 9.2.4 Latent layers

Writing latent vector $Z_t = (\nu_t, \pi_t^1, \pi_t^2)$ and observation $y_t = (y_{1,t}, y_{5,t}, y_{6,t})$ to cope with the notations of Fig. 1.13, we are now in search of a hierarchical structure to express that, to some extent, years may look like one another. The natural choice for the latent distributions are the Beta distribution (see Eq. (2.6), page 54) for $\pi_t^1$ and for $\pi_t^2$

$$\begin{cases} \pi_t^1 & \sim Beta(a_1, b_1) \\ \pi_t^2 & \sim Beta(a_2, b_2) \end{cases} \tag{9.8}$$

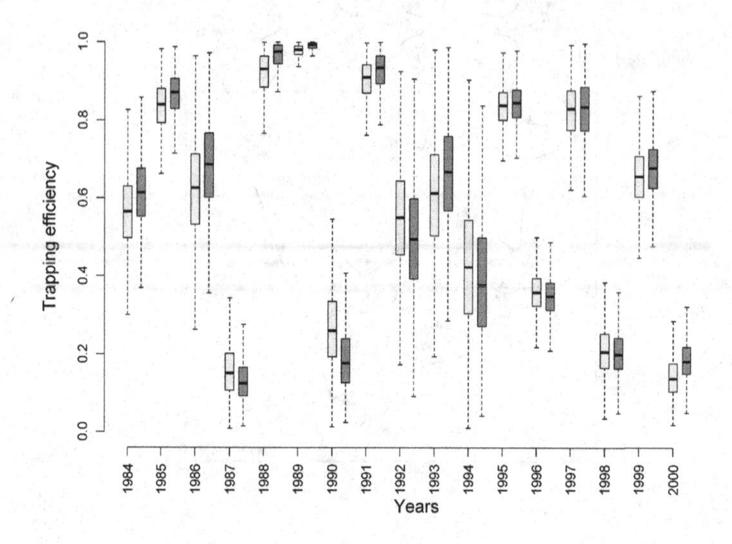


**FIGURE 9.3**: Box and whisker plots of a size-15,000 MCMC sample from the marginal posterior distribution of the trapping efficiency at the Cerisel trap by migration year obtained under two model configurations, the model in which years are treated independently (light gray) and the hierarchical model (dark gray). The boxes indicate the interquartile range and the median.

and the Negative Binomial distribution (already met at Eq. (4.11)) for $\nu_t$

$$\nu_t \sim NegBinomial(c, d) \tag{9.9}$$

Figure 9.2 shows a DAG representation of the exchangeable hierarchical model for the joint modeling of the capture-mark-recapture for the 17 years. The higher level parameters $(a_1, a_2, b_1, b_2, c, d)$, denoted $\theta$ consistently with notations of Fig. 1.13, are generally assigned a diffuse prior distribution to reflect some ignorance about them. A common practice is to set a prior on some appropriate one-to-one transformed parameters and then to go back to the original parameters via the inverse transformation. Most often, the transformation recovers the mean and variance because of their well-understood meaning. Sometimes, the mean can be assigned a rather informative prior but it is generally not the case of the variance (that describes between years variability), an uppermost unknown quantity.

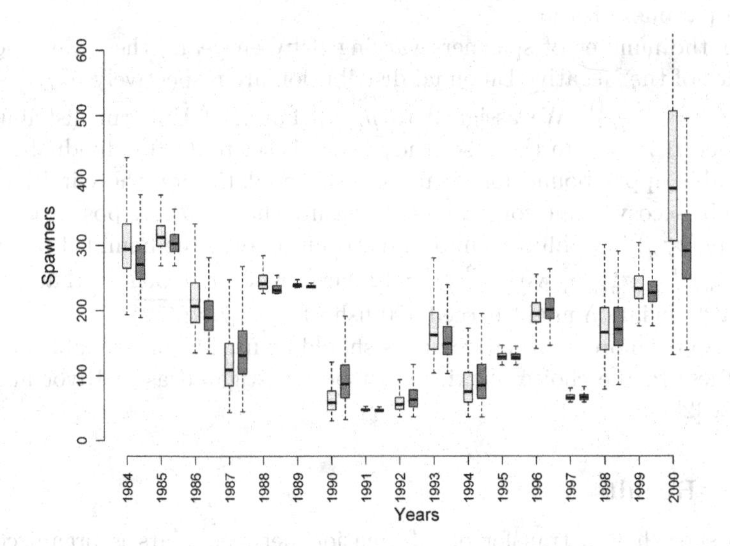


**FIGURE 9.4**: Box and whisker plots of a size-15,000 MCMC sample from the marginal posterior distribution of the number of spawners by migration year obtained under two model configurations, the model in which years are treated independently (light gray) and the hierarchical model (dark gray). The boxes indicate the interquartile range and the median.

For the capture and recapture efficiencies, a diffuse (noninformative) prior can be put on the mean $\mu_{a,b} = \frac{a}{a+b}$ and variance $\sigma^2_{a,b} = \frac{ab}{(a+b)^2(a+b+1)}$ of the Beta distributions. But this transformation introduces some unnecessary difficulty in order to ensure that both parameters $a$ and $b$ are positive, the variance must satisfy the constraint

$$\sigma^2_{a,b} < \mu_{ab}(1 - \mu_{ab})$$

We used a rather simpler transformation by considering the transformation $\mu_{a,b} = \frac{a}{a+b}$ and $u_{a,b} = (a+b)$ and by drawing $\mu_{a,b}$ in a diffuse prior distribution (we took a $Beta(1.5, 1.5)$) and $log(u_{a,b})$ in a $Uniform(0, 10)$ distribution. Keeping in mind the interpretation of Beta coefficients $(a, b)$ as prior distributions for Binomial trials (see Chapter 2), $a$ and $b$ are to be interpreted as prior number of success and failures, respectively. Then, $u = (a+b)$ is interpreted as a prior sample size that scales the variance of the Beta prior distribution, and a Uniform distribution on the log-scale is appropriate for a diffuse prior (see also [187] for another utilization of such a parameterization).

For the number of spawners varying between years, the mean and variance of the Negative Binomial distribution are respectively $\mu_{c,d} = \frac{c}{d}$ and $\sigma^2_{c,d} = \frac{c(d+1)}{d^2}$. We assigned to $\mu_{c,d}$ a bounded Uniform distribution over $]0, \mu_{max}]$. In the case study, $\mu_{max}$ is set to 3000 individuals, a reasonable upper bound for a salmon fish population on the Oir River due to bio-ecosystemic constraints. To ensure that $(c, d)$ are positive, we draw $log(\sigma^2_{c,d})$ in a diffuse Uniform distribution over the bounded range $[log(\mu_{c,d}), log(\sigma^2_{max})]$ with $\sigma^2_{max} = 12$ since we do not believe that the standard deviation might exceed 400 fish ($400 \approx \sqrt{\exp(12)}$).

Of course more informative priors should be used when available and robustness to the choice of prior must be investigated as in Rivot and Prévost [255].

### 9.2.5 Results

To show how a transfer of information between years is organized by the hierarchical model, we compare its results with the model assuming independence between years. For the models with independence, independent prior distributions with known parameters were set on $(\nu_t, \pi^1_t, \pi^2_t)$:

$$\begin{cases} \pi^1_t \sim Beta(1.5, 1.5) \\ \pi^2_t \sim Beta(1.5, 1.5) \\ \nu_t \sim Uniform(1, 3000) \end{cases} \tag{9.10}$$

Inference has been performed via WinBUGS (see the supplementary material available from the book's website *hbm-for-ecology.org*).

Results highlight that hierarchical modeling has no effect on the inferences on the capture efficiencies (Fig. 9.3), but greatly improves posterior inferences for the number of spawners migrating back to the Oir River (Fig. 9.4).

Posterior mean values of the capture probabilities $\pi^1_t$ do not seem to shrink much toward their overall grand mean (Fig. 9.3) and the recapture probabilities $\pi^2_t$'s (not shown) are only slightly subjected to the shrinkage effect. In Fig. 9.3, there remains a lot of between-year variability in the experimental conditions at the Cerisel trapping facility.

Conversely, the hierarchical structure hypothesized on $\nu_t$'s strongly reduces the skewness and uncertainty in the estimation of the number of spawners. The grey boxplots of Fig. 9.4 clearly point out that the most precise inferences are obtained under the hierarchical model, especially for the years with sparse CMR data, *i.e.*, low number of marked released

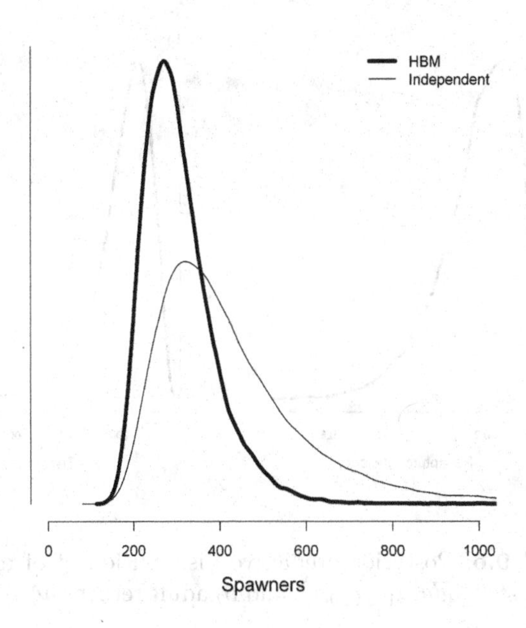


**FIGURE 9.5**: Marginal posterior distribution of the number of spawners $\nu_{2000}$ obtained under two model configurations, the model in which years are treated independently (thin line) and the hierarchical model (bold line).

or, more importantly, low number of recaptures of previously marked fish yield (*e.g.*, years 1987, 1990, 1994 and 2000). For this latter year, the upper bounds of the 95% Bayesian credibility intervals obtained with the model assuming independence between years appears unrealistically high given the size of the Oir River and the available knowledge on the biology and ecology of Atlantic salmon as exemplified for year 2000 in Fig. 9.5.

A straightforward result of the hierarchical model are the posterior predictive distributions of the trapping or recapture efficiencies and of the number of returns , denoted $[\pi^{1,new}|data_{1984:2000}]$ and $[\pi^{2,new}|data_{1984:2000}]$, and $[\nu^{new}|data_{1984:2000}]$, respectively. The posterior predictive of the trapping efficiency is an informative distribution with a mean value 0.124 and 95% of its density in the range $[0.016, 0.567]$ (Fig. 9.6). The posterior predictive of the returns has a mean value around 230 fish and 95% of its density in the range $[40, 610]$ (Fig. 9.6). Thus, the data of all years combined allow discarding *a priori* the pos-

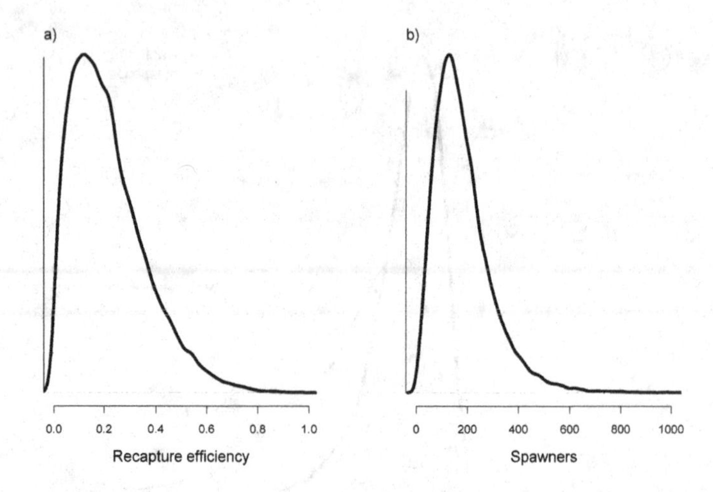


**FIGURE 9.6**: Posterior predictive distribution of of a) the trapping efficiency $[\pi^{1,new}|data_{1984:2000}]$, and b) adult returns $[\nu^{new}|data_{1984:2000}]$.

sibility of very high trapping efficiency (*i.e.*, greater than 0.5) or high spawner population size (*i.e.*, greater than a thousand) in any additional year.

In addition, the posterior inferences derived under the hierarchical model are rather insensitive to changing priors on $\mu_{a,b}$, whilst the model assuming independence is not, especially for the spawner stock of the sparse data years $\nu_{1987}, \nu_{1990}$, and $\nu_{1994}$ (results not shown). In-depth sensitivity analyses and another observation model for the recaptures can be found in Rivot and Prévost [255].

## 9.3 Hierarchical stock-recruitment analysis

This section develops a hierarchical extension of the Ricker Stock Recruitment model introduced in Chapter 7. The data and models are based on a published paper by Prévost et al. [239]; but see also [238].

We show that the hierarchical assemblage of several salmon populations (the biological analogs of baseball players from Appendix E) which we model as exchangeable units appears once again as the working solution to transfer information and borrow strength from data-rich to data-poor situations. The main difference with the previous models is that the data structure gets more sophisticated with Normal latent vectors (instead of Beta and Negative Binomial ones as in the previous section), and that the probabilistic structures have to be designed conditionally on some available covariates, namely the latitude and the riverine wetted area accessible to salmon, following Fig. 1.12.

### 9.3.1 Data

As already explained in Chapter 7, the analysis of stock and recruitment (SR) relationships is the most widely used approach for deriving *Biological Reference Points* in fisheries sciences. It is particularly well suited for anadromous salmonid species for which the recruitment of juveniles in freshwater is more easily measured than the recruitment of marine species. SR relationships developed in Chapter 7 are critical for setting reference points for the management of salmon populations, such as the spawning target $S^*$, a biological reference point for the number of spawners which are necessary to guarantee an optimal sustainable exploitation, or the maximum sustainable exploitation rate $h^*$.

| River | Country | Latitude (°N) | Riverine wetted area accessible to salmon ($m^2$) | Number of SR observations (years) |
|---|---|---|---|---|
| Nivelle | France | 43 | 320995 | 12 |
| Oir | France | 48.5 | 48000 | 14 |
| Frome | England | 50.5 | 876420 | 12 |
| Dee | England | 53 | 6170000 | 9 |
| Burrishoole | Ireland | 54 | 155000 | 12 |
| Lune | England | 54.5 | 4230000 | 7 |
| Bush | N. Ireland | 55 | 845500 | 13 |
| Mourne | N. Ireland | 55 | 10360560 | 13 |
| Faughan | N. Ireland | 55 | 882380 | 11 |
| Girnock Burn | Scotland | 57 | 58764 | 12 |
| North Esk | Scotland | 57 | 2100000 | 6 |
| Laerdalselva | Norway | 61 | 704000 | 8 |
| Ellidaar | Iceland | 64 | 199711 | 10 |

**TABLE 9.2**: Location, size and SR time series length of the 13 monitored Atlantic salmon index rivers (from [239]; see also map in Fig. 9.7).

There are several hundreds of salmon stocks in the northeast Atlantic area, each having its own characteristics with regard to the size and

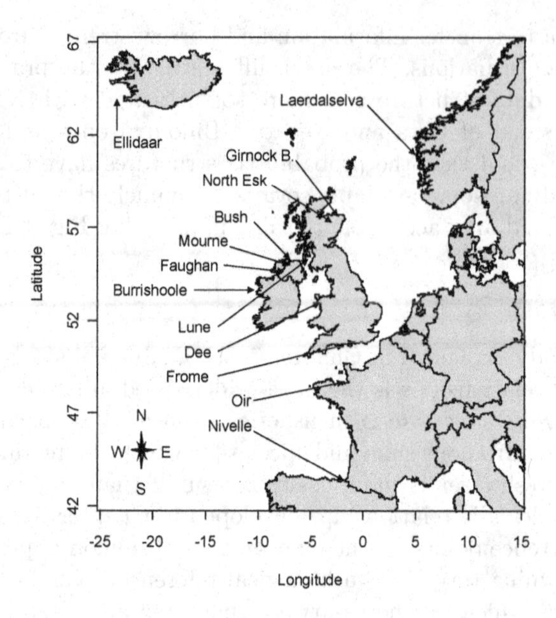


**FIGURE 9.7**: Location of the 13 index rivers with available stock-recruitment datasets for A. salmon (from [239]).

productivity of the salmon populations. But resources to collect SR data are limited and suitable SR series (both in terms of length and reliability of observations) such as the ones in Table 9.2 (see also map in Fig. 9.7) are only available for a handful of monitored rivers spread throughout the European area of distribution of the species. These so-called *index rivers* are a representative sample from the salmon rivers located in western Europe and under the influence of the Gulf Stream. This sample covers a broad area including Spain, France, UK, Ireland, Norway, the western coast of Sweden and the southwestern coast of Iceland (Fig. 9.7). The collection and pre-processing procedures used to obtain the data ready for SR analysis presented in Table 9.2 and Fig. 9.8 are described in detail in [77] or [239].

Here, hierarchical modeling will be used to address two questions:

- How is the SR information transferred from the monitored data-rich rivers to set Biological Reference Points for other sparse-data salmon rivers, while accounting for the major sources of uncertainty?
- How can the joint analysis of the SR relationship for the 13 in-

dex rivers be used to forecast biological reference points for a new river without any SR data but for which relevant covariates are available?

## 9.3.2 Model assuming independence (and no covariate)

Latitude and riverine wetted surface area accessible to salmon are relatively easily obtained from map based measurements. These variables given in Table 9.2 are candidate covariates for explaining variations in SR parameters among rivers. Because the eggs represent the end product of a generation and the starting point of the next, both stock and recruitment are expressed in terms of eggs for each of the 13 rivers. When standardized by riverine wetted surface area accessible to salmon, they provide meaningful measurements expressed in $eggs/m^2$ with regards to Atlantic salmon ecology (mainly territorial behavior and competition for resources spatially limited at both juvenile and adult stages).

### 9.3.2.1 Likelihood

Let us denote $k = 1, ..., 13$ the indices for the 13 rivers as ordered in Table 9.2. Within a river $k$, the recruitment process is modeled by means of a Ricker function with independent logNormal process errors such as in Eq. (7.4). We use the reformulation with management parameters $(S^*, h^* = \frac{R^*-S^*}{R^*})$ defined by Eq. (7.7) as in [274]. For river $i$, one relates the recruitment $R_{k,t}$ of the cohort born in year $t$ to the associated spawning stock $S_{k,t}$:

$$\begin{cases} log(R_{k,t}) = h_k^* + log(\frac{S_{k,t}}{1-h_k^*}) - \frac{h_k^*}{S_k^*} S_{k,t} + \varepsilon_{k,t} \\ \varepsilon_{k,t} \overset{iid}{\sim} Normal(0, \sigma_k^2) \end{cases} \tag{9.11}$$

where $\sigma_k$ is the standard deviation of the Normal distribution of $log(R_{k,t})$, $S_k^*$ and $h_k^*$ are respectively the stock which are necessary to guarantee an optimal sustainable exploitation and the associated optimal exploitation rate for the river $k$.

### 9.3.2.2 Prior

The informative priors of Chapter 7 defined in Table 7.3 encoded the available knowledge about the productivity of the Margaree River Salmon population. In the present situation, much fewer is known about $h^*$ and $S^*$ apart that 200 eggs/$m^2$ is definitively an upper limit for salmon egg density in any river. Therefore we adopt the following diffuse

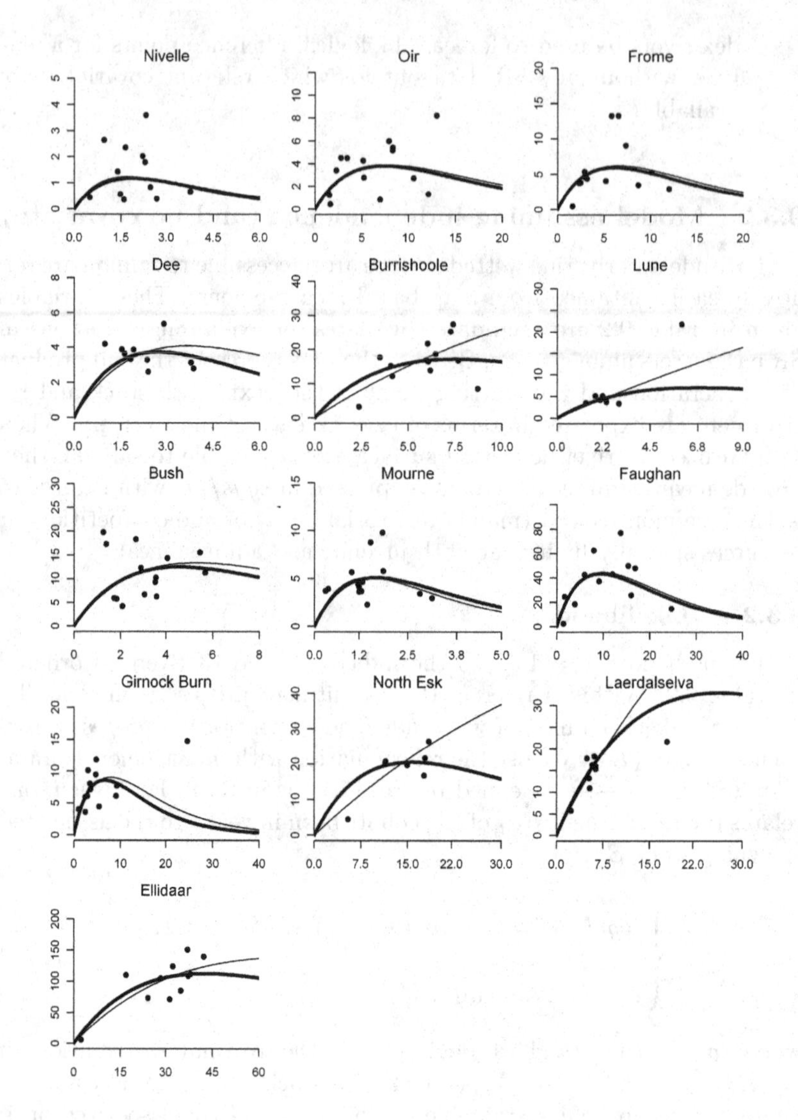


**FIGURE 9.8**: The Atlantic salmon SR series on the 13 index rivers and fitted SR relationships. $S$ ($x$-axis) and $R$ ($y$-axis) are the stock and recruitment variables after standardization for river size expressed in eggs per $m^2$ of riverine wetted area accessible to salmon. The SR Ricker curves are graphed for two model configurations, the model assuming independence between rivers (thin line) and the hierarchical model (bold line). SR curves are graphed with parameters $(S^*, h^*)$ set at the median of their marginal posterior distributions.

priors, independently for each river $k$:

$$\begin{cases} h_k^* & \sim Beta(1,1) \\ S_k^* & \sim Uniforn(0,200) \end{cases} \tag{9.12}$$

A more refined elicitation of $S^*$ would specify the expectation to be about $\mu_{S^*} = 40$ eggs per $m^2$ with a standard deviation of the same order ($\sigma_{S^*} = 40$ and $CV_{S^*} = \sigma_{S^*}/\mu_{S^*} = 1$), thus leading to a Normal prior distribution with mean $\mu_{S^*} = 40$ and variance $\sigma^2_{S^*} = 1600$. Equivalently but taking into account the positiveness of $S^*$, one may prefer a Gamma pdf with shape parameter $a$ and scale parameter $b$ such that $\mu_{S^*} = \frac{a}{b}$, $CV_{S^*} = \frac{1}{\sqrt{a}}$ and constrained to the range $]0, 200]$ :

$$\begin{cases} \mu_{S^*} = 40\, eggs/m^2;\ CV_{S^*} = 1 \\ a = \dfrac{1}{CV^2_{S^*}};\ b = \dfrac{1}{\mu_{S^*} \times CV^2_{S^*}} \\ S_k^* \sim Gamma(a,b) 1_{S_k^* < 200} \end{cases} \tag{9.13}$$

The full prior structure is completely specified by adding a prior on the variance of the environmental noise $\sigma_k^2$ for each river $k$. We assume $\sigma_k^2$ is constant across all the rivers in the study, and a Gamma prior distribution was assigned on the precision:

$$\begin{cases} \forall k,\ \sigma_k = \sigma \\ \sigma^{-2} \sim Gamma(p,q) \end{cases} \tag{9.14}$$

The Gamma distribution $Gamma(p,q)$ is commonly chosen for the precision $\sigma^{-2}$. Since information conveyed in all the datasets will be used to make posterior inference about $\sigma$ (instead of possibly short series if we had kept different $\sigma_k$ for each location), we can here afford to use diffuse prior by letting $p$ and $q$ being very small.

#### 9.3.2.3 Posterior inferences on the model assuming independence between rivers

Bayesian inference of the monitored rivers SR series was first performed on a river by river basis according to likelihood (Eq. (9.11)) and priors (Eqs. (9.12)-(9.14)). The fitted SR relationships drawn for each river with the point estimates of parameters (set to their posterior median) reported in Fig. 9.8 (thin line) show a large variety of shapes.

The marginal distributions of parameters $S^*$ and $h^*$ are reported in Figs. 9.9 and 9.10 (light gray). All distributions exhibit quite heavy tails, but depending on the river, the number of observations, and on the contrast between the $S$ values in the observation sample, uncertainty range

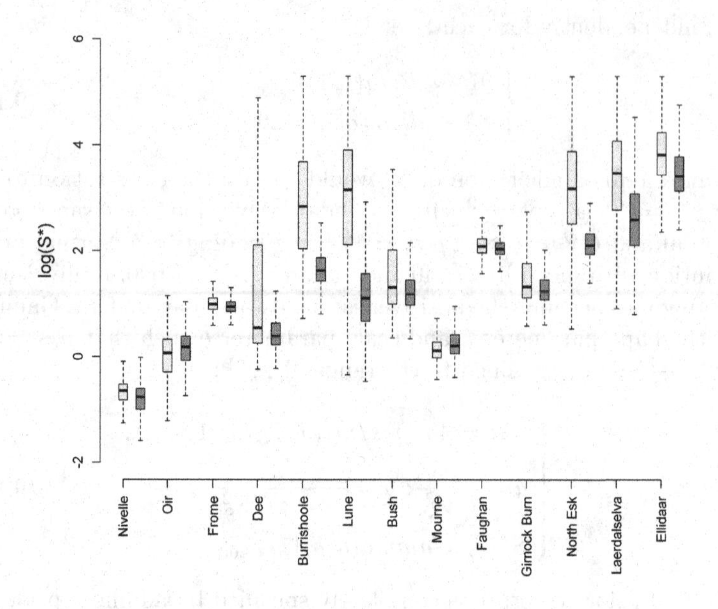


**FIGURE 9.9**: Marginal posterior distribution of $log(S^*)$ (in eggs per $m^2$) for the 13 index rivers obtained under two model configurations: the model assuming independence between rivers (light gray) and the hierarchical model (dark gray).

in the posterior inferences may differ from several orders of magnitude. The boxplots of the parameters' posterior pdfs on Figs. 9.9 and 9.10 are drawn in the same order than in Table 9.2, *i.e.*, by increasing latitude. They reveal an increasing latitudinal gradient in the $S_k^*$'s. Although less evident, the same pattern seems to exist for the exploitation rate $h_k^*$'s.

### 9.3.3 Hierarchical model with partial exchangeability

Figure 9.11 gives a Directed Acyclic Graph representation of a hierarchical structure for the joint modeling of stock-recruitment relationships for the 13 rivers. The hierarchical structure is designed to improve the estimation of parameters $S^*$ and $h^*$ for data-poor rivers by borrowing strength from data-rich to data-poor rivers. Second, it is designed to capture the between-rivers variability of the parameters $(S^*, h^*)$ conditionally on the latitude.

The gradient on $S_k$ with increasing latitude can be incorporated in the model by writing a log-linear relationship between the expectation of

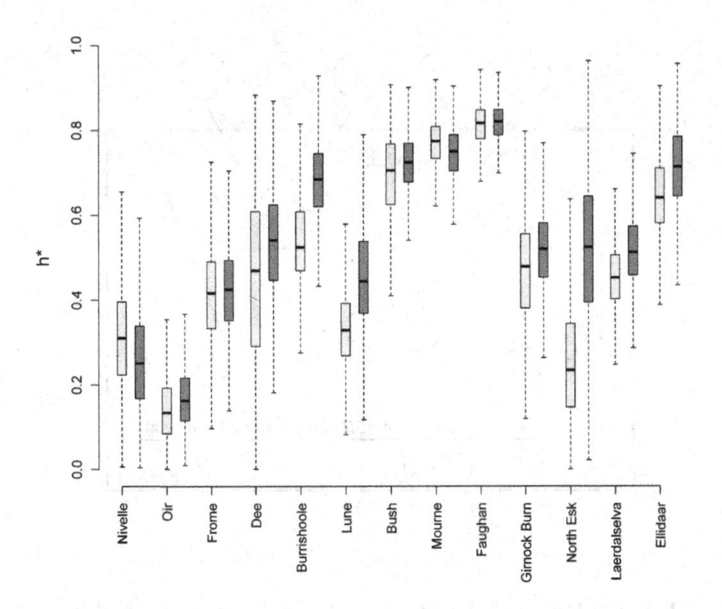


**FIGURE 9.10**: Marginal posterior distribution of $h^*$ for the 13 index rivers obtained under two model configurations: the model assuming independence between rivers (light gray) and the hierarchical model (dark gray).

$S_k^*$ and the latitude of river $k$ denoted $x_k$. As no information is available on the slope ($\alpha$) and intercept ($\beta$) of the linear regression in the log scale, they will be given a flat prior, for instance uniform pdfs with large bounds:

$$\begin{cases} log(\mu_{S_k^*}) = \alpha \times x_k + \beta \\ \alpha \sim Uniform(-5, 5) \\ \beta \sim Uniform(-50, 50) \end{cases} \tag{9.15}$$

Given the expected mean $\mu_{S_k^*}$, parameter $S_k^*$ for the river $k$ is drawn a priori in a Gamma distribution as in Eq. (9.13), but with parameters

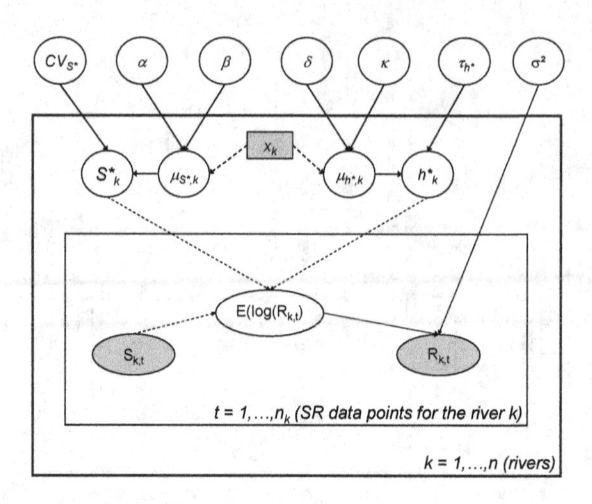


**FIGURE 9.11**: Directed Acyclic Graph representation of a hierarchical structure for the joint modeling of stock-recruitment relationships for the 13 rivers designed to capture the between-rivers variability of the parameters $(S^*, h^*)$ conditionally on the latitude $x_k$.

$a_k$ and $b_k$ depending upon the latitude $x_k$:

$$\begin{cases} CV_{S^*} & \sim Uniform(0,20) \\ a_k & = \dfrac{1}{CV_{S^*}^2} \\ b_k & = \dfrac{1}{\mu_{S_k^*} \times CV_{S^*}^2} \\ S_k^* & \sim Gamma(a_k, b_k) 1_{S_k^* < 200} \end{cases} \tag{9.16}$$

Because $h_k^*$ varies between 0 and 1, the $logit(\cdot)$ transform of $h$ is used to introduce some covariates (see also Chapter 8). The $logit(\cdot)$ transformation $((logit(h) = log\frac{h}{1-h}))$ will output values ranging from $-\infty$ to $\infty$. It is therefore convenient to express the gradient with increasing latitude as a linear regression on $\mu_{h^*}$ in the logit scale. Given the expected mean, a specific river contribution is modeled as a Normal distribution with variance $\tau_{h^*}^2$ that depicts the residual degree of similarity between rivers once the latitude gradient is accounted for:

$$\begin{cases} logit(\mu_{h_k^*}) = \delta \times x_k + \kappa \\ \delta \sim Uniform(-5, 5) \\ \kappa \sim Uniform(-50, 50) \end{cases} \tag{9.17}$$

And given the expected mean in the logit scale, $logit(h_k^*)$ for each river $k$ is drawn in a Normal distribution with expected mean $logit(\mu_{h_k^*})$ and precision $\tau_{h^*}^{-2}$. A diffuse prior was set on the precision:

$$\begin{cases} logit(h_k^*) \sim Normal(logit(\mu_{h_k^*}), \tau^2) \\ \tau_{h^*}^{-2} \sim Gamma(0.001, 0.001) \end{cases} \tag{9.18}$$

### 9.3.4 Results from the hierarchical approach

Salient features of the marginal posterior parameter pdfs represented in Fig. 9.12 are that the posterior probability $P(\alpha < 0|data)$ is null while that of $\delta < 0$ is 0.05, which indicates that the covariate latitude offers a good statistical explanation of positive variations between rivers in both $h^*$ and $S^*$. A more formal model comparison could be set (computing Bayes factors to test the present model versus the one with $\alpha = 0$ and $\delta = 0$) to definitively validate this choice.

Figures 9.9 and 9.10 give boxplot representations of what is known for the Biological Reference Points given data and the hierarchical model. The posterior distributions of $S^*$ and $h^*$ for the monitored 13 rivers reveal:

- Considerable within-river uncertainty in some cases despite SR data being available (*e.g.*, the Lune R. and the Laerdalselva R.);
- Significant variations among rivers, even within a relatively narrow latitudinal range (*e.g.*, the Bush R., the Mourne R. and the Faughan R., all located in Northern Ireland);
- An increasing trend with latitude.

$h^*$ is often poorly estimated (Fig. 9.10) and the increasing latitudinal gradient is less evident that for $S^*$. However, hierarchical modeling allows to improve data-poor rivers' inference as seen when comparing the inferences of the hierarchical model with those obtained under the model with independent rivers (Figs. 9.9 and 9.10). The reduction in posterior uncertainty is particularly visible for rivers with very poor SR data, such as the Laerdalselva River (with only 8 observations). Thanks to the transfer of information through the hierarchical structure, a more

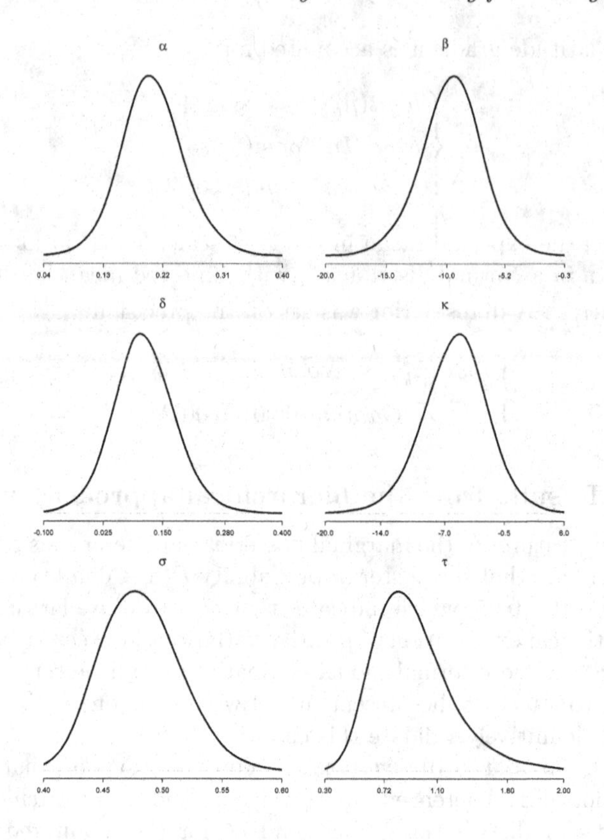


**FIGURE 9.12**: Marginal posterior probability shapes of the parameters $\alpha$, $\beta$, $\delta$, $\kappa$, $\sigma$ and $\tau$ from the hierarchical model.

precise knowledge of the biological reference points is gained with a more acceptable expected shape of the Ricker relationship (Fig. 9.8). As a consequence, a sensible diagnosis on the past working conditions of the river with regards to the ecological conservation limits (probable overexploitation) can be re-assessed.

### 9.3.5 Prediction of $S^*$ and $h^*$ given the latitude

The distributions of ultimate interest are the posterior predictive distributions, which represent our uncertainty/knowledge for sparse-data rivers without SR observations. The marginal posterior predictive distribution of $h^*_{new}$ and $S^*_{new}$ at various latitudes covering the salmon distribution range in the northeast Atlantic area (46°, 52°, 59° and 63° north) have been added to Figs. 9.13 and 9.14. They indicate that when moving north, salmon stocks can sustain higher exploitation rates $h^*$, can produce higher recruitment at MSY ($R^* = \frac{S^*}{1-h^*}$), but at the same time should be set at higher conservation limits $S^*$. However all the posterior predictive distributions are rather wide, suggesting that there remains great uncertainty in the spawning stock, the recruitment and the exploitation rates at MSY for a sparse data river; important residual variations stem from other explanatory covariates than the riverine wetted area and the latitude.

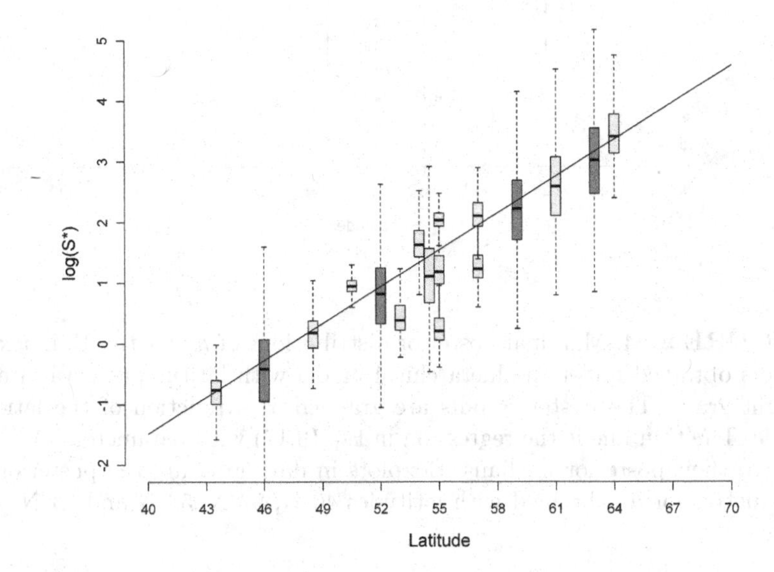


**FIGURE 9.13**: Marginal posterior distributions of $log(S^*)$ for the 13 index rivers obtained under the hierarchical model with latitude as covariate (light gray). The posterior pdfs are graphed as a function of the latitude. The thin line is the regression in Eq. (9.15) with parameters $(\alpha, \beta)$ set to their posterior medians. Boxplots in dark gray are the posterior predictive for $log(S^*)$ obtained with latitudes 46°N, 52°N, 59°N and 63°N.

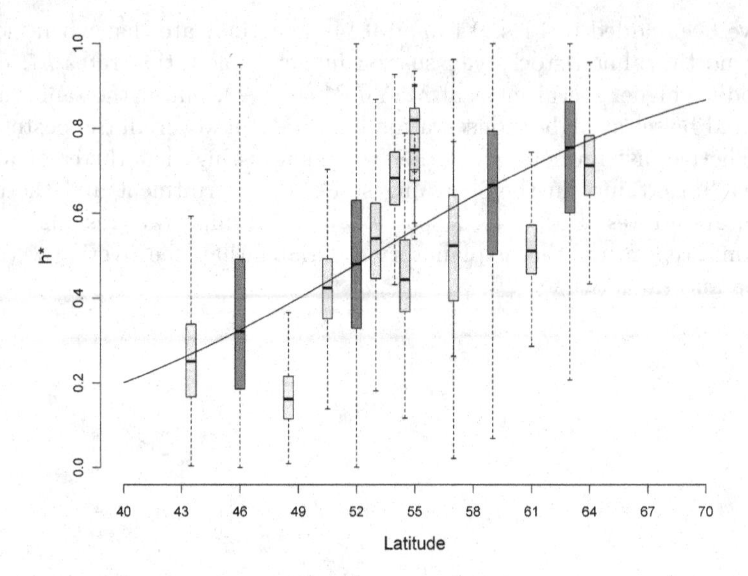


**FIGURE 9.14**: Marginal posterior distributions of $h^*$ for the 13 index rivers obtained under the hierarchical model with latitude as covariate (light gray). The posterior pdfs are graphed as a function of the latitude. The thin line is the regression in Eq. (9.17) with parameters $(\delta, \kappa)$ set to their posterior medians. Boxplots in dark gray are the posterior predictive for $h^*$ obtained with latitudes 46°N, 52°N, 59°N and 63°N.

---

## 9.4 Further Bayesian comments on exchangeability

An infinite sequence $Y_1, ..., Y_n, ...$ of random variables is said to be exchangeable if for all $n$, the joint pdf $[Y_1, ..., Y_n]$ remains invariant under any permutation of $\{1, 2, ..n\}$. If $Y_1, ..., Y_n, ...$ are independent and identically distributed, they are exchangeable, but not conversely. Latent variables and observables from hierarchical structures with a DAG such as the one of Fig. 1.13 (or Fig. E.2 for the baseball players' example in Appendix E) are exchangeable random variables (see [30]). Reciprocally, De Finetti [82] shows that for all exchangeable binary sequences $Y_1, ..., Y_n$, there exists a probability distribution $F$ on $[0, 1]$ such that

$$[Y_1, ..Y_n] = \int_{\theta=0}^{1} \theta^{\sum_{i=1}^{n} y_n} \times (1-\theta)^{n-\sum_{i=1}^{n} y_n} : dF(\theta)$$

When the probability distribution $F$ is diffuse, with pdf $[\theta]$ (we can then use for instance parametric models like the Beta pdf), one recovers the standard Bayesian predictive equation such as in Eq. (1.26) in Chapter 1. Hewitt and Savage [134] have generalized de Finetti's results: For all exchangeable sequences $Z_1, ..., Z_n$, their joint distribution can be written as a fully exchangeable hierarchical structure such as Eq. (9.2) (or some generalization if the distribution function of their empirical measure does not admit a pdf).

The concept of exchangeability stands at the core of statistical modeling, because it provides a marginal interpretation of conditional independence. For Bayesians, in addition to uncertainties by ignorance (as in priors) and by essence (as in likelihood), it simply brings a third type of uncertainty (by resemblance) to be modeled by probability distributions: using a biological metaphor, the basic idea is that if *phenotypes* $Y_{kj}$ and $Y_{kj'}$ of unit $k$ look alike, they must share something in common $(Z_k)$, to be drawn from an urn of *genotypes*.

One should not overlook the very important frequentist literature on *mixed effect* models, the frequentist appellation for exchangeable hierarchical models. The required statistical basic background for linear mixed model methodology with some extensions can be found in books such as [70], [200] and [305]. Fahrmeir and Tutz [101] cover multivariate statistical approaches. Since Rao and Toutenburg [249] gave a theoretical treatment of linear models including extensions to GLM (binary data) and missing data, there is a recent flourishing literature about mixed effect models from a Bayesian viewpoint ([89]; [281]).

# Chapter 10

## Hierarchical Bayesian Modeling II: Piling up simple layers


### Summary

In this chapter, we keep on playing with *LEGO* bricks to build more and more complex hierarchical models by piling up several simple layers. We show that hierarchical structures are fruitful to integrate multiple and various sources of data so as to learn from characteristics of ecological systems.

The first example of such *cocktail* models is inspired from Rivot *et al.* [257]. The example shows that Hierarchical Bayesian Modeling (HBM) is flexible and effective for the treatment of successive removal catch data introduced in Chapter 4. We perform the estimation of the population of Atlantic salmon juveniles in the Oir River (France) with ten inventory sites sampled by one or two removals over the period 1985-2005 given in Rivot *et al.* [257]. We show how to develop a general model to assess the effects of temporal variations and habitat type, on two latent quantities of interest: the density of fish and the probability of capture. Ecological expertise and Bayesian model comparison techniques are used to keep the model credible, parsimonious and realistic. Predictions of the total number of 0+ juveniles in the entire river reach can be derived, while accounting for all sources of uncertainty involved in this extrapolation. Finally, we show how to pile up the HBM built for estimating the 0+ population size with a rough observation model for the number of smolts (of the same cohort) migrating to the sea in order to estimate a survival rate between the 0+ juveniles and the smolts stage.

The second example is inspired from Brun *et al.* [39]. We develop a three-pass successive removal hierarchical model to estimate A. salmon juvenile abundance in the Nivelle River, and extend the model by connecting a second observation module depicting a rapid sampling technique as a possible alternative to the costly present successive removal procedure. We showed how simultaneously integrating these two different observation models within a single HBM can contribute to improve estimates of the total 0+ juvenile abundance.

---

## 10.1 Hierarchical model for successive removal data with habitat and time covariates

### 10.1.1 Why sampling freshwater juveniles?

The estimation of salmonid juvenile abundance matters both for the analysis of stock (*i.e.*, egg deposition by spawners) and for recruitment (*i.e.*, juvenile production of the subsequent generation) relationships. The transition from the egg to the 0+ juvenile (see Chapters 1 and 11) is a major bottleneck for Atlantic salmon populations ([99]; [207]). From a management perspective of salmon stocks, the freshwater juvenile is the earliest, and the most widely used, but often the sole, development stage which can be monitored prior to, and independently from, fisheries (see Chapters 7 and 9 for an estimation of biological reference points from stock-recruitment analysis and Chapter 12 for some insights on salmon stock management policies).

This chapter falls within the perspective of building a detailed observation sub-model of a general state-space model for A. Salmon life cycle. In a state-space life cycle model like the one introduced at the beginning of this book in Chapter 1 and developed further in Chapter 11, linking successive removal data to parr abundance can be used as an observation process to update the hidden population renewal process.

Two steps are required to estimate riverine fish population size:

1. Estimate population size (or density) at the sampling site level;
2. Predict the population on the whole river stretch relying on these sampling sites estimates. But only a small proportion of the wetted area is generally sampled. As a result, the uncertainty due to the extrapolation process may represent the most important part of the overall uncertainty ([130]; [209]).

This chapter revisits the HBM approach of successive removal data via electrofishing of Rivot *et al.* [257] to estimate salmon juvenile population size (Fig. 10.1 illustrates an electrofishing campaign of the 0+ juvenile population from the Oir River, Lower Normandy, France). Similar modeling approaches can be found in recent papers ([39]; [81]; [316]; [317]). There are conceptual difficulties very specific to the case study:

**FIGURE 10.1**: Electrofishing of 0+ A. salmon juveniles in the Oir River.

1. The available data set may seem large: 7 to 10 sites are sampled over a series of 20 years (1985 to 2005; year 1990 is missing) with typically two successive removals by site. Yet, it is a sparse dataset, as the number of fish caught is often very low with many missing data in the second pass.

2. Special care must be taken to check that the hierarchical structure of the proposed models depicts satisfactorily the multiple sources of variability (*e.g.*, between sites, between years). Variable selection is needed to make sure that habitat and sector covariates explain both the probability of capture and the fish density. Model comparison will discard unduly sophisticated models. Posterior checking will help exploring the dispersion allowed by the various model structures.

3. Predicting the population size of the entire river reach relies on extrapolation; the targeted values are blurred by many interfering sources of uncertainty.

### 10.1.2 Study site, sampling design and data

The survey covering all the areas colonized by A. Salmon in the Oir River network, has already been presented in Fig. 4.1 in Chapter 4. With

| Years (i) | Habitat type (h) | | | | | | | | | |
|---|---|---|---|---|---|---|---|---|---|---|
| | h=1 | | | | h=2 | | | | | |
| | Site (k) | | | | Site (k) | | | | | |
| | 1 | 2 | 3 | 4 | 1 | 2 | 3 | 4 | 5 | 6 |
| 1985 | 316 | 355 | 666 | 360 | 342 | - | - | 286 | - | 398 |
| 1986 | 316 | 355 | 666 | 360 | 342 | - | - | 286 | - | 398 |
| ... | | | | | | | | | | |
| 2005 | 340 | 355 | 172 | - | 137 | 131 | 111 | 262 | 154 | - |

**TABLE 10.1**: Surface area ($m^2$) of the sampling sites. Habitat type 1: Rapid/riffle; 2: Run. The symbol – indicates that the site was not sampled. Note that year 1990 is missing.

regard to the ecology of A. salmon, habitat was classified into three categories, known as rapids/riffles, runs and pools ([16]). Pools (only 3.5% of the water surface area) were neglected owing to the evidenced absence of 0+ salmon in this type of habitat. Rapids/riffles and runs are identified based on a combination of depth ($< 25cm$, between 25 and $60cm$, respectively), water velocity ($> 40cm{\cdot}s^{-1}$, between 20 and $40cm{\cdot}s^{-1}$, respectively) and bottom substrate (a mixture of sand, gravel, and pebbles with a higher proportion of coarse material in the rapid/riffles compared to the runs). Owing to the habitat preferences of A. salmon ([19]), this classification has been shown to explain a significant part of the spatial variability of the 0+ juveniles density in French rivers ([15]).

Since 1985, the 0+ juvenile production of the main stream is surveyed every year in autumn over a 12.3-km-long stretch extending from the trapping facility to an impassable dam (see Fig. 4.1). The data are collected according to a two-stage sampling scheme:

- 1*st* stage: Depending upon the year, 7 to 10 sampling sites (inventory sites, also called experimental repetitions in the following) were selected within the area of interest (Table 10.1). Their location and habitat remain the same every year during the study period, but their surface area may vary between years. Each site is a section of the river associated with a unique habitat type, *i.e.*, rapid/riffle or run. Each sampling unit is identified by three indices $(i, h, k)$ : $i = 1, ..., 20$ for the year (1985 to 2005; year 1990 is missing); $h = 1, 2$ for the habitat type (rapid/riffle, run) respectively; $k$ stands for the repetition per stratum $(i, h)$ (the number of repetitions varies among strata).

- 2*nd* stage: In each unit $(i, h, k)$, the 0+ salmon population was sampled by electrofishing with two successive removals. From 1985

| Years (i) | Habitat type (h) | | | | | |
|---|---|---|---|---|---|---|
| | h=1 | | | h=2 | | |
| | 1 | ... | 4 | 1 | .... | 6 |
| 1985 | 7/0 | ... | 73/5 | 25/3 | ... | 16/0 |
| 1986 | 35/2 | ... | 22/3 | 15/- | ... | 3/- |
| 1987 | 3/0 | ... | 20/1 | 0/0 | ... | 0/0 |
| 1988 | 20/5 | ... | 30/2 | 14/1 | ... | 0/0 |
| 1989 | 19/1 | ... | 57/6 | 11/1 | ... | 0/0 |
| ... | | | | | | |
| 2001 | 74/27 | ... | 49/- | 15/3 | ... | -/- |
| 2002 | 21/8 | ... | -/- | 0/0 | ... | -/- |
| 2003 | 31/4 | ... | 62/12 | 4/- | ... | 2/- |
| 2004 | 24/5 | ... | 0/0 | 12/1 | ... | 0/- |
| 2005 | 40/9 | ... | 22/1 | 6/- | ... | 5/- |

**TABLE 10.2**: Number of fish captured at the first and second passes ($C^1/C^2$). /−: Second pass not completed. Symbol −/− indicates that the site was not sampled at all. Habitat type: 1 = rapid/riffle; 2 = run. Note that year 1990 is missing.

to 2005, the survey was conducted according to a similar operating protocol (the one described in Section 4.2) and with essentially the same staff. The $2^{nd}$ pass was always realized shortly after the first one. The complete dataset consists of the numbers of fish captured at the first and the second pass for 190 sampling units (Table 10.2).

There is quite a lot of missing data for the second pass. The sampling rate (measured as the proportion of the wetted area sampled) in each stratum $(i, h)$ varies between 2.9% and 12.2% (see Table 10.4).

### 10.1.3 General model formulation

#### 10.1.3.1 Observation model for successive removals

We stick to the classical assumptions of the successive removal method ([45]), already detailed in Chapter 4. Keeping the same notations, we denote, for each unit $(i, h, k)$, $C^1_{i,h,k}$ and $C^2_{i,h,k}$ the catch data of the first and the second pass, respectively, $\pi_{i,h,k}$ the probability of capture, $\nu_{i,h,k}$ the initial population size. There are only two passes and the sampling distributions of the catch data are given by the following Binomial equations:

$$\begin{cases} C^1_{i,h,k} \sim Binomial(\nu_{i,h,k}, \pi_{i,h,k}) \\ C^2_{i,h,k} \sim Binomial(\nu_{i,h,k} - C^1_{i,h,k}, \pi_{i,h,k}) \end{cases} \quad (10.1)$$

#### 10.1.3.2 Role of the sampling area on fish number

The population size $\nu_{i,h,k}$ depends on the expected fish density $\delta_{i,h,k}$ (fish per m$^2$) and on the surface area $S_{i,h,k}$ of the sampling sites. As in Eq. (4.8) and for the same reasons, we assume that $\nu_{i,h,k}$ is Poisson-distributed with parameter scaled by the sampling area $\delta_{i,h,k} \times S_{i,h,k}$:

$$\nu_{i,h,k} \sim Poisson(\delta_{i,h,k} \times S_{i,h,k}) \tag{10.2}$$

#### 10.1.3.3 A Normal hierarchical structure for the latent variables $\delta$ and $\pi$

The density $\delta$ and the catchability $\pi$ are latent variables depending on the year $i$, the habitat category $h$ and the sites sharing the same habitat conditions (repetition $k$). Because they are convenient to work with real numbers, $log(\cdot)$ and $logit(\cdot)$ transformations on the density and catchability respectively were used. A general model formulation would decompose the various sources of variability for the latent variables by introducing year and habitat effects. The influence of these covariates is introduced via additive effects on the expected mean of $log(\cdot)$ and $logit(\cdot)$ transformations of the density and catchability, respectively:

$$\begin{cases} \mathbb{E}(log(\delta_{i,h,k})) = \mu_\delta + \alpha_{\delta_i} + \beta_{\delta_h} \\ \mathbb{E}(logit(\pi_{i,h,k})) = \mu_\pi + \alpha_{\pi_i} + \beta_{\pi_h} \end{cases} \tag{10.3}$$

$\mu_\delta$ and $\mu_\pi$ are the overall mean of log-density and logit-catchability respectively, $\alpha_{\delta_i}$ and $\alpha_{\pi_i}$ are the effects of year $i$, and $\beta_{\delta_h}$ and $\beta_{\pi_h}$ are the effects of habitat type $h$.

Two sites sharing the same habitat conditions and recorded the same year may still differ due to uncontrolled conditions. Because the density and the catchability are likely to be correlated, a bivariate Normal distribution on $\begin{pmatrix} log(\delta_{i,h,k}) \\ logit(\pi_{i,h,k}) \end{pmatrix}$ was used to capture the residual variability once the systematic effects of years and habitat have been accounted for:

$$\begin{pmatrix} log(\delta_{i,h,k}) \\ logit(\pi_{i,h,k}) \end{pmatrix} \sim Normal_2\left(\begin{pmatrix} \mathbb{E}(log(\delta_{i,h,k})) \\ \mathbb{E}(logit(\pi_{i,h,k})) \end{pmatrix}, \Sigma_{\delta\pi}\right) \tag{10.4}$$

with a variance-covariance matrix $\Sigma_{\delta\pi}$:

$$\Sigma_{\delta\pi} = \begin{pmatrix} \sigma_\delta^2 & \rho_{\delta\pi}\sigma_\delta\sigma_\pi \\ \rho_{\delta\pi}\sigma_\delta\sigma_\pi & \sigma_\pi^2 \end{pmatrix}$$

### 10.1.4 From a general mathematical formulation to a model that makes ecological sense

#### 10.1.4.1 Correlation between $\delta$ and $\pi$

No spatial dependence between adjacent sites is introduced in this model and we interpret the variance-covariance matrix $\Sigma_{\delta\pi}$ as a degree of similarity between sites (after subtracting the effects of year and habitat conditions). For instance, $\sigma_\delta = 0$ would mean that the density at two sites sampled the same year with the same habitat would be strictly the same. The role of a correlation $\rho_{\delta\pi} \neq 0$ is much more questionable. $\rho_{\delta\pi} > 0$ would mean that sites with fish density estimated above the average are also sites with a better estimated probability of capture, which would not really match the independence assumptions of the successive removal method (see Section 4.2 in Chapter 4 ).

#### 10.1.4.2 Effects of year and habitat on the density

There are potentially many factors causing yearly variations in fish recruitment but they are mostly not identified, uncontrolled and not monitored ([99]; [159]; [207]). Within a given year, the spatial variability of the density is high and correlated with riverine physical habitat ([6]; [19]). Ecological expertise would therefore deny considering models without year and site effects on fish density.

Linear additive effects of Eq. (10.3) for the logarithm of the density turn into multiplicative effects when getting back to the original variable. This should cause no trouble since recruitment of 0+ juveniles is known to be highly variable between years.

The effect of years are considered as a sample drawn from an infinite set of years that are exchangeable regarding their effect on density and catchability. Thus, the $\alpha_{\delta_i}$'s are modeled as random effects, drawn from a common distribution. A Normal pdf with mean= 0 and variance $\sigma^2_{\alpha_\delta}$ is the generic choice for representing the occurrences of these 20 latent variables (one vector for each year):

$$\alpha_{\delta_i} \sim Normal(0, \sigma^2_{\alpha_\delta}) \tag{10.5}$$

In contrast with years, the habitat types are two mutually exclusive categories and the $\beta_{\delta_h}$'s are modeled as fixed effects. In French rivers, the 0+ juveniles salmon densities in autumn are notably higher in riffles/rapids compared to runs ([15]; [17]; [97]). The $\beta_\delta(h)$'s must *a priori* keep to the constraint $\beta_{\delta_1} > \beta_{\delta_1}$. To meet this requirement, $\beta_{\delta_1}$ can be drawn from a half-positive Normal prior with large variance, say $V = 100$, and a sum-to-zero constraint is imposed to avoid confusion with the overall mean $\mu_\delta$ :

$$\begin{cases} \beta_{\delta_1} & \sim & Normal(0, V = 100) I(0, +\infty) \\ \beta_{\delta_2} & = & -\beta_{\delta_1} \end{cases} \tag{10.6}$$

#### 10.1.4.3 Model comparison for testing the effects of year and habitat on the probability of capture

The probability of capture $\pi_{i,h,k}$ can also vary among units. In stream-dwelling salmonids, many influential factors have been identified (*e.g.*, river width, depth, water velocity, temperature, water conductivity, fish size, habitat complexity), but the results are not fully consistent across studies ([229]; [264]; [282]).

Sticking back to the terms of the general structure in Eqs. (10.3) and (10.4), we do not want to consider $\sigma_\pi = 0$. Yet, it is not clear whether the year or the habitat type would *a priori* cause systematic variations of the probability of capture. Consequently, four models denoted $M_0$, $M_1$, $M_2$, and $M_3$ can be designed as indicated in Table 10.5. They can be compared according to Bayes Factors or Deviance Information Criteria (see Appendix B).

All models have year (random) and habitat (fixed) effects on the density. The simplest model $M_0$ is called the "reference model" and has no year and habitat effect on the probability of capture. $M_1$ considers year effect on the probability of capture, but no habitat effect. The year effect $\alpha_{\pi_i}$'s are also modeled as random effect:

$$\alpha_{\pi_i} \sim Normal(0, \sigma^2_{\alpha_\pi}) \tag{10.7}$$

$M_2$ considers habitat effect on the probability of capture but no year effect. The $\beta_{\pi_h}$'s are modeled as fixed effects. But conversely to habitat effect on the density in Eq. (10.6), no order would be introduced *a priori* (the prior on $\beta_{\pi_1}$ is not restricted to positive values):

$$\begin{cases} \beta_{\pi_1} & \sim & Normal(0, V = 100) \\ \beta_{\pi_2} & = & -\beta_{\pi_1} \end{cases} \tag{10.8}$$

All models are nested in the full model $M_3$ that considers both year effects (as in Eq. (10.7)) and habitat effects (as in Eq. (10.8)) on the probability of capture.

### 10.1.5 Directed Acyclic Graph and prior specification on hyper-parameters

Figure 10.2 provides a Directed Acyclic Graph representation of the baseline model $M_0$ with the conventions of Chapter 1. All the free pa-

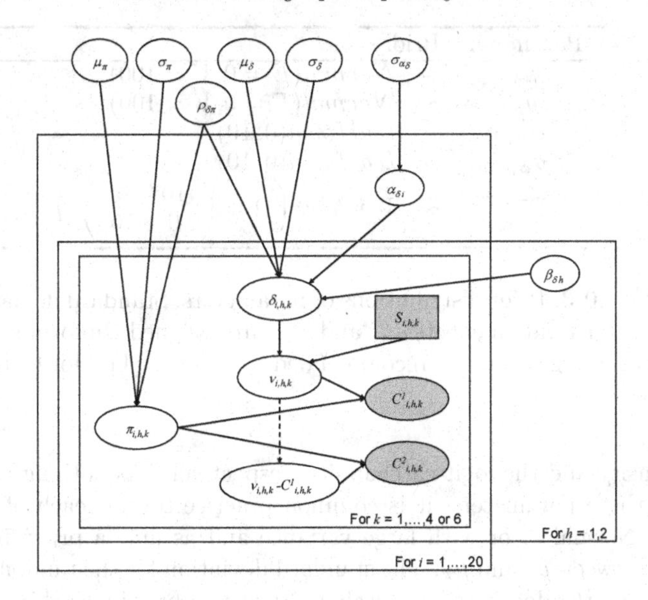


**FIGURE 10.2**: Directed Acyclic Graph of the baseline model $M_0$ (year and habitat effects on the density only); shaded rectangles: known constant; shaded ellipse: observed probabilistic nodes; white ellipse: nonobserved probabilistic nodes; arrow: conditional dependencies between nodes; solid arrow: probabilistic dependency; broken arrow: logical dependency; frame: repetition of structure over units.

rameters of the model, *i.e.*, those not conditioned by any quantity, are assigned weakly informative and independent prior pdfs (Table 10.3). In agreement with Gelman's statement ([116]), we follow the approach of Bernardo ([29]) and consider so-called noninformative priors as "reference priors" to be used as a standard of comparison or starting point in place of the proper, informative prior distributions that would be appropriate for a full Bayesian analysis.

The first idea for setting a prior on the three parameters $\sigma_\delta$, $\sigma_\pi$ and $\rho_{\delta\pi}$ would be to independently pick a $Uniform(-1, 1)$ for $\rho_{\delta\pi}$ and vague Gamma distributions for the precisions $\sigma_\delta^{-2}$ and $\sigma_\pi^{-2}$. The Wishart distribution, a multivariate generalization of the Gamma distribution, could also be considered as a prior pdf for the $2 \times 2$ precision matrix $\Sigma_{\delta\pi}^{-1}$. Advanced mathematical properties justify the classical recourse to this sophisticated prior ([140], page 257; [164], page 373). It is the conjugate prior for the inverse of a variance-covariance matrix when working with a multivariate Normal likelihood. $\mu_\delta$ and $\mu_\pi$ are the mean effects for the

| Parameter | | Prior |
|---|---|---|
| $\mu_\delta$ | $\sim$ | $Normal(E=0, V=100)$ |
| $\mu_\pi$ | $\sim$ | $Normal(E=0, V=100)$ |
| $\sigma_{\alpha_\delta}$ | $\sim$ | $Uniform(0, 10)$ |
| $\sigma_{\alpha_\pi}$ | $\sim$ | $Uniform(0, 10)$ |
| $\Sigma_{\delta\pi}^{-1}$ | $\sim$ | $Wishart\left(\Omega = \begin{pmatrix} 10 & 0 \\ 0 & 10 \end{pmatrix}, p=2\right)$ |

**TABLE 10.3**: Prior distributions of parameters. Standard deviation parameters for random effects $\sigma_{\alpha_\delta}$ and $\sigma_{\alpha_\pi}$ are assigned Uniform prior on a sufficiently large range, as recommended by Gelman [116] for hierarchical models.

log-density and the logit-catchability, respectively. As nothing is known about these parameters, it is common practice to give each of them a diffuse Normal prior with large variance and assume a priori independence between $\mu_\delta$ and $\mu_\pi$. The standard deviation for the random effects of years on the densities and catchability were a priori picked in Uniform distributions.

### 10.1.6 Extrapolation to the whole river stretch

| Years | Hab. type | | | | | | | | | |
|---|---|---|---|---|---|---|---|---|---|---|
| | 1 | 2 | 1 | 2 | 1 | 2 | 1 | 2 | 1 | 2 |
| | $S_{s_{i,h}}$ | | $S_{s_{i,h}}$ | | $S_{e_{i,h}}$ | | $n_{e_{i,h}}$ | | $S_{e_{i,h}}$ | |
| 1986 | 1697 | 1026 | 424 | 342 | 12181 | 26078 | 29 | 76 | 420 | 343 |
| ... | | | | | | | | | | |
| 2005 | 1031 | 860 | 258 | 143 | 12847 | 26244 | 50 | 183 | 257 | 143 |

**TABLE 10.4**: Sampled and not sampled (extrapolation) surface area ($m^2$) by habitat type. $S_{s_{i,h}}$: total sampled area in year $i$ and habitat type $h$. $\bar{S}_{s_{i,h}}$: mean area of sampling sites. $S_{e_{i,h}}$: total extrapolation area. $n_{e_{i,h}}$: number of sites for the extrapolation (see text). $\bar{S}_{e_{i,h}}$: mean area of the extrapolation sites. Habitat type: 1 = rapids/riffles; 2 = runs. Total surface area ($m^2$) of the river by habitat type: Type 1: 13878; type 2: 27104.

The ultimate goal of the study is to assess the total 0+ juvenile population in the surveyed section of the Oir River (Figure 4.1 in Chapter 4) for each year $i$, denoted $\nu_i$. It is the sum of the population size $\nu_{s_i}$ estimated on the sampling sites with the population size $\nu_{e_i}$ predicted on the non-sampled area, where $\nu_{s_i}$ and $\nu_{e_i}$ are calculated as a sum over the habitat types:

$$\nu_i = \sum_{h=1}^{2} \nu_{s_{i,h}} + \sum_{h=1}^{2} \nu_{e_{i,h}} \tag{10.9}$$

The population sizes $\nu_{e_{i,h}}$ have to be extrapolated for the $(i, h)$ strata. It can definitively not be computed simply as $\nu_{e_{i,h}} = \nu_{s_{i,h}} \times \frac{S_{s_{i,h}}}{S_{e_{i,h}}}$ since the between sampling unit variability $\sigma_\delta$ is of the same order of magnitude of the mean site effect $\mu_\delta$ (see Table 10.6 in the Results section) and thus the many unidentified factors in addition to the habitat and year cannot be neglected at all!

To mimic the spatial variability of the original experiment, we assume that the nonsampled surface area $S_{e_{i,h}}$ (Table 10.4) is divided into $n_{e_{i,h}}$ of sites of equal surface area $\bar{S}_{e_{i,h}}$, as close as possible to the mean surface of the sampled sites in the habitat $h$ during year $i$ and consider virtual repetitions $k' = 1, ..., n_{e_{i,h}}$ for the nonsampled section. Two sources of uncertainty are attached to the random variable $\nu_i$:

1. One source of uncertainty is the partial knowledge of the parameters, which is encoded in the posterior distribution of the unknowns of the model of interest for fish density, *i.e.*, $(\sigma_\delta, \mu_\delta, \alpha_{\delta_i}, \beta_{\delta_h}, \nu_{s_{i,h}})$;
2. Another part of uncertainty arises from the fact that we have no data to update the nonsampled units and we must take into account the between sampling unit variability: for each of the virtual repetitions $k' = 1, ..., n_{e_{i,h}}$, we first draw the density $\delta'_{i,h,k}$ in its posterior predictive:

$$\begin{cases} \mu_{\delta'_{i,h,k}} = \mu_\delta + \alpha_{\delta_i} + \beta_{\delta_h} \\ log(\delta'_{i,h,k}) \sim Normal(\mu_{\delta'_{i,h,k}}, \sigma^2_\delta) \end{cases} \tag{10.10}$$

and finally generate a Poisson number of fish:

$$\nu'_{e_{i,h,k}} \sim Poisson(\bar{S}_{e_{i,h}} \times \delta'_{i,h,k}) \tag{10.11}$$

The latter source of uncertainty must not be forgotten in the extrapolation term $\nu_{e_i} = \sum_{h,k'} \nu_{e_{i,h,k'}})$. It depends on the sampling rate and on the spatial variability $\sigma_\delta$ of the density. If the sampling rate is low and the density is highly variable, it can represent the greatest source of uncertainty in the estimation of $\nu_i$.

### 10.1.7 Linking the 0+ population size to the smolts run

The estimates of the total 0+ juveniles population size in the Oir River obtained from the procedure defined above in Section 10.1.6 can

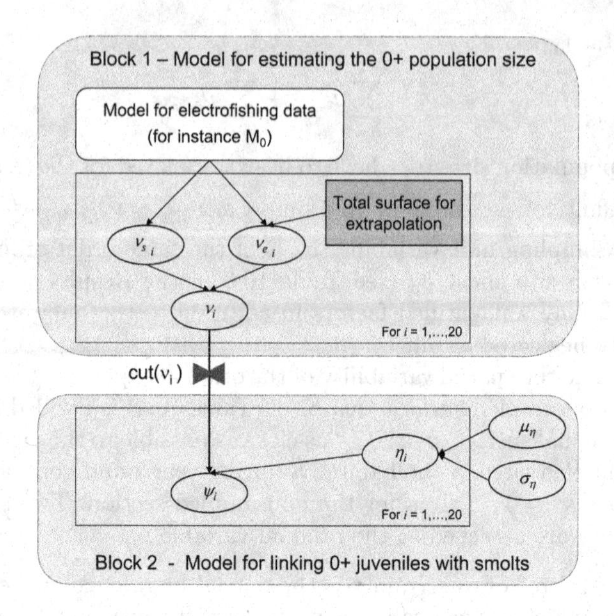


**FIGURE 10.3**: Directed Acyclic graph of the model combining the two blocks. The first block is the model built to estimate the 0+ density and to extrapolate the total 0+ population size $\nu_i$. The second block is designed to estimate the 0+ $\rightarrow$ smolts survival rated $\eta_i$. The $cut(\cdot)$ function between the two blocks allows the information for flowing downward to estimate the $\eta_i$'s, but information conveyed by the smolts estimates $\psi_i$ is prevented from flowing upward model structure to update the number of 0+ juveniles.

be compared with those of the smolts production of the corresponding cohorts. Linking the 0+ number with the smolts number of the corresponding cohort would be a first step toward building a population dynamic model as it would enable us to estimate the 0+ to smolts survival rate for each cohort for which data are available. As a first approach to estimate the 0+ $\rightarrow$ smolts survival rate, one could simply stick to the ratio of points estimates of the number of 0+ and smolts. However, we would then neglect the uncertainty attached to the estimates of the number of 0+ juveniles (see the results of this chapter in Section 10.1.12) and smolts (see Section 4.1 in Chapter 4 for a description of the capture-mark-recapture model built to infer the smolts run).

To achieve such a transfer of information and estimate the survival rate within a full Bayesian scheme, one has to devise an additional layer in the hierarchical structure described in Fig. 10.2 to link the juvenile population size $\nu_i$ at year $i$ to the corresponding smolts run, denoted $\psi_i$ in the following. The trick is to consider the whole hierarchical electrofishing model detailed in Fig. 10.2 which provides a prior for the population size $\nu_i$ before it is linked to the smolts run $\psi_i$, and then to propose a conditional model for the number of smolts conditioned by $\nu_i$. Figure 10.3 presents a simplified DAG for such a model.

Let us denote $\eta_i$ the survival rate from 0+ juveniles to smolts for the cohort $i$. The $\eta_i$'s could be taken a priori from a Beta distribution. From the capture-mark-recapture model built in Chapter 4, it is possible to obtain a posterior distribution of the number of smolts $\psi_i$ for each cohort $i$. As a realistic synthesis of the results, one could consider that the posterior distribution of the $\psi_i$'s are log-Normally distributed with mean indicated in Fig. 10.9 and an average coefficient of variation $CV_\psi = 0.1$ (supposed constant for all cohorts $i$). Hence, conditionally upon $\eta_i$ and $\nu_i$, a simple probabilistic model for the survival including observation error on the total number of smolts (which magnitude is scaled by $CV_\psi$) would be:

$$log(\psi_i) \sim Normal(\eta_i \times \nu_i - \frac{1}{2}\sigma_\psi^2, \sigma_\psi^2) \tag{10.12}$$

with $\sigma_\psi^2 = log(CV_\psi^2 + 1)$. $\frac{1}{2}\sigma_\psi^2$ is a correction factor ensuring that the expected mean of $\psi_i$ is $\eta_i \times \nu_i$.

To complete this additional block added to the full model, an exchangeable hierarchical prior structure is set up on the $\eta_i$'s to capture the between years variability of the survival rate. Instead of considering the $\eta_i$'s as a priori independently sampled from conditional Beta distribution with parameters common to all years (as detailed in Section 9.2 of Chapter 9), the *logit*-transform of the $\eta_i$'s are a priori drawn in a Normal distribution with mean and variance parameters $\mu_\eta$ and $\sigma_\eta$ shared between years.

$$logit(\eta_i) \sim Normal(\mu_\eta, \sigma_\eta^2) \tag{10.13}$$

The $\eta_i$'s are then calculated by the inverse transformation $logit^{-1}(\cdot)$. $\mu_\eta$ and $\sigma_\eta$ are a priori drawn in a $Normal(0, 100)$ and $Uniform(0, 10)$ prior distribution, respectively.

### 10.1.8 Bayesian computations

#### 10.1.8.1 Missing data

Some of the $C_{i,h,k}^2$ catch data are missing. The missing data mechanism can be realistically assumed to be ignorable, in the sense defined by Gelman [117]. The *Missing At Random* conditions can be checked:

- The missing process parameter priors must be independent from those ruling the observation mechanism,
- The probability that a data is missing should not depend on the value of the missing data to be observed.

When missing at random, the treatment of missing data in the Bayesian setting is straightforward and can be done by data augmentation techniques (see Gelman [117] for more details and various examples about the Bayesian treatment of missing data and Rivot *et al.* [257] for a discussion in the case of electrofishing data). They are then considered as any other unknown quantity of a model. Posterior inference about any set of unknowns of interest is obtained by integrating over the posterior pdf of the missing data. Such an integration is painless when MCMC sampling methods are used to derive the posterior inferences.

#### 10.1.8.2 Controlling the flow of information in the DAG

In the full model defined in Fig. 10.3, we do not wish that the estimates of the smolt numbers contribute to the estimation of unknown variables defined in the first block (the one defined in Fig. 10.2). We rather want to use the first block of the model to form a prior distribution for the number of 0+ juveniles, and then in the second block to combine this prior information with estimates of the smolts run to estimate the 0+ $\rightarrow$ Smolts survival rate, but we do not want any information *feedback* from the second block to the first one. To achieve such a unidirectional flow of information, we use the $cut(\cdot)$ function of OpenBUGS® that forms a kind of one-way pipe in the model. Instead of Eq. (10.12), we rather write

$$log(\psi_i) \sim Normal(\eta_i \times cut(\nu_i) - \frac{1}{2}\sigma_\psi^2, \sigma_\psi^2) \quad (10.14)$$

In Eq. (10.14) that uses $cut(\nu_i)$ instead of $\nu_i$, information is allowed to flow *downward* through the cut to estimate the $\eta_i$'s, but information conveyed by Eq. (10.14) is prevented from flowing upward model structure to update the number of 0+ juveniles (see the OpenBUGS manual for more details).

#### 10.1.8.3 MCMC sampling

As in the previous chapters, all the posterior inferences are performed by means of the OpenBUGS software. Convergence of the MCMC chains is checked via the Gelman-Rubin (GR) diagnostics as implemented by OpenBUGS. We used three independent chains, and the first 10,000 iterations are discarded as an initial burn-in period. Then, 100,000 further iterations are performed. Inferences are based on a 30,000 MCMC sample obtained from pooling the three chains after thinning by a factor of 10 to reduce autocorrelation in the final sample.

### 10.1.9 Results of the comparison between $M_0$,...,$M_3$

Based on the guidelines proposed by Kass and Raftery [157], both models $M_2$ and $M_3$ are clearly discarded by the *BF*(and also by the *DIC*, see Spiegelhalter *et al.* [283] and Appendix B), while the more parsimonious models $M_0$ and $M_1$ appear as good candidates (see Table 10.5).

In what follows, we focus on the baseline model $M_0$, the most parsimonious and most credible one. When necessary, the robustness of our findings can be checked against the results provided by model $M_1$ as a possible alternative.

| Models | Effect on the density | | Effect on the prob. of capture | | DIC | $BF M_0 vs M_i$ (log scale) |
|---|---|---|---|---|---|---|
| | Year | Habitat | Year | Habitat | | |
| $M_0$ | + | + | - | - | 924 | 0 |
| $M_1$ | + | + | + | - | 919 | 1.2 |
| $M_2$ | + | + | - | + | 982 | 3.7 |
| $M_3$ | + | + | + | + | 972 | 5.6 |

**TABLE 10.5**: Alternative competing models differing by the way year and habitat effects are introduced in the probability of capture, and criteria for model selection. Values are computed with the prior configuration described in the Material and Methods section; "+": The effect of the corresponding covariate is introduced; "-": No effect is introduced; *DIC*: Deviance Information Criterion. *BF*: Bayes Factors. DIC and BF were calculated as explained in Appendix B.

### 10.1.10 Estimation with model $M_0$

The estimation of the overall mean density in the log scale ($\mu_d$) is fairly precise (Table 10.6). As shown in Fig. 10.6, the posterior distribution of the correlation between the density and the probability does not indicate a strong negative or positive correlation between $\pi$ and $\delta$. Results are then compatible with the common sense that $\delta$ and $\pi$ do not vary with a strong systematic positive or negative correlation.

The posterior pdfs of the $\alpha_{\delta_i}$'s (Fig. 10.4) point out the high between years variability of the 0+ salmon density without any particular trend.

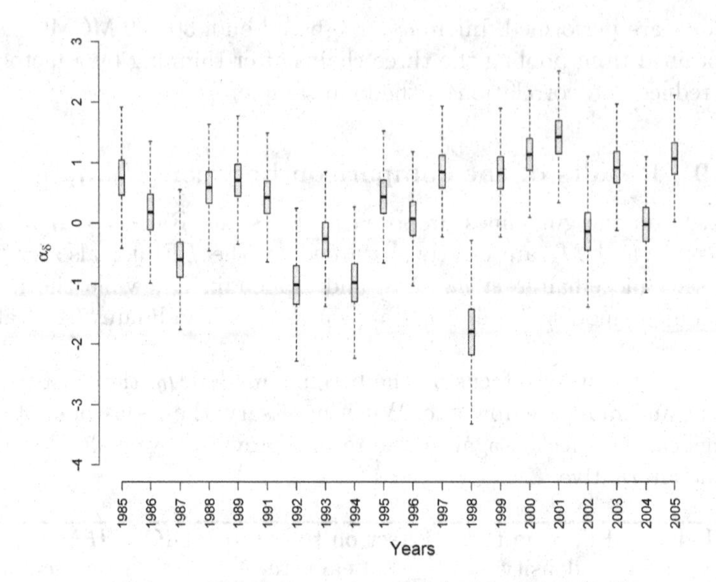


**FIGURE 10.4**: Box and whisker plots of a size-30,000 MCMC sample from the marginal posterior pdfs of the $\alpha_{\delta_i}$'s in model $M_0$ (year and habitat effect on the density only).

Year 1998 (see data in Table 10.2 and estimates of $\alpha_{\delta_{1998}}$ in Fig. 10.4) has been noticeably poor. When getting back to the 1998 data, the total number of captures for the ten locations at the first pass were only six individuals and none was caught during the second pass.

The variability between the $\alpha_{\delta_i}$'s (the mean of the posterior *pdf* of $\sigma_{\alpha_\delta}$ is 0.97; Table 10.6) seems much higher than the within-year uncertainty of estimated effects (the standard deviation of the posterior *pdf* of the $\alpha_{\delta_i}$'s range from 0.42 to 0.58 according to the year). In the natural scale of the density, the mean of the posterior predictive pdfs of the density on a rapid/riffle habitat in a given year ranges from 2 fish per 100 $m^2$ (year 1998) to 38 fish per 100 $m^2$ (year 2001).

The effect of the habitat type on the density is very strong (Fig. 10.5). The mean of the posterior *pdf* of the density ratio in the habitat types 1 versus 2, calculated as $e^{\beta_{\delta 1}}/e^{\beta_{\delta 2}}$, is 6.5. Even after accounting for the effects of the year and the habitat type, the between-site variability of the density, $\sigma_\delta$, remains high when compared to the overall mean $\mu_\delta$ or the between-year standard deviation $\sigma_{\alpha_\delta}$ (Table 10.6).

The overall mean probability of capture in the *logit* scale, $\mu_\pi$, is

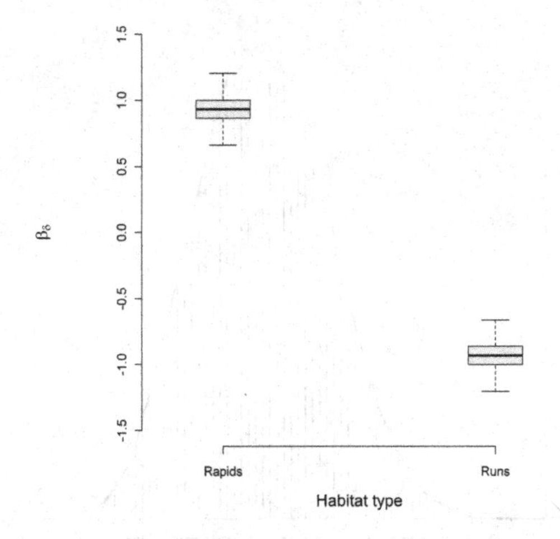


**FIGURE 10.5**: Box and whisker plots of a size-30,000 MCMC sample from the marginal posterior pdfs of the $\beta_{\delta_h}$'s under model $M_0$ (year and habitat effect on the density only).

estimated with a good precision (Table 10.6). In the natural scale, the mean of the posterior predictive *pdf* of the probability of capture is 0.82, confirming that a good job is done by the fishing team! The between-units standard deviation of the probability of capture ($\sigma_\pi$) is rather low compared to the overall mean $\mu_\pi$ (Table 10.6). Consistently, even for units with missing or little informative data, the probability of capture (not shown) is rather precisely estimated (the $\pi_{i,h,k}$'s have dome-shaped posterior densities close to 0 around 0 and 1) and appear rather high (the mean of the posterior *pdf* of the probability of capture ranges from 0.64 to 0.90 according to the sampling unit). The results obtained under model $M_1$ as summed in Table 10.6 are quite comparable to the ones of $M_0$.

### 10.1.11 Posterior checking

Many sources of variability were hypothesized: fish densities vary according to sampling units and years, so do probabilities of capture. This creates $2 \times (190+20) = 420$ latent variables for models $M_3$ and $M_2$, and 400 ones for models $M_1$ and $M_0$. Given the density on a sampling site, an additional Poisson stochastic behavior entails the number of fish, still a latent variable. The database contains 190 records for the first pass

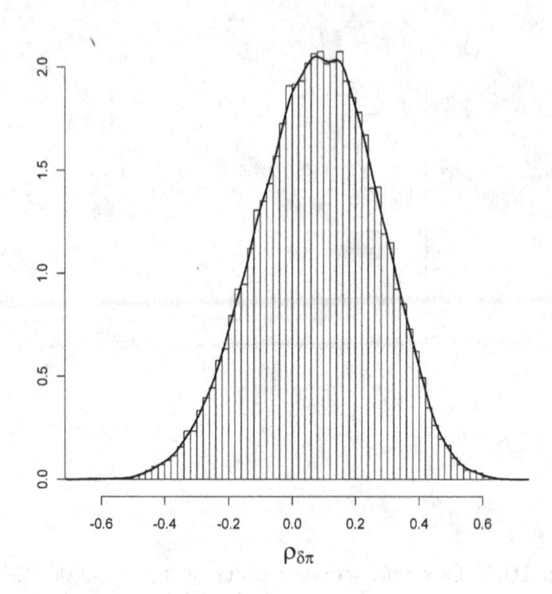


**FIGURE 10.6**: Marginal posterior distribution of $\rho_{\delta\pi}$ under model $M_0$.

and 125 records for the second one. Once Bayesian inference has been performed, what means do we have to check that the variability observed in the data has been correctly represented in the deeper layers of the *HBM*? Bayesian statisticians have devised various statistical tools.

Replicate the data and work out posterior predictive checks as advocated as an efficient method to track for inconsistency in hierarchical models ([74]; [117]). Gelman *et al.* [117] suggest to use the omnibus $\chi^2$-discrepancy as a summary statistics to measure the consistency between the model and data. As the experiments are Binomial, we evaluate for each sampling unit $(i, h, k)$, the following omnibus $\chi^2$-discrepancies:

$$\begin{cases} \chi^2_{(1)}(C^1, \pi, \nu) = \left( \dfrac{C^1 - \nu\pi}{\sqrt{\nu\pi(1-\pi)}} \right)^2 \\ \chi^2_{(2)}(C^1, C^2, \pi, \nu) = \left( \dfrac{C^2 - (\nu - C^1)\pi}{\sqrt{(\nu - C^1)\pi(1-\pi)}} \right)^2 \end{cases} \tag{10.15}$$

and sum these quantities on all sampling units (with $\chi^2_{(2)} = 0$ when the second pass is missing). We did the same work by replacing $C^1$ and $C^2$ with replicates $C^{1,rep}$ and $C^{2,rep}$ drawn from their predictive distri-

| | Mean | Sd | 2.5% | 5% | 25% | med | 75% | 95% | 97.5% |
|---|---|---|---|---|---|---|---|---|---|
| $M_0$ | | | | | | | | | |
| $\mu_\delta$ | 0.87 | 0.25 | 0.37 | 0.46 | 0.71 | 0.87 | 1.03 | 1.27 | 1.35 |
| $\sigma_\delta$ | 1.09 | 0.09 | 0.93 | 0.95 | 1.03 | 1.09 | 1.15 | 1.25 | 1.29 |
| $\mu_\pi$ | 1.62 | 0.11 | 1.41 | 1.44 | 1.55 | 1.62 | 1.70 | 1.81 | 1.85 |
| $\sigma_\pi$ | 0.61 | 0.13 | 0.37 | 0.41 | 0.52 | 0.60 | 0.69 | 0.84 | 0.89 |
| $\sigma_{\alpha_\delta}$ | 0.97 | 0.22 | 0.63 | 0.67 | 0.82 | 0.95 | 1.10 | 1.36 | 1.47 |
| $M_1$ | | | | | | | | | |
| $\mu_\delta$ | 0.61 | 0.12 | 0.36 | 0.41 | 0.53 | 0.61 | 0.69 | 0.81 | 0.84 |
| $\sigma_\delta$ | 1.09 | 0.09 | 0.93 | 0.95 | 1.03 | 1.09 | 1.15 | 1.25 | 1.29 |
| $\mu_\pi$ | 1.55 | 0.17 | 1.26 | 1.31 | 1.42 | 1.52 | 1.64 | 1.85 | 1.93 |
| $\sigma_\pi$ | 0.26 | 0.14 | 0.06 | 0.06 | 0.12 | 0.25 | 0.35 | 0.51 | 0.67 |
| $\sigma_{\alpha_\delta}$ | 0.93 | 0.20 | 0.60 | 0.64 | 0.78 | 0.90 | 1.03 | 1.28 | 1.38 |
| $\sigma_{\alpha_\pi}$ | 0.79 | 0.15 | 0.55 | 0.58 | 0.68 | 0.77 | 0.87 | 1.06 | 1.12 |
| $\rho_{\alpha\delta}$ | -0.01 | 0.3 | -0.56 | -0.48 | -0.23 | -0.02 | 0.19 | 0.47 | 0.54 |

**TABLE 10.6**: Main statistics of the marginal posterior pdfs of the free hyperparameters in model $M_0$ (year and habitat effect on the density only) and $M_1$ (year and habitat effect on the density and year effect on the probability of capture).

bution conditional to $(\pi, \nu)$. The Bayesian $p-$values are the probability that $\chi^2_{(1)}(C^{1,rep}, \pi, \nu) > \chi^2_{(1)}(C^1, \pi, \nu)$ and $\chi^2_{(2)}(C^{1,rep}, C^{2,rep}, \pi, \nu) > \chi^2_{(2)}(C^1, C^2, \pi, \nu)$. If the model fits appropriately, replicated data should look similar to observed ones; consequently the $\chi^2$-discrepancy calculated with replicated data should not be too different from the ones calculated with observed data, and the $p$-values should not be too different from 0.5. When $p$ is markedly too small or too large, serious warnings are cast about the consistency between model and data since the amplitudes of their possible variations do not match.

The two panels of Fig. 10.7 show reasonable behavior of model $M_0$: $p = 0.42$ for the $\chi^2_{(1)}$-discrepancy and $p = 0.57$ for the $\chi^2_{(2)}$-discrepancy.

### 10.1.12 Estimation of the total number of 0+ juveniles

Results of the extrapolation method are given in Figure 10.8. The posterior predictive pdfs of the $\nu_i$'s are almost identical both in terms of expected values and precision, whether model $M_0$ or $M_1$ is used (not shown).

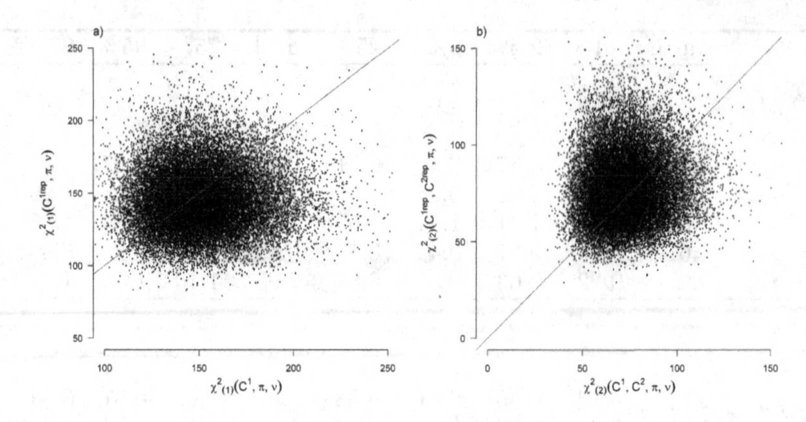


**FIGURE 10.7**: Posterior predictive checks of (a) omnibus $\chi^2_{(2)}(C^1, \pi, \nu)$ discrepancy ($p$-value=0.42), and (b) omnibus $\chi^2_{(2)}(C^1, C^2, \pi, \nu)$ discrepancy ($p$-value=0.57), both obtained under the model configuration $M_0$ (year and habitat effects on the density only).

The between-year variability of the estimated population size is high with no particular trend. The best estimates (posterior mean) for $\nu_i$ ranges from 310 (year 1998) to 6760 (year 2001). The contrast in the total uncertainty about the yearly population size estimates is large as well. As expected when using log-Normal distributions for the densities, the uncertainty about the population size estimates increases with their means.

As the fish density is extrapolated over 95% of the study area with high spatial variability of the density, a high price is paid in terms of uncertainty for the very low sampling rate: the population size $\nu_i$ does not have good precision (although it may vary from one year to the other). In addition, Fig. 10.8 helps to visually disentangle the uncertainty stemming from the between-site variability alone (the inner grey area gives an 80% confidence interval assuming no parameter uncertainty but various perturbations and Poisson draws for the sites where extrapolation is needed) and the total predictive uncertainty stemming from all unknowns (the larger bounds).

### 10.1.13 Linking 0+ juveniles to smolts run to estimate survival rates

The estimates of the $0^+$ population size (posterior means of the $\nu_i$'s, years 1985-2005 obtained from model $M_0$) can be compared with those

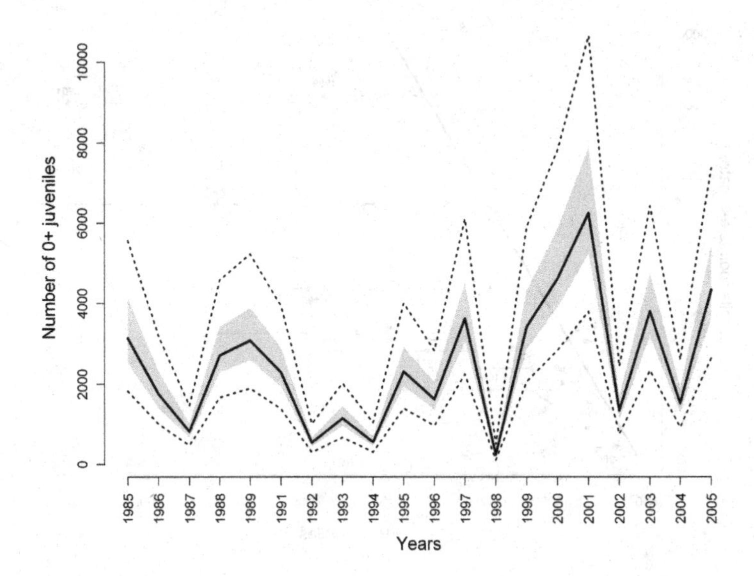


**FIGURE 10.8**: 80% confidence intervals of the posterior predictive distribution of the total $0^+$ juveniles population size obtained under the model configuration $M_0$ (note that year 1990 is missing). The shaded inner area points out the 80% confidence intervals assuming only between sites variability but no parameter uncertainty. The wider uncertainty area (dotted line) cumulates all sources of uncertainty.

of the corresponding smolt run (from capture-mark-recapture data) provided by Baglinière *et al.* [17] (Fig. 10.9). Resulting estimates of the survival rates from 0+ juveniles to smolts of the corresponding cohort are shown in Fig. 10.10.

Results are consistent with the previous knowledge about the value of the 0+-to-smolt survival rates in French rivers. The solid line in Fig. 10.9 shows that a linear model between the two estimates might be consistent with moderate yearly fluctuations of the 0+-smolt survival rate in the Oir River (around 50% in average).

Figure 10.9 confirms that the 0+ population size in 1998 was exceptionally small because the corresponding smolts run was also very small. The thick dotted line corresponds to a survival rate of 100%, which casts special attention on the results associated with the 1994 cohort. Can unusual measurement errors explain that the estimated number of smolts exceeds the estimated number of juveniles? Although some part of the

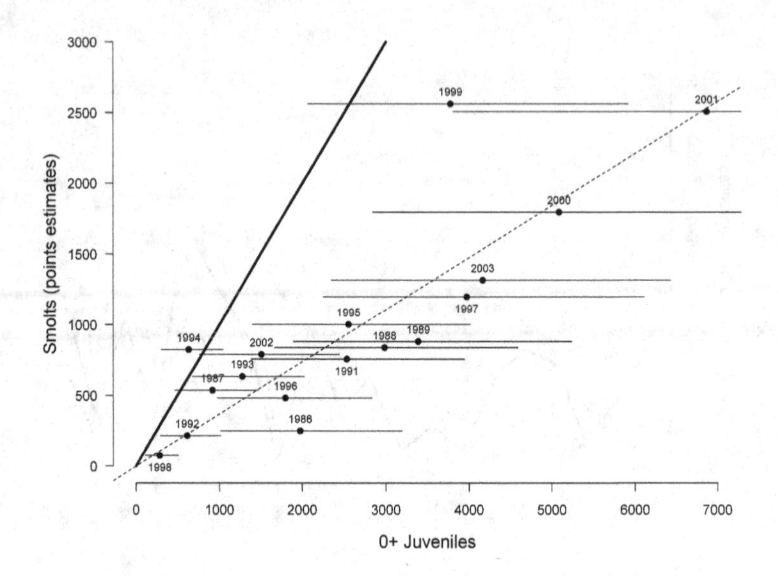


**FIGURE 10.9**: Estimates of the smolts run size by cohort (from capture-mark-recapture data [17]) plotted against the corresponding 0+ juveniles population size (posterior means of the $\nu_i$'s, years 1985-2005) estimates.

80% confidence interval lies below the limits, the corresponding survival rate is surprisingly assessed very close to 100%.

### 10.1.14 Discussion

#### 10.1.14.1 Changing prior assumptions?

No sensitivity analysis to the choice of priors is provided in this chapter but the interested reader will find in Rivot *et al.* [257] that the outcomes of the model comparison and the posterior estimation are robust to the choice of the prior configuration.

Nonconstant capture probability between successive pass is likely to cause underestimation of the population size ([187]; [229]; [254]; [264]), but also underestimation of the uncertainty and overestimation of spatio-temporal variability ([187]). In the case study, the probability of capture is high (on average from 0.68 to 0.9 according to the year; not shown). In addition, the 0+ salmon juveniles are mainly found in open, shallow waters in the middle of the river channel and their size range is limited.

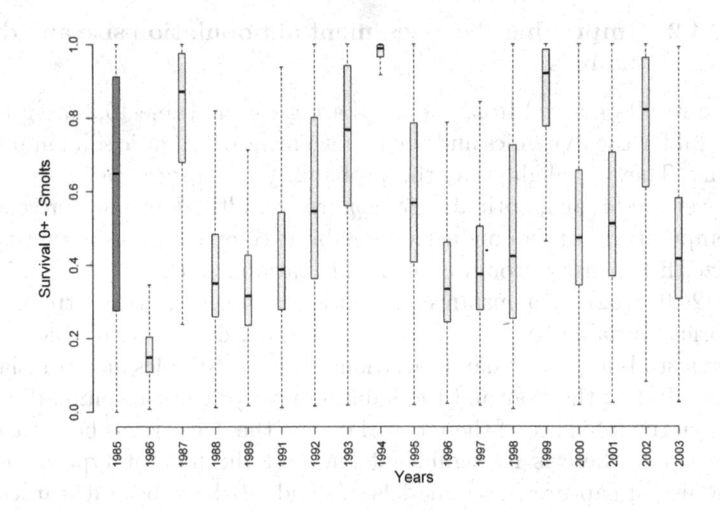


**FIGURE 10.10**: Box and whisker plots of a size-30,000 MCMC sample from the marginal posterior pdfs of the survival rate from 0+ juveniles to smolts by cohort ($\eta_i$) under model $M_0$ (year and habitat effect on the density only).

Such conditions favor robustness against the violation of the constant catchability hypothesis that we adopted.

Fish densities have been assumed to be log-Normally distributed. We did not formally test for the relevance of this modeling hypothesis, which is both convenient and classical. It allows to introduce multiplicative effects of covariates on the density by means of standard linear-normal modeling of the log-density. Log-Normal distribution as a common model for fish population abundance has been critically examined by Halley and Inchausti [126] that plead for the Gamma *pdf* as the most valuable alternative, not difficult to implement under the HBM framework (see Brun *et al.* [39] for a recent example; see also the hierarchical stock-recruitment model in Section 9.3 of Chapter 9 for an application using the Gamma distribution within a hierarchical context). No spatial dependence between densities at adjacent sites was introduced in this chapter. Further development of spatially structured models is certainly warranted as it has been demonstrated that juvenile salmon densities are spatially auto-correlated ([86]). The spatially explicit model coupled with a GIS developed by Wyatt [317], is an interesting development in that direction.

#### 10.1.14.2 Improving the assessment of population size and dynamics

The results derived from the database given in Table 10.2 are quite insightful for the dynamics and the management of Atlantic salmon populations. They highlight that the probability of capture varies among units. Several factors, both abiotic, *e.g.*, river width, depth, water velocity, temperature, water conductivity, habitat complexity, or biotic (*e.g.*, fish size, fish density) could explain this variability ([229]; [248]; [254]; [263]; [264]; [282]). For instance, the location along the river stream (a categorical variable coded as upstream, middle, downstream) may also be worth studying. Systematic variations of these variables may translate into an effect of the year and the habitat type. But our results indicate that once the influence of these covariates on the density has been taken into account, there is no significant effect of the habitat type on the probability of capture (since models $M_2$ and $M_3$ have been discarded). This can be explained by two facts: (*i*) In spite of the variability of the salmon habitat, many physical characteristics of the stream remain relatively homogeneous along the stretch of interest; (*ii*) the electrofishing crew is trained to operate in a consistent manner in order to maximize the capture efficiency whatever the habitat type. The conclusions with regard to the effect of the year on the probability of capture are less clear-cut, but as far as prediction of the total population size is concerned, models $M_0$ and $M_1$ cannot be distinguished. Variations of the hydrological conditions between years during the electrofishing campaigns may also explain part of the observed pattern. The 0+ salmon density fluctuates widely between years as seen in Fig. 10.8. This is a typical result for juvenile salmonid recruitment which is highly sensitive to environmental fluctuations ([17]; [99]; [207]). On average, the density of 0+ juveniles on rapid/riffle habitat is 6.5 times higher than on the run habitat. These figures are in agreement with previous findings on the habitat preferences of Atlantic salmon ([6]; [19]) and with the inter-habitat ratios established for French rivers ([17]). In conclusion, HBM is a flexible step-by-step methodology: it can accommodate an additional observation submodel to the model of Chapter 10 devised for the successive removal data. The Bayesian learning procedure progressively improves the estimation of demographic parameters of the Salmon life cycle and the predictions about the future of the population. But the *HBM* framework for the estimation of 0+ population size can also serve to incorporate information into a more complex age-structured population dynamic model as illustrated in Section 11.3 of Chapter 11.

## 10.2 Combining different observations processes of the same unknown quantity

### 10.2.1 Hierarchical model for electrofishing Salmon juveniles with successive removal technique

In this section, we turn again to the successive removal Salmon data from the Nivelle, a 40-km river flowing from the Spanish Pyrenées to the bay of Saint Jean de Luz. Details about the fishing procedure have been already given in Section 4.2 from Chapter 4, page 83. Similarly to the previous part of this chapter, we work with a complete dataset involving 3 years (2003-2005) with 11 sites located from the Nivelle River mouth to impassable upstream dams on the main stream and one of its tributaries. For each of the 33 experiments the number of fish caught were recorded for the first ($C^1$), second ($C^2$) and eventually third pass ($C^3$) of the electrofishing experiment (Fig. 10.11). In addition, Table 10.7 provides the area $S$ in $m^2$ that was fished each time.

**FIGURE 10.11**: Electrofishing of 0+ A. salmon juveniles.

It makes sense to share information between experiments about the probability of capture as well as the densities of juveniles, therefore leading to a hierarchical version of the successive removal model without changing the core of the model already composed of the equations (Eqs. (4.7)+(4.8)+(4.10)). The statistical status of the unknown $\pi$ and $\delta$ changes from parameter to latent variable (and from scalar to vector

| Site | S | $C^1$ | $C^2$ | $C^3$ |
|---|---|---|---|---|
| 1 | 608 | 23 | 14 | 3 |
| 2 | 294 | 35 | 7 | NA |
| 3 | 565 | 38 | 17 | NA |
| 4 | 564 | 41 | 28 | 15 |
| 5 | 229 | 17 | 8 | NA |
| ... | | | | |
| 29 | 667 | 73 | 10 | NA |
| 30 | 606 | 104 | 28 | NA |
| 31 | 712 | 190 | 43 | NA |
| 32 | 341 | 16 | 2 | NA |
| 33 | 632 | 17 | 4 | NA |

**TABLE 10.7**: Three passes during successive removal electrofishing on the Nivelle River (11 sites, years 2003-2005; NA = Not Available).

as far as their numerical status is concerned). It is tempting to recycle part of the previous model prior structure as a hierarchical additional layer that ties the experiments together (*i.e.*, the intersite variability submodel). We will assume that for each site $i$ ranging from 1 to 33:

$$\begin{cases} \pi_i \sim Beta(a, b) \\ \delta_i \sim Gamma(c, d) \end{cases}$$

As nodes without parents in the DAG (see Fig. 10.12), $a, b, c$ and $d$ are the parameters of the hierarchical model: they are unknown and must be assigned prior distributions. How to design such priors?

The prior used in Section 9.2.4 can be recycled for the parameters $(a, b)$. We used a rather simpler transformation by considering the transformation $\mu_{a,b} = \frac{a}{a+b}$ and $u_{a,b} = (a + b)$ and by drawing $\mu_{a,b}$ in a diffuse prior distribution (we took a $Beta(1.5, 1.5)$) and $log(u_{a,b})$ in a $Uniform(0, 10)$ distribution. Keeping in mind the interpretation of Beta coefficients $(a, b)$ as prior distributions for Binomial trials (see Chapter 2), $a$ and $b$ are to be interpreted as prior number of success and failures, respectively. Then, $u = (a + b)$ is interpreted as a prior sample size that scales the variance of the Beta prior distribution, and a Uniform distribution on the log-scale is appropriate for a diffuse prior.

Following the same line of reasoning, a change of variables (mean $\mu_\delta$ and standard deviation $\sigma_\delta$) is worth considering for parameters $c$ and $d$

$$\begin{cases} \mu_\delta = \dfrac{c}{d} \\ \sigma_\delta = \dfrac{\sqrt{c}}{d} \end{cases}$$

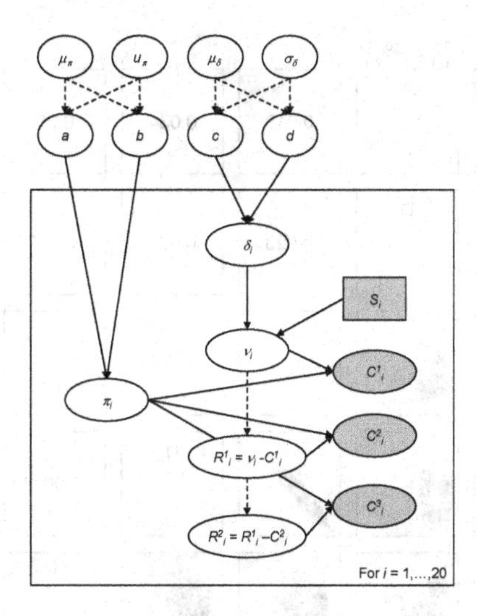

**FIGURE 10.12**: A hierarchical model for successive removals.

The mean density of juveniles $\mu_\delta$ could be assigned a noninformative gamma distribution as in Eq. (4.10) or an informative prior. Prior knowledge about juvenile Salmon density suggests a Gamma(1,1): most of the occurrences of this latter distribution adequately lies between 0.025 and 3.6 individuals per $m^2$. As no definitive idea can be given about the standard deviation, a quasi-uniform Gamma($1,10^{-3}$) can be a reasonable choice for $\sigma_\delta$.

$$\begin{cases} \mu_\delta & \sim \; Gamma(1,1) \\ \sigma_\delta & \sim \; Gamma(1,10^{-3}) \end{cases}$$

We therefore obtain the hierarchical successive removal model depicted by the DAG given in Fig. 10.12.

We have to point out that missing data occurring during the third capture in many sites (see Table 10.7) should bother the analyst. As in Section 10.1.8.1, it is important to note that the *Missing At Random*conditions ([117]) can be realistically assumed. When these conditions are fulfilled, no further modeling of the missing data process is needed: the missing data can be considered as latent variables whose posterior distribution is to be directly inferred without changing the DAG of Fig. 10.12. The posterior results from the Bayesian analysis are graphed in Fig. 10.13.

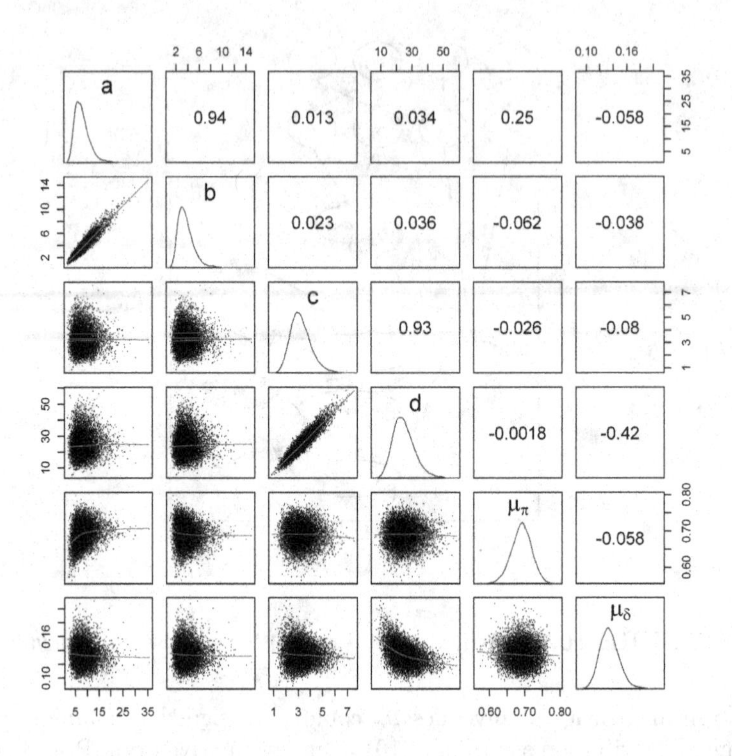


**FIGURE 10.13**: Posterior distributions of parameters for the hierarchical successive removal model.

The expected efficiency of capture $\mu_\pi$ lies around 0.7 and the $\pi$'s are dispersed among sites approximately like a $Beta(7,3)$. We get a good posterior knowledge about the expected density: $\mu_\delta$ is about 13 individuals per $100m^2$. In addition, one obtains a rather small standard deviation $\sigma_\delta$ of fish densities among sites but, locally, the densities $\delta_i$ 's are generally not precisely estimated. Missing data are assigned a predictive posterior distribution; for instance, had a third pass been achieved on site 26 where $C^1 = 142$ and $C^2 = 81$ , one would have expected to catch around 38 fish (with a standard deviation of 10). Figure 10.14 compares the distribution of fish densities between the hierarchical (grey boxplots) and the independent models (white boxplots).

Figure 10.14 makes clear the shrinkage effect due to hierarchical modeling: with the hierarchical model, the densities are less dispersed around the grand mean. At the same time, sharing information between sites

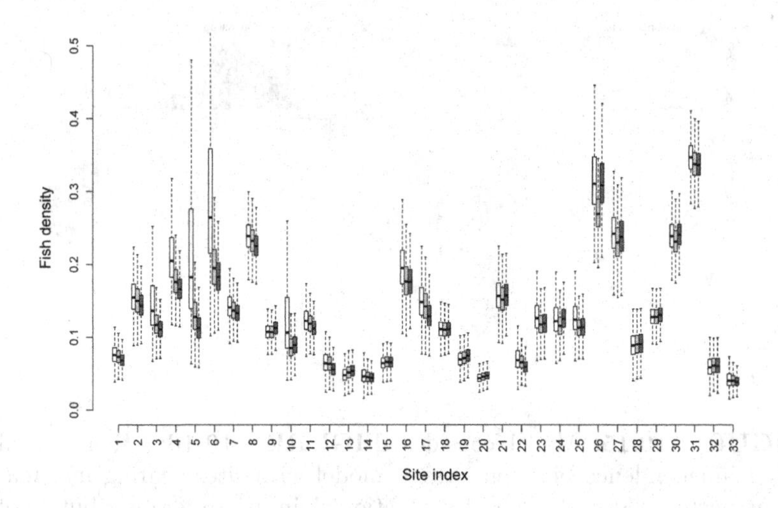


**FIGURE 10.14**: Marginal posterior distributions of salmon densities. White boxplots are for the independent successive removal model, grey ones for hierarchical model. Black boxplots are obtained when taking into account the information from abundance indices.

allows for more precise estimations of the local fish densities. In addition, the hierarchical model borrows strength from neighbors. Consequently the problem of banana-shaped joint distribution between $\pi$ and $d$ (already met in Fig. 4.8, page 94) is reduced when recourse is made to hierarchical modeling. As an example, when considered solely, site 5 exhibits poor determination of $\pi$ and $\delta$. Figure 10.15 shows that the data at site 5 alone are insufficient to opt for a large capture efficiency $\pi$ associated a small population size or for a small capture efficiency with a high local population density $\delta$. Under the hierarchical model though, sufficient transfer of information between sites helps specify the probable joint domain where $\pi$ and $\delta$ are to be expected (see Fig. 10.16).

### 10.2.2 Combining successive removals and a rapid sampling technique

Removal sampling by electric fishing is time and manpower consuming, because it is difficult to store the removed fish and to go thrice in the field for a large number of sites. To increase the number of sites sampled with limited budget and diminishing human resources, recourse can be made to rapid electric fishing assessment techniques. A 5-mn

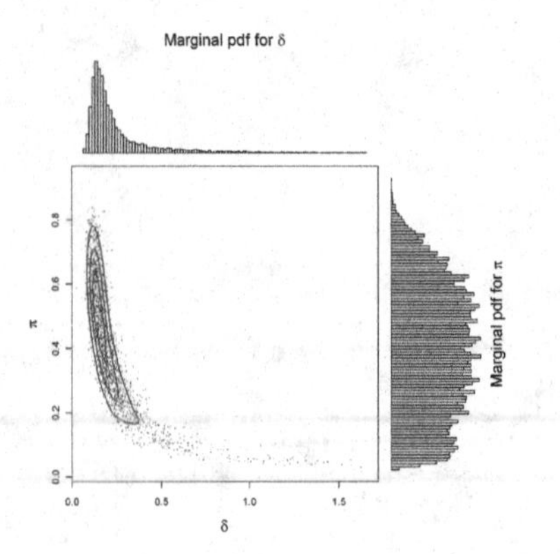


**FIGURE 10.15**: Model considering independence between sites. Joint posterior distribution of fish density and capture efficiency at site 5.

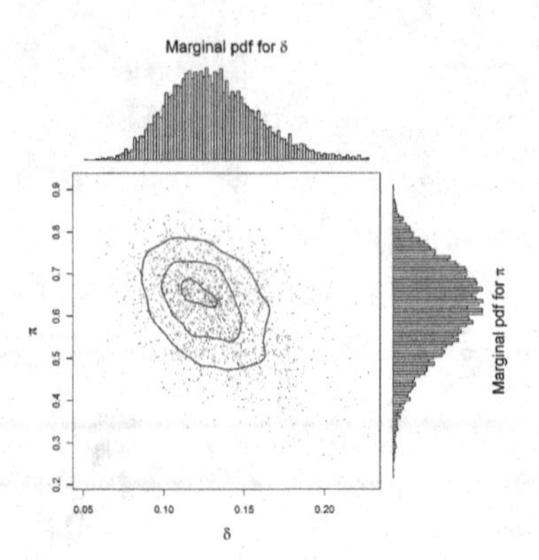


**FIGURE 10.16**: Hierarchical model with sites sharing information. Joint posterior distribution of fish density and capture efficiency at site 5.

sampling technique requires a single-timed sample for each site under a lighter protocol but only provides relative abundance measures. To be used for absolute population abundance estimation, abundance indices ($AI$) need to be first calibrated with another method of population size estimation. Table 10.8 gives such $AI$'s performed on the same sites with the successive removal sampling data already depicted in Table 10.7.

| Site | 1 | 2 | 3 | 4 | 5 | ... | 6 | 7 | 8 | 9 | 10 |
|---|---|---|---|---|---|---|---|---|---|---|---|
| AI | 6 | 17 | 12 | 17 | 11 | ... | 21 | 41 | 44 | NA | 5 |

**TABLE 10.8**: Abundance indices on the Nivelle River (11 sites, years 2003-2005).

To perform such a calibration in a consistent Bayesian framework, we need to model jointly successive removals and $AI$'s. Figure 10.17 shows that there is a strong link by plotting these observed abundance indices versus the mean of the posterior pdf of $\delta$ that we take as an estimate of the fish density from each site: one can figure out a linear trend but with an increasing dispersion. Conversely to the homoscedastic behavior of the linear regression model, the variance should depend on the mean. Keeping these clues in mind, we relate the vector of Salmon juveniles' density $\delta_i$'s to the vector of abundance indices $IA_i$'s as follows:

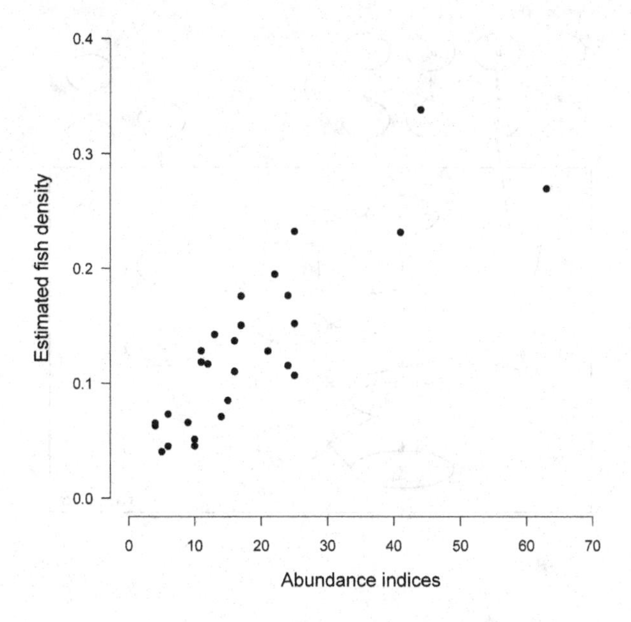


**FIGURE 10.17**: Posterior means of Salmon juvenile density (under the HBM model) versus observed Abundance Indices.

- As we are counting events, we hypothesize for each site $i$, a Poisson distribution for AI around a mean value $\mu_{AI}$ :

$$AI_i \sim Poisson(\mu_{AI_i})$$

- In turn, this true mean value $\mu_{AI}$ will be drawn from a Gamma distribution.

$$\mu_{AI_i} \sim Gamma(\alpha_{AI_i}, \beta_{AI_i})$$

Such a choice can be hypothesized for two reasons:

1. The convolution of a Poisson by a Gamma distribution makes a negative Binomial pdf, a common model for overdispersion in a counting experiment;
2. By setting $\beta_{AI_i} = f$, $\alpha_{AIi} = k \times f \times \delta_i$, one can describe a variance

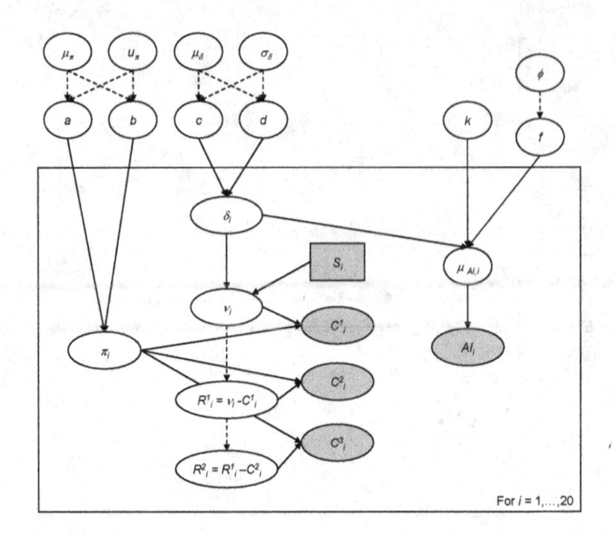


**FIGURE 10.18**: DAG of the hierarchical model for successive removals including abundance indices as an additional source of information for the 0+ juveniles density.

increasing (linearly) with the mean since:

$$\begin{cases} \mathbb{E}(\mu_{AI_i}) = \dfrac{\alpha_{AI_i}}{\beta_{AI_i}} = k\delta_i \\ \mathbb{V}(\mu_{AI_i}) = \dfrac{\alpha_{AI_i}}{\beta^2_{AI_i}} = \dfrac{\mathbb{E}(\mu_{AI_i})}{f} \end{cases}$$

The additional parameters $k$ and $f$ must be assigned a prior pdf. Parameter $k$ is given a flat noninformative gamma prior:

$$k \sim Gamma(1, 10^{-3})$$

It is convenient to work with $\phi$ such that $F = \frac{\phi}{1-\phi}$. $\phi$ is the probability parameter from the Negative Binomial and we naturally opt for a uniform pdf:

$$\phi \sim Beta(1, 1)$$

The complete model with the additional brick from the $AI$ observation submodel is sketched in Fig. 10.18 (to be compared with Fig. 10.12).

The statistical Bayesian learning machinery provides informative posterior distribution for $\phi$ and $k$ (see the lower panels of Fig. 10.19):

- $\phi$ is far from uniform, its posterior pdf can be approximated by a $Beta(3.81, 2.07)$
- $k$ varies close to 140 with a small standard deviation (a good approximation is a $Gamma(235.7, 1.7)$)

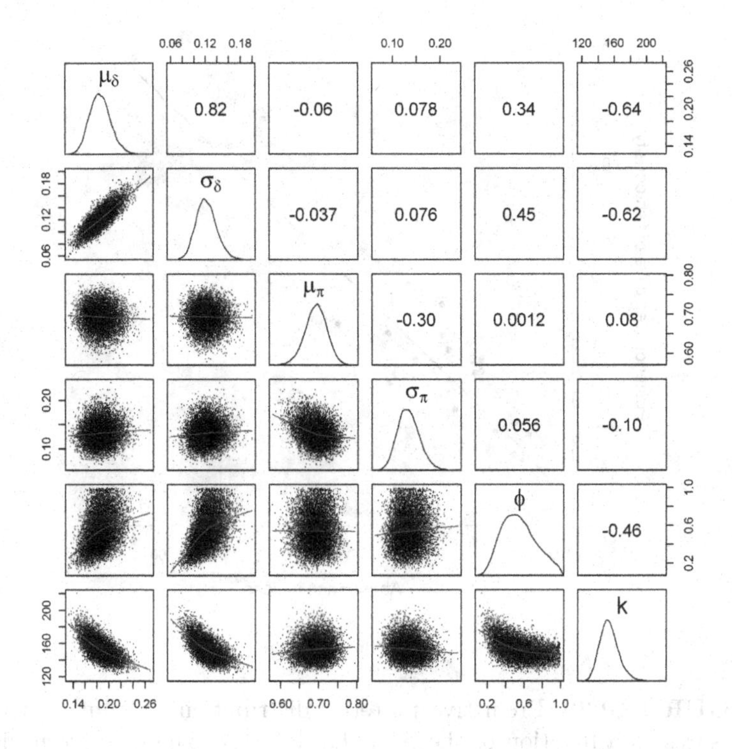


**FIGURE 10.19**: Posterior distributions of parameters for the full AI + successive removal model.

Figure 10.19 provides posterior estimates of the parameters in the hierarchical model combining successive removals and rapid sampling techniques. The black boxplots from Fig. 10.14 depict the posterior pdf of fish density $\delta$ when modeling jointly the successive removals and the $AI$'s. When compared to the previous model without taking into account the information conveyed by the abundance index (grey boxplots), we see that there is only a small shrinkage effect: although the 50% credible intervals are generally narrower, *i.e.*, we learn something from the *AI* sample but not that markedly. Site 26 bears a noticeable exception that can be explained by contradictory information brought by successive removals and by AI. If we were to trust the successive removal model only, the juvenile density for site 26 should be around 0.3 and, as $k$ is close to 140, we should expect a value of $140 \times 0.3 = 42$ individuals for the abundance index, but we actually observed 63 juveniles in the *AI* data!

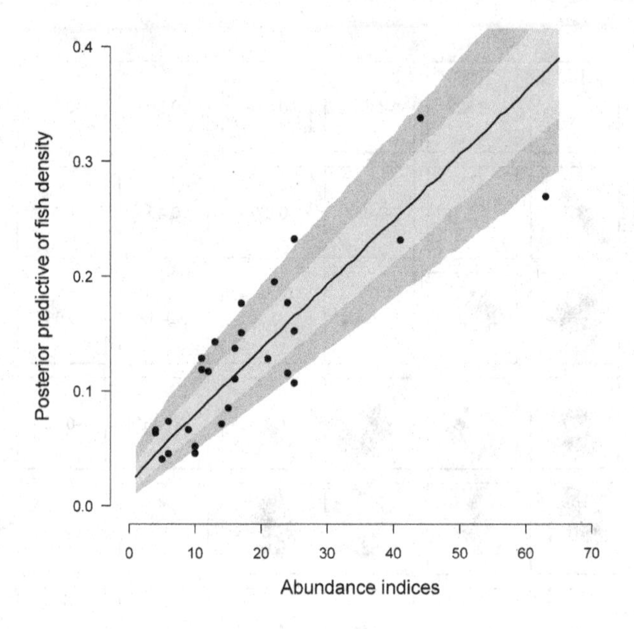


**FIGURE 10.20**: Predictive posterior distribution of Salmon juvenile densities as a function of the Abundance Index. The grey (resp. light-grey) polygon gives a 95% (resp. 50%) credible interval.

### 10.2.3 Predictive efficiency of the 5-mn standardized sampling technique

Modeling jointly *AI* and successive removals allows for the calibration of the 5-mn standardized sampling technique. Based on posterior knowledge from the joint calibration, predictive estimates of $0^+$ juvenile salmon density can now be obtained from abundance index data alone.

From consideration of Fig. 10.19, $\phi$ and $k$ have already been respectively approximated by Beta and Gamma independent distributions. Unfortunately $\mu_\delta$ and $\sigma_\delta$ exhibit strong dependence a posteriori but $\mu_\delta$ and $\frac{\mu_\delta}{\sigma_\delta}$ do not. Their posterior can be approximated by Gamma distributions with coefficients $(89, 668)$ and $(60, 104)$, respectively. These distributions are used to design a prior for 0+ juvenile salmon density, that one can update by the observation of various abundance index from the 5-mn standardized sampling protocol. As a result, Fig. 10.20 shows that the index of abundance alone allows for differentiation between contrasting levels of fish density without recourse to the successive removal technique in the future. Yet fairly imprecise, this rapid sampling technique also provides the corresponding confidence bounds that match the cloud made by the posterior means of Salmon juvenile density (under the HBM model) and the Abundance Indices observed in the dataset.

# Chapter 11

# Hierarchical Bayesian Modeling III: State-space models


## Summary

This chapter is devoted to state-space models, *i.e.*, models with dynamic state transition in the latent layer of the hierarchical structure. Conversely to the examples presented in the previous chapters, unknown quantities of interest (*e.g.*, the number of fish, the biomass of a fish stock) evolve with time while observables only give a noisy piece of information about these latent variables. From a graphical modeling perspective, state-space modeling consists of adding arrows between variables within the hidden layer of a DAG so as to create the temporal link between these variables. To go one step further, state-space modeling consists of defining two key equations: the process equation with process noise that captures the stochastic dynamics of the hidden state variables, and the observation equation that relates the data at hand to the state variables, which may involve some observation noise. Surprisingly enough, statistical estimation under the Bayesian setting of (even complex) state-space models remains easily tractable. The flexibility of Bayesian analysis of state-space models is exemplified through two examples of growing complexities; both present educational qualities for illustrating the strengths and limits of the Bayesian analysis of state-space models. The first example sketches the dynamics of the biomass of a fish stock under fishing pressure. The model is used to derive estimates of key management parameters and to forecast changes in biomass under different management scenarios. The second example is an aged-structured population model for A. salmon. The model mimics the salmon life cycle with all its development stages, represented through a multidimensional state-space model.

## 11.1 Introduction

Modeling the system state dynamics is a critical issue encountered in many ecological applications such as population dynamics. Identifying the factors that control the system dynamics and being able to forecast its future evolution are key issues for the modeler.

In a state-space model, transition equations between state variables are used to sketch the dynamics of the system, and observation equations link the state variables to some observables. Such models with dynamic state transition in the latent layer of the hierarchical structure are encompassed in the general family of state-space models sometimes also referred to as hidden Markov models ([40]; [41]; [65]; [75]; [136]; [259]). From a graphical modeling perspective, the sophistication is a simple matter of adding arrows between variables within a hidden layer of a DAG to sketch the temporal link between the variables (see Figs. 1.12 and 1.14 in Chapter 1). Surprisingly enough, estimation under the Bayesian setting of (even complex) state-space models remains easily tractable.

To go further into details, state-space modeling consists of ascertaining two key equations (or set of equations): the process equation with parameters $\theta_1$ and identically independently distributed *process noise* $\epsilon(t)$ that captures the stochastic dynamics of the hidden (not observed) state variables, $Z_t$, and the observation equation that relates the data at hand $y_t$ to the state variables $Z_t$ through an observation function involving parameters $\theta_2$ and eventually some *iid observation noise* $\omega(t)$ (see also Chapter 1):

$$\begin{cases} Z_{t+1} = f(Z_t, \theta_1, \epsilon(t)) \\ y_t = g(Z_t, \theta_2, \omega(t)) \end{cases} \tag{11.1}$$

Using the convenient bracket notation of Chapter 1, Eq. (11.1) can also be written as:

$$\begin{cases} [Z_{t+1}|Z_t, \theta_1] \\ [y_t|Z_t, \theta_2] \end{cases} \tag{11.2}$$

where $[Z_{t+1}|Z_t, \theta_1]$ denotes the conditional pdf of state vector at time step $t+1$ given the state vector at time step $t$ and parameters $\theta_1$, and $[y_t|Z_t, \theta_2]$ denotes the conditional distribution of observation $y_t$ given the state vector at time $t$ and parameters $\theta_2$.

As shown in Chapter 1 (Eq. (1.28)), once a prior $[Z_1]$ is specified for the state vector at the first time step, the joint prior distribution of $\theta_1$ and $Z_{1:n} = (Z_1, ..., Z_n)$ can be written as in Eq. (11.3) below, thus emphasizing the Markovian property in the state dynamics which defines the prior structure in the latent layer $Z$:

$$[Z_{1:n}, \theta_1] = [\theta_1] \times [Z_1] \times \prod_{t=1}^{n-1} [Z_{t+1}|Z_t, \theta_1] \tag{11.3}$$

Conditionally upon states $Z_t$ and parameters $\theta_2$, observations $y_t$ are mutually independent and the observation equation also factorizes:

$$[y_{1:n}|Z_{1:n}, \theta_2] = \prod_{t=1}^{n} [Y_t = y_t|Z_t, \theta_2] \tag{11.4}$$

Following the general factorization of the joint posterior distribution for a Bayesian hierarchical model given in Eq. (1.25), the joint distribution of all state variables and parameters $\theta = (\theta_1, \theta_2)$ is straightforwardly obtained:

$$\begin{aligned}[Z_{1:n}, \theta|Y_{1:n} = y_{1:n}] \propto [\theta] \times [Z_1] \times \prod_{t=1}^{n-1} [Z_{t+1}|Z_t, \theta_1] \\ \times \prod_{t=1}^{t=n} [Y_t = y_t|Z_t, \theta_2]\end{aligned} \tag{11.5}$$

## 11.2 State-space modeling of a Biomass Production Model

### 11.2.1 Motivating example: The Namibian hake fishery

Let us consider as a first example the data from the Namibian hake fishery. Two hake species (*Merlucius capensis* and *Merlucius paradoxus*) are targeted by this fishery. The data analyzed here concern the fishery operating in zones 1.3 and 1.4 of the International Commission for the South-East Atlantic Fisheries (ICSEAF) from 1965 to 1988. For further details about the fishery shown in Fig. 11.1, we refer to the report from the International Commission for Southeast Atlantic Fisheries ([144]) or to [136] and [196].

The catch-effort data are presented in Table 11.1. The two targeted species are pooled in the dataset. The catches concern the total annual commercial catches of hakes (in thousand tons) realized by large ocean-going trawlers operating in the ICSEAF zones 1.3 and 1.4. The catches per unit effort data (CPUEs) are the catches per hours of fishing for a

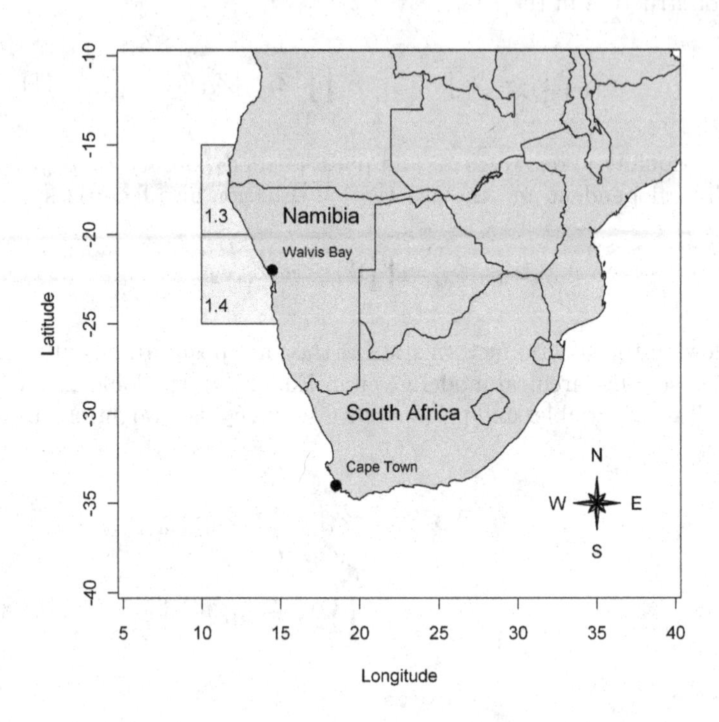


**FIGURE 11.1**: Location of the ICSEAF fishery areas 1.3 and 1.4 in the Southeast coast of Africa.

specific class of Spanish trawlers. As the CPUEs are standardized, they are considered here as a reliable index of abundance for the Namibian hake stock.

We rely on these data to model the dynamics of hake stock biomass through dynamic Biomass Production Models (BPM) (see also Figure 1.2, page 10). A dynamic BPM is a voluntarily crude but useful simplification of some harvested fish population dynamics ([137]; [244]) that only aims at helping fisheries scientists to interpret the data (*e.g.*, catches and abundance indices in Table 11.1). Analyzing data through BPMs allows to assess how the fishery pressure has impacted the biomass. For instance, it provides answers to questions such as:

- What is the maximum sustainable yield ($C_{MSY}$) and what are the past and current levels of yield sustainable with regards to the $C_{MSY}$?

| Years | Catches | CPUE |
|---|---|---|
| 1964 | 1.8 | NA |
| 1965 | 93.5 | 1.78 |
| 1966 | 212.4 | 1.31 |
| 1967 | 195.0 | 0.91 |
| 1968 | 382.7 | 0.96 |
| ... | | |
| 1984 | 228.7 | 0.64 |
| 1985 | 212.2 | 0.66 |
| 1986 | 231.2 | 0.65 |
| 1987 | 136.9 | 0.61 |
| 1988 | 212.0 | 0.63 |

**TABLE 11.1**: Catches and abundance indices for the Namibian Hake fishery in the ICSEAF Divisions 1.3 and 1.4 ([144]). Catches are in thousand tons. Catches per unit effort (CPUE) are in tons per standardized trawler hours. The data are reproduced after McAllister and Kirkwood [196].

- How large was the abundance in year 1988 (the last year for the dataset) with reference to its level when the fishery began?
- Could biomass level increase and yield be improved if more restrictive fishing quotas are imposed in the future?

The latter question is particularly important to test the performance of alternative management scenarios when efforts are being made to control the level of catches and promote sustainable harvest.

### 11.2.2 A state-space model for the biomass surplus production

The backbone of state-space modeling (SSM) of a BPM is to consider the time series of abundance indices as noisy observations of an underlying hidden process which mimics the dynamics of a fish stock biomass. Designing a SSM involves four steps:

1. Propose a mathematical model for the biomass dynamics including fishery removals;
2. Propose a model that sketches the noisy observation process;
3. Link these two components by a conditional structure;
4. Derive inferences using inverse (*i.e.*, Bayesian) reasoning.

#### 11.2.2.1 Process equation for the underlying dynamics of the Biomass

The process equation models the underlying dynamics of the Biomass. The variable of interest is the total biomass in the population at each time $t$, denoted $B_t$. In BPMs with continuous time, the instantaneous change in the biomass level is modeled thanks to a differential equation which, when no exploitation occurs, can be written as:

$$\frac{dB_t}{dt} = h(B_t) \tag{11.6}$$

where $h(B_t)$ is the *production function.* It quantifies the balance between recruitment (arrival of new individuals in the stock biomass), growth (weight), natural mortality, and eventually emigration-immigration.

Maybe the most classical choice for the production function is the *logistic* equation, first proposed as a population model by P. F. Verhulst in 1938:

$$h(B_t) = r \times B_t \times (1 - \frac{B_t}{K}) \tag{11.7}$$

For simplifications of Eq. (11.7), the two parameters – the population intrinsic growth rate $r$ and the carrying capacity $K$ – are generally considered constant over time. The ecological theory behind such a model is that the production rate $\frac{1}{B_t} \times \frac{dB_t}{dt}$ will be maximum when the population stands at a very low level, and will decrease continuously (*e.g.*, because of intra-specific competition for the resource) when the size of the population increases. As no production occurs when $B = K$, $K$ stands for the carrying capacity of the habitat. The dynamic BPM with the logistic production function is also known as the Schaefer biomass production model.

The dynamics can be modeled in discrete time, most often on a year-to-year basis. When fishing occurs, the biomass at the beginning of time step $t + 1$, denoted $B_{t+1}$, is obtained from $B_t$ through a rather simple budget equation:

$$B_{t+1} = B_t + h(B_t) - c_t \tag{11.8}$$

where $c_t$ is the observed harvest (in weight) between $t$ and $t + 1$. When no catch occurs, it is easy to see that the biomass will stabilize at the virgin equilibrium level $B = K$.

A LogNormal random noise term is generally added to capture the biological variability due to (unpredictable) environmental variations. Hence, the stochastic version of the deterministic Eq. (11.8) is:

$$B_{t+1} = (B_t + h(B_t) - c_t) \times e^{\epsilon_{t+1}} \tag{11.9}$$

with $\epsilon_{t+1}$ a Normally distributed $N(0, {\sigma_p}^2)$ random term standing for the environmental noise (process error) with variance ${\sigma_p}^2$ in the log scale.

In a first approach, catches can be considered as observed without errors, and entered into the process equation (Eq. (11.9)) as observed covariates (hence the notation $c_t$ with a lowercase letter). But catches could more realistically be considered as observed with some errors, or landings statistics can be systematically lower than true catches because many fish are discarded, landed illegally or the catches are simply misreported (see Hammond and Trenkel [128] for an example of a biomass surplus production model accounting for censored catches).

Equation 11.9 defines a stochastic Markovian transition which can alternatively be written as the probability distribution of the state variable $B_{t+1}$ conditionally upon the previous state $B_t$, some parameters $\theta_1 = (r, K, \sigma_p)$ and the observed catches $c_t$. Following the notation used in Chapter 1, this Markovian transition will be written as:

$$[B_{t+1}|B_t, \theta_1; c_t] \tag{11.10}$$

nothing more than the general process Eq. (11.2) adapted to the biomass dynamics example.

To reduce the number of unknowns in the model and ensure statistical identifiability, one must set a constraint for the initial condition $B_1$. It is often assumed that the biomass at the beginning of the time series is a known proportion of the carrying capacity $K$. In the hake fishery, because catches were very low before the year 1964, it was assumed that the stock was not fished at the beginning of the time series. To be consistent with Eq. (11.9), the biomass of the first year was considered as LogNormally distributed around the carrying capacity $K$:

$$B_1 = K \times e^{\epsilon_1} \tag{11.11}$$

Let $t = 1, ..., n$ denote the time series for which observations are available (in what follows, we will often use *1:n* to denote the indices of the time series $t = 1, ..., n$). Conditionally upon the parameters $\theta_1 = (r, K, \sigma_p)$ (*e.g.*, those involved in the process) and upon the catches $c_{1:n}$, the sequence of unknown states $B_{1:n}$ follows a first-order Markov chain. The chain is initialized by Eq. (11.11). The transition kernel of the Markov process is defined by the dynamic process Eq. (11.10). Thanks to the conditional independence property, one can split the whole joint pdf into the product of single unit time steps. Once a prior distribution is specified for the parameters $\theta_1$ and for the first state $B_1$ (conditionally upon $\theta_1$, the prior $[B_1|\theta_1]$ is fully defined from Eq. (11.11)), the process equation can be factorized as Eq. (11.12) which is a specialized reformulation of the general process Eq. (11.3):

$$[B_{1:n}, \theta_1] = [\theta_1] \times [B_1|\theta_1] \times \prod_{t=1}^{t=n-1} [B_{t+1}|B_t, \theta_1; c_t] \tag{11.12}$$

#### 11.2.2.2 Observation equation to link the data to the hidden process

The abundance indices $i_{t=1:n}$ are often assumed proportional to the current biomass $i_t = q \times B_t, \forall t \in \{1, ..., n\}$ with a catchability parameter $q$, considered constant over time. A common, although simplifying, supplementary assumption is that the relative abundance index of each year $t$ is related to the unobserved biomass through a stochastic observation model:

$$i_t = q \times B_t \times e^{\omega_t} \tag{11.13}$$

with $\omega_t$ a Normally distributed $N(0, {\sigma_o}^2)$ random term representing the uncertainty in observed abundance indices due to measurement and sampling error (observation error). In addition, the $\omega_t$ are considered as mutually independent.

The observation Eq. (11.13) links the available data to the underlying dynamics. Following the notation used in Chapter 1, it defines the probability distribution of any abundance indice $i_t$ conditionally upon the biomass $B_t$ and the parameters $\theta_2 = (q, \sigma_o)$:

$$[i_t | B_t, \theta_2] \tag{11.14}$$

Equation 11.14 also defines the likelihood function, which gives the probability of the series of observations $i_{1:n}$ conditionally on the actual states $B_{1:n}$ and on the parameters $\theta_2$. Conditionally on state $B_t$ and parameters $\theta_2$, the observations $i_t$ are mutually independent, and the likelihood can be factorized as already shown in the general case in Eq. (11.4):

$$[i_{1:n} | B_{1:n}, \theta_2] = \prod_{t=1}^{t=n} [i_t | B_t, \theta_2] \tag{11.15}$$

It is not rare that fisheries scientists get more than one series of abundance indices to fit the model. For instance, as an additional source of information to the catch-per-unit-effort data in Table 11.1, other fishery-independent abundance indices might be available such as standardized CPUE from scientific surveys or abundance indices from acoustic scientific surveys. The information provided by additional abundance indices $i'_{1:n}$ can be easily integrated in the analysis by supplementing the observation Eq. (11.15) with conditional distributions $[i'_t | B_t, \theta'_2]$ associated with abundance indices $i'_{1:n}$ (see [195] or [197] for examples of SSM fitted to several abundance indices).

### 11.2.3 Deriving inferences and predictions

Combining the prior on the parameters $[\theta] = [\theta_1, \theta_2]$ with the process and the observation Eqs. (11.12) and (11.15) yields the full joint distribution of the model. One recognizes the general factorization of a hierarchical model as explained in Eqs. (1.17) and (1.18):

$$\begin{aligned}[Parameters, Process, Observables] \\ &= [Parameters] \\ &\times [Process|Parameters] \\ &\times [Observables|Process, Parameters]\end{aligned}$$

Applying Bayes' rule, the full joint posterior distribution of all unknowns is decomposed under the previous three terms as follows:

$$\begin{aligned}[B_{1:n}, \theta|i_{1:n}; c_{1:n}] \\ &\propto [\theta] \\ &\times [B_1] \times \prod_{t=1}^{t=n-1} [B_{t+1}|B_t, \theta_1; c_t] \\ &\times \prod_{t=1}^{t=n} [i_t|B_t, \theta_2]\end{aligned} \tag{11.16}$$

The DAG of the model in Eq. (11.16) is given in Fig. 11.2. A sample of the full joint posterior distribution (Eq. (11.16)) can be easily obtained from MCMC sampling using WinBUGS software. The posterior distribution can be used to answer the following questions:

- What are the best guesses for the parameters $(r, K)$ and the associated uncertainty?
- What are the credible values for the historical trajectory of the biomass $B_{1:n}$ and what is the level of the Biomass depletion over the time series $\frac{B_n}{B_1}$?
- What are the estimates for the management reference points $C_{MSY}$ and $B_{MSY}$ (directly derived from the parameters $(r, K)$ following Eq. (1.5)) and their associated uncertainty?

Posterior predictive distributions can also be used to derive predictions of future trajectory of the biomass and catches over a time series $t = n+1, ..., n+k$ under alternative *management scenarios*. Simulations typically aim at comparing different harvest control rules, *e.g.*, different level of catches $c_{n+1:n+k}$.

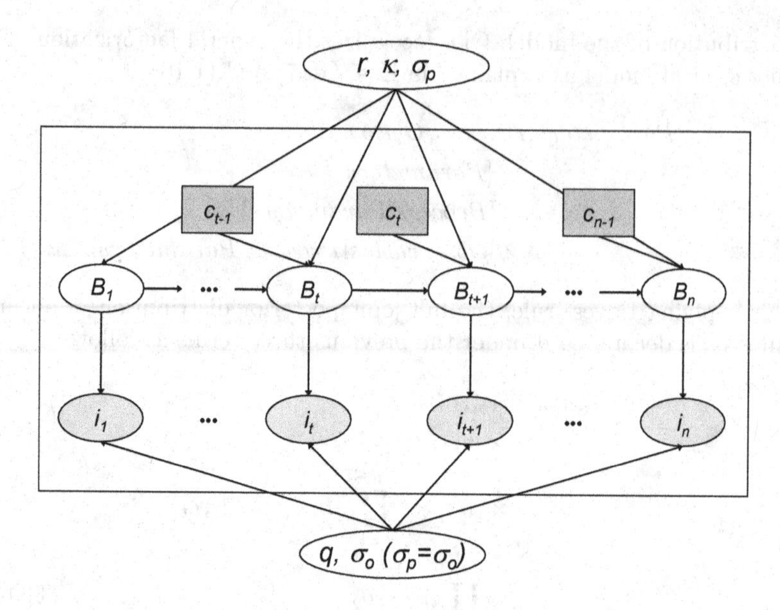


**FIGURE 11.2**: Directed acyclic graph of the state-space biomass production model.

### 11.2.4 Application to the dynamics of the Namibian hake stock

#### 11.2.4.1 Additional hypotheses, priors and computational details

Back to the application to the Namibian Hake fishery, we now detail the additional hypotheses and some technical tricks that were made to fit the model.

To ensure that all parameters can be estimated, an additional hypothesis was formulated: the variance of the process and the observation errors were set to be equal. Indeed, the magnitude of process and observation errors $\sigma_p$ and $\sigma_o$ can hardly be specified a priori. But in the absence of any prior information on $\sigma_p$ and $\sigma_o$, uncertainty in the data can be all transferred either in the process noise or to the observation noise, making it difficult to identify both variances precisely if they are totally set free a priori. Fixing the ratio of variances $\lambda = \frac{\sigma_p}{\sigma_o}$ is a classical additional trick used to reduce the number of unknown parameters. In the Bayesian setting $\lambda$ would be equipped with a reasonably informative prior. However, it is worth noting that inferences in SSM are seldom sensitive to the value of $\lambda$, unless it strongly favors observation or process error ([162]; [240]; [275]; [276]). Therefore, in the absence of any particular information on the relative magnitude of process versus observation error in the case of the Namibian hake fishery, $\lambda = 1$ was used as a default choice.

Fixing the process error variance $\sigma_p^2$ to 0 yields to the so-called *observation error model*, whereas fixing the observation error variance $\sigma_o^2$ to 0 yields to the so-called *process error model* ([240]; [275]).

As the data look informative, rather diffuse priors were set on the parameters $\theta = (r, K, q, \sigma^2)$ (Table 11.2).

| Parameter | Prior |
|---|---|
| $r$ | $\sim Uniform(0.01, 3)$ |
| $K$ | $\sim Uniform(100, 15000)$ |
| $log(q)$ | $\sim Uniform(-20, 20)$ |
| $log(\sigma^2)$ | $\sim Uniform(-20, 20)$ |

**TABLE 11.2**: Prior distributions on parameters of the biomass production state-space model applied to the Namibian hake fishery. The unit of the carrying capacity is in thousand tons. Following Millar [205], diffuse priors were assigned on $q$ and $\sigma^2$.

Following Meyer and Millar [203], the dynamic state-space Eqs. (11.10) and (11.11) were reparameterized using the state variable $P_t = \frac{B_t}{K}$. Indeed, the parameter $K$ and the latent state variables $B_t$ are unknown but are in the same scale, *i.e.*, the scale of the absolute stock size. The constraint $B_1 = K \times \epsilon_1$ in Eq. (11.11) shows that conditionally upon a value of $K$, the scale of the whole time series of the biomass $B_{1:n}$ is fully determined. This dynamical structure induces a strong dependency between $K$ and the whole biomass trajectory, which impedes an efficient exploration of the support of the posterior distribution by the Gibbs sampler. As shown by [202] and [203], using the new parameterization $P_t = \frac{B_t}{K}$ drastically improves the efficiency of the Gibbs sampler and the mixing speed of the MCMC chains.

For all estimations, three MCMC independent chains with dispersed initialization points were used. For each chain, the first 50,000 iterations were discarded. After this "burn-in" period, only one in 10 iteration steps (thinning) was kept to reduce the MCMC sampling autocorrelation. Inferences were then derived from a sample of 30,000 iterations obtained from three chains of 10,000 iterations. All the modeling results have undergone tests to assess convergence of MCMC chains. It is therefore assumed that the reported probability density functions are representative of the underlying stationary distributions, *i.e.*, the posterior pdf.

#### 11.2.4.2 Inferences using the Schaefer-type production function

The posterior distributions of the key parameters are shown in Fig. 11.3 along with their main statistics in Table 11.3.

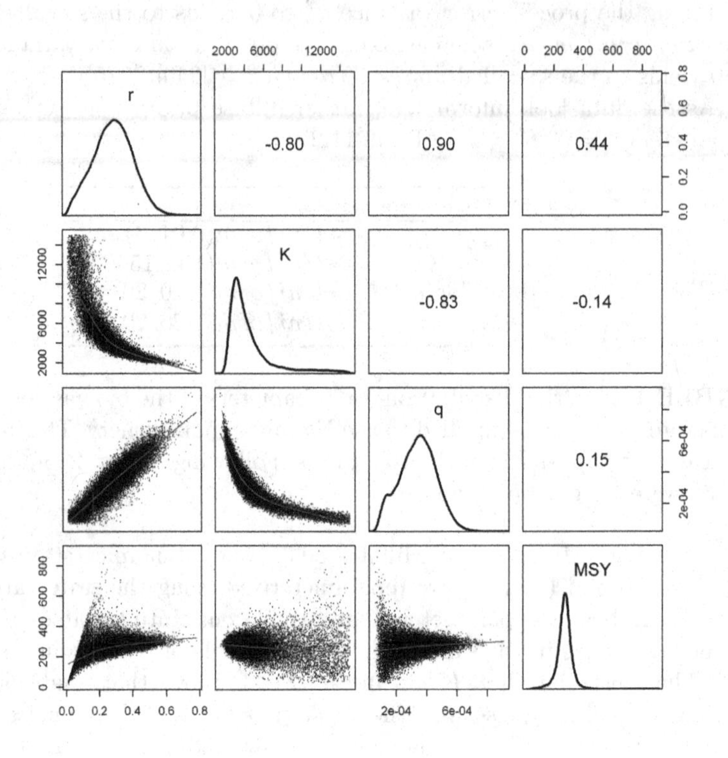


**FIGURE 11.3**: Posterior distributions of the key parameters obtained using the Schaefer-type production function. The marginal distributions are shown in the diagonal. Pairwise MCMC plots are shown in the lower part. The upper part shows the linear correlation between the MCMC draws.

The annual intrinsic growth rate is about 0.3 (Table 11.3), which is rather high and indicates that the stock should exhibit a high resilience to exploitation. The posterior median of $C_{MSY}$ is about 268 (thousand tons), thus indicating that the stock was exploited above its optimal sustainable harvest rate during 10 years between 1968 and 1977.

Parameters $(K, q)$ and $(r, K)$ are highly negatively correlated, as shown by the joint MCMC samples in Fig. 11.3. The negative corre-

| Parameter | Mean | Sd | 2.5% pct. | Median | 97.5% pct. |
|---|---|---|---|---|---|
| $r$ | 0.29 | 0.12 | 0.066 | 0.29 | 0.52 |
| $K$ | 4500 | 2343 | 2225 | 3729 | 11790 |
| $q$ | 0.00034 | 0.00012 | 0.00012 | 0.00034 | 0.00058 |
| $\sigma^2$ | 0.0093 | 0.0032 | 0.0049 | 0.0087 | 0.017 |
| $C_{MSY}$ | 266 | 53 | 152 | 268 | 364 |
| $B_{MSY}$ | 2250 | 1172 | 1112 | 1864 | 5894 |
| $F_{MSY}$ | 0.14 | 0.058 | 0.033 | 0.14 | 0.26 |

**TABLE 11.3**: Main statistics of the marginal posterior distributions of the key parameters obtained using the Schaefer-type production function.

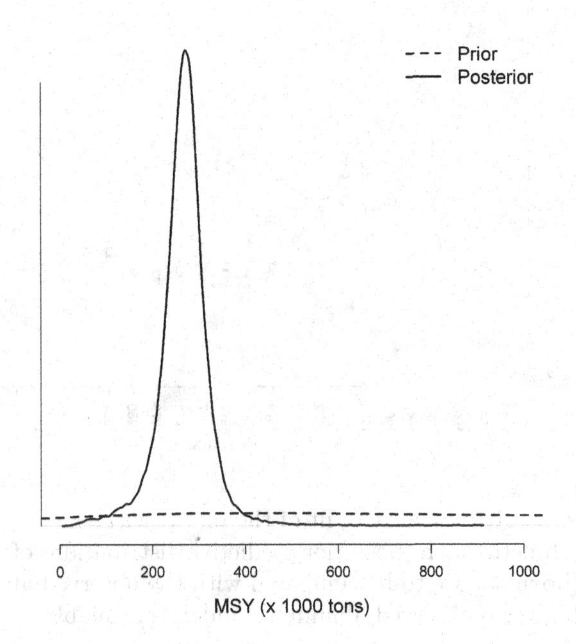


**FIGURE 11.4**: Prior and marginal posterior distribution of the Maximum Sustainable Yield ($C_{MSY}$) estimated using the Schaefer-type production function.

lation between $K$ and $q$ results from the structure of the observation Eq. (11.13). As the abundance indices $i_{1:n}$ are known, the data convey information about the product $q \times B_t$ (for the whole biomass trajectory $B_{1:n}$). As the biomass trajectory $B_{1:n}$ is scaled by parameter $K$, the parameters $q$ and $K$ can hardly be identified individually. Had an informative prior been available for $q$ or $K$, better inferences would have been obtained.

The negative correlation between $r$ and $K$ indicates that the data alone do not enable to clearly disentangle a very abundant population (large $K$) with a rather low growth rate $r$ from one with a lower $K$ but a higher $r$. However, the posterior distribution of the $C_{MSY}$ ($C_{MSY}$ is calculated from the product $r \times K$) is clearly updated by contrast with the prior (Fig. 11.4).

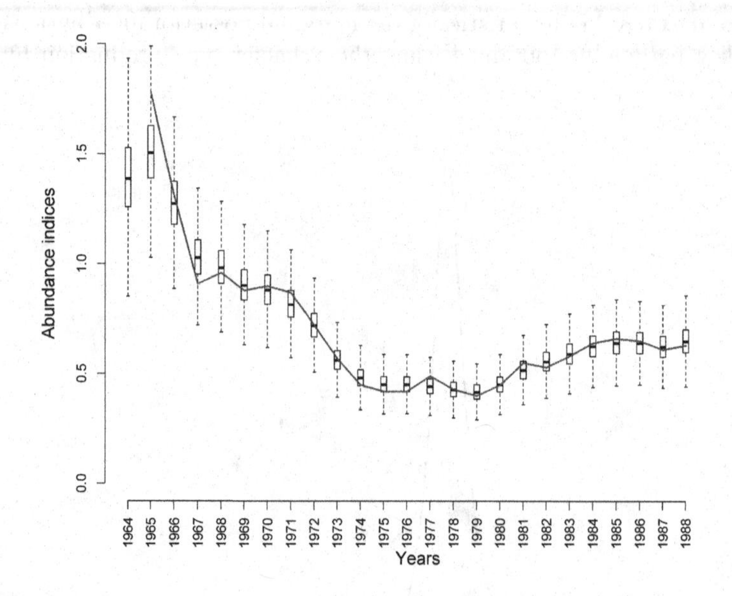


**FIGURE 11.5**: Quality of fit of the model with the Schaefer-type production function. The posterior predictive distributions of the abundance index (boxplot) are to be compared with the observed abundance index (solid line). No observed abundance index is available for year 1964.

To check whether the model reasonably fits the data, the posterior predictive of the abundance index for the whole time series (denoted $[\tilde{\imath}_{1:n}|data]$) were plotted together with the observed series of abundance index (Fig. 11.5). The joint posterior predictive of the abundance index was computed as:

$$[\tilde{\imath}_{1:n}|data] = \int_{\theta, B_{1:n}} [\tilde{\imath}_{1:n}|\theta, B_{1:n}] \times [\theta, B_{1:n}|data] \times d(\theta, B_{1:n}) \quad (11.17)$$

with

$$[\tilde{i}_{1:n}|\theta, B_{1:n}] = \prod_{t=1}^{t=n} [\tilde{i}_t|\theta, B_t] \tag{11.18}$$

and

$$log(\tilde{i}_t)|\theta, B_t \sim Normal(q \times B_t, \sigma^2) \tag{11.19}$$

As highlighted in Figure 11.5, the fit is pretty good. The fluctuations of the observed abundance indices over the time series are well captured by the predicted abundance indices. For the whole time series, the observed abundance indices are almost always contained in the predictive 50% Bayesian credibility intervals.

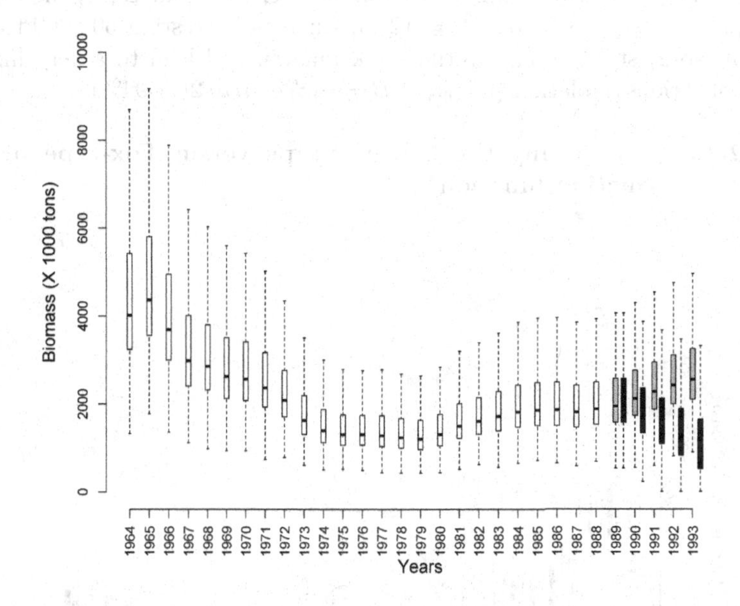


**FIGURE 11.6**: Marginal posterior distributions of the Biomass estimated using the Schaefer-type production function. Boxplots in gray fonts show the posterior predictive of the Biomass assuming a 100,000 tons constant quotas management option during 5 years after the year 1988, whereas boxplots in dark-gray are for quotas of 500,000 tons.

As a very interesting result, the model also provides estimates of the whole time series of the latent biomasses together with an appreciation of their uncertainty (Fig. 11.6). As previously shown, the shape of the time series of abundance indices is well reproduced. In 1964 (first year), the estimated biomass was about 4 million tons. From this time to the year 1972, the catches increased up to 606,000 tons in 1972, and remained higher than the estimated maximum sustainable yield up to 1977. Consequently, the Biomass decreases by a factor 2 and was estimated as less than 2 million tons in 1978. By this time, the catches were reduced by more restrictive management rules and the biomass started to rebuild.

As an additional result, two illustrative scenarios were run which enable to forecast the evolution of the biomass during 5 years (after 1988, the last year for which data are available) under two contrasted harvest rules (Fig. 11.6). The first one mimics constant catches of 100 thousand tons during 5 years, and the second one sketches constant catches of 500,000 tons. Forecasting indicate that harvesting 100,000 tons during 5 years will allow the biomass to increase again and lead to a very low risk of being at a biomass level less than $B_{MSY}$ in 1993 ($[B_{1993} < B_{MSY} | Scenario\ 1] = 12\%$), whereas harvesting 500,000 tons (scenario 2) strongly affects the stock renewal and lead to a very high risk of serious depletion ($[B_{1993} < B_{MSY} | Scenario\ 2] = 94\%$).

#### 11.2.4.3 Comparing the Schaefer-type versus Fox-type production function

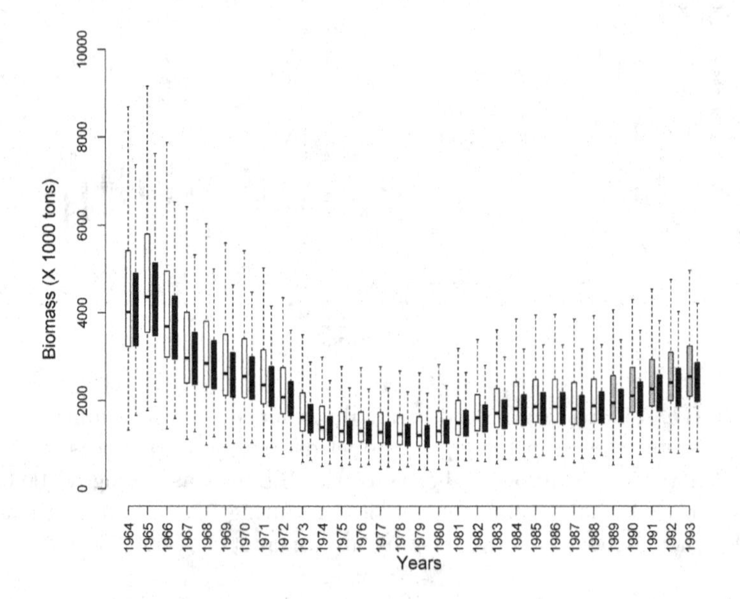


**FIGURE 11.7**: Marginal posterior distributions of the Biomass estimated using the Schaefer-type production function (gray font) and using the Fox-type production function (dark gray). The last 5 years are obtained assuming a 100, 000 tons constant quotas management option.

Consider now an alternative form for the production $h(B_t)$, the so-called Fox model ([137]; [244]):

$$h(B_t) = r \times B_t \times (1 - \frac{log(B_t)}{log(K)}) \tag{11.20}$$

and its associated fisheries management reference points:

$$\begin{cases} C_{MSY} = \dfrac{r \times K}{e^1 \times log(K)} \\ B_{MSY} = \dfrac{K}{e^1} \end{cases} \tag{11.21}$$

Figure 11.7 shows how the model behaves when the Fox-type production function is used instead of the Schaefer-type. The overall shape of the estimated biomass trajectory only differs slightly; the biomass estimated with the Fox production function being systematically lower than with the Schaefer one.

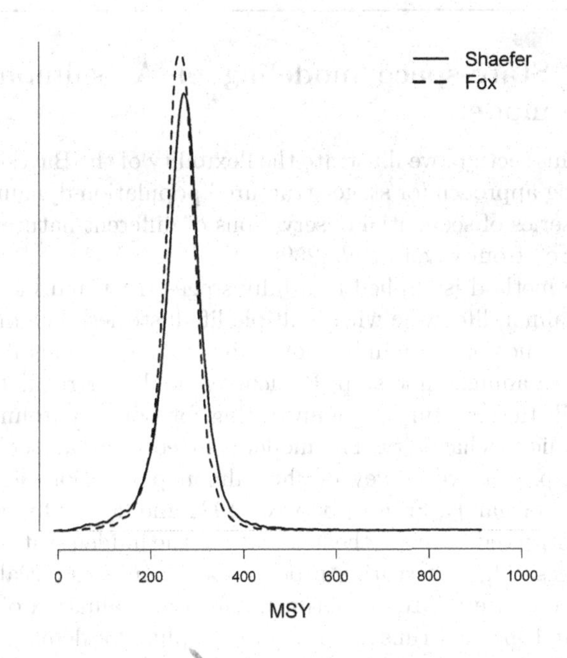


**FIGURE 11.8**: Marginal posterior distribution of the Maximum Sustainable Yield ($C_{MSY}$) estimated using the Schaefer-type (solid line) and the Fox-type (dotted line) production function.

As shown in Figure 11.8, both production functions lead to minor differences in the marginal posterior distributions of the maximum sustainable yield.

It is worth noting that both the Schaefer and the Fox production functions are particular cases of a more general equation with 3 parameters known as the *Pella-Tomlinson* production function ([244]):

$$h(B_t) = r \times B_t \times (1 - \frac{B_t}{K}^{m-1}) \tag{11.22}$$

Setting $m = 2$ yields to the Schaefer type production function (Eq. (11.7)), and $m = 1$ yields to the Fox production (Eq. (11.20)). However, only poor information generally exists in the dataset to estimate the parameter $m$, and the particular forms (Eqs. (11.7) or (11.20)) are generally preferred to Eq. (11.22).

---

## 11.3 State-space modeling of A. salmon life cycle model

In this section, we illustrate the flexibility of the Bayesian state-space modeling approach for stage-structured population dynamics models fitted to series of sequential observations of different nature. The example is inspired from Rivot *et al.* [259].

The method is applied to a fully stage-structured model for the Atlantic salmon life cycle with multiple life histories. The model describes the dynamics of the numbers of individuals at various life stages, with a discrete annual time step. It includes nonlinear regulation and has a probabilistic structure accommodating for both environmental and demographic stochasticity. The model is fitted to a dataset resulting from the comprehensive survey of the salmon population of the Oir River (Lower Normandy, France) between 1984 and 2001. Observation models are constructed to relate the field data to the hidden states at the various life stages. The observation process corresponds essentially to capture-mark-recapture (CMR) experiments for the evaluation of migrating juvenile and spawner runs and random sampling for demographic features. The ecological significance of the inferences is discussed at the end of the section.

Here, we largely rely on the A. salmon population ecology described in Chapter 1, but we take into consideration a more complete description of life histories: fish from the two smolt age classes (1+ Smolts and 2+

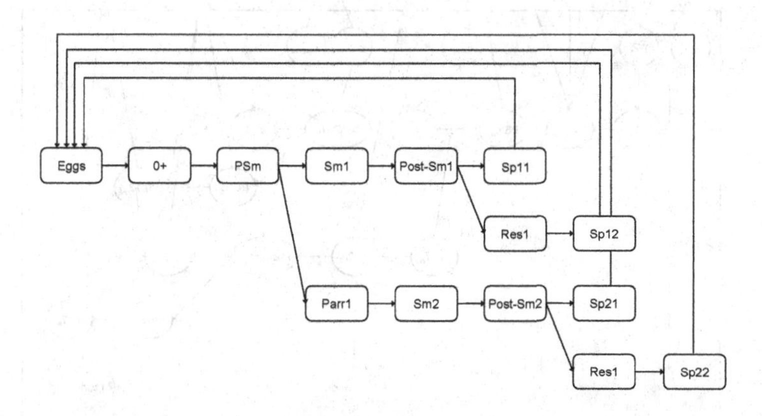


**FIGURE 11.9**: Atlantic salmon life cycle model including the four main life histories encountered in French populations. Eggs spawned at year $t$ give rise to spawners $Sp11$ returning year $t+3$, $Sp12$ and $Sp21$ returning year $t+4$ and $Sp22$ returning year $t+5$.

Smolts) can either spend one or two winters at sea (1SW and 2SW in the following) before returning as spawners (Fig. 11.9). One and two sea winter spawners resulting from 1+ Smolts are denoted $Sp11$ and $Sp12$, respectively, and those issued from 2+ Smolts are denoted $Sp21$ and $Sp22$, respectively. Both $Sp11$ and $Sp21$ are 1SW fish but with different smolt-ages, and $Sp12$ and $Sp22$ are 2SW fish with different smolt-ages.

### 11.3.1 Process equations describing the hidden population dynamics

In such population dynamics state-space models, the general process Eq. (11.2) giving $[Z_{t+1}|Z_t, \theta]$ is fully defined by the probabilistic demographic transitions used to describe the population dynamics. The stage-structured A. salmon life cycle model can be viewed as a conditional model. The following Eqs. (11.23)-(11.33) mimic the transition of the number of individuals from one development stage to the next one, on a discrete, yearly basis time step. Most of the transitions are built on a probabilistic rationale to capture demographic or environmental stochasticity. For the sake of clarity, equations are written with time

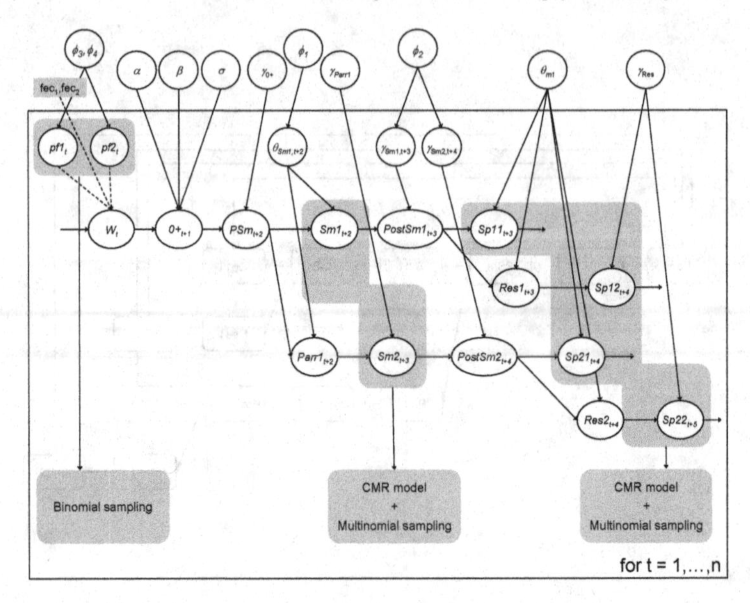


**FIGURE 11.10**: DAG of the hidden dynamics for the Atlantic salmon life cycle including the observation process on both the smolt and spawner runs. All probabilistic equations for the population dynamics are given in Eqs. (11.23)-(11.33). For the sake of clarity, the observtion process is represented by shaded boxes and the observation Eqs. (11.37)-(11.42) are not detailed in the figure.

indices following the fate of a cohort issued from eggs spawned in year $t$. The DAG of the model presenting the hidden population dynamics is shown in Fig. 11.10.

1. *Spawners → Eggs*

   Each year, four age classes of spawners return from the sea and reproduce in the river. The number of eggs laid by females spawning in year $t$, $W_t$, is a deterministic function of the number of spawners in the two sea-age classes, the proportion of females in 1SW and 2SW fish, denoted $p_{f1,t}$ and $p_{f2,t}$ respectively, and of the mean fecundity of these females denoted $fec_1$ and $fec_2$, both considered as known and constant over time:

$$\begin{aligned} W_t = & (Sp11_t + Sp21_t - y^{Sp}_{2,t}) \times p_{f1,t} \times fec_1 \\ & + (Sp12_t + Sp22_t - y^{Sp}_{3,t}) \times p_{f2,t} \times fec_2 \end{aligned} \tag{11.23}$$

   $y^{Sp}_2$ and $y^{Sp}_3$ are known numbers of fish (defined in the dataset

| Model component | Parameter | Prior |
|---|---|---|
| **Population dynamics** | | |
| Proportion of females | $\mu_{pf1}$ | $\sim Normal(0, 100)$ |
| | $\sigma_{pf1}$ | $\sim Uniform(0, 5)$ |
| | $\mu_{pf2}$ | $\sim Normal(0, 100)$ |
| | $\sigma_{pf2}$ | $\sim Uniform(0, 5)$ |
| Eggs $\rightarrow$ 0+ Juveniles (Ricker) | $log(\alpha)$ | $\sim Uniform(-10, 0)$ |
| | $\beta$ | $\sim Normal(0, 100)$ |
| | $log(\sigma^2)$ | $\sim Uniform(-20, 20)$ |
| 0+ Juveniles $\rightarrow$ Pre-smolts | $\gamma_{0+}$ | $\sim Beta(15, 15)$ |
| Probability to smoltify as 1+ Smolts | $\mu_{\theta_{Sm1}}$ | $\sim Normal(0, 100)$ |
| | $\sigma_{\theta_{Sm1}}$ | $\sim Uniform(0, 5)$ |
| 1+ Parrs $\rightarrow$ 2+ Smolts | $\gamma_{Parr1}$ | $\sim Beta(20, 10)$ |
| Survival 1+ Smolts | $\mu_{\gamma_{Sm1}}$ | $\sim Normal(0, 100)$ |
| | $\sigma_{\gamma_{Sm1}}$ | $\sim Uniform(0, 5)$ |
| Survival 2+ Smolts | $\delta_{\gamma}$ | $\sim Uniform(0, 10)$ |
| Probability to mature as 1SW | $\theta_{m1}$ | $\sim Beta(3, 2)$ |
| Survival of nonmaturing adults at sea | $\gamma_{Res}$ | $\sim Beta(3, 2)$ |
| Initial states variables | $0+^*_{t=1}$ | $\sim Uniform(0, 1)$ |
| | $Psm1_{t=1}$ | $\sim Uniform(1, 10000)$ |
| | $Sm2_{t=1}$ | $\sim Uniform(1, 300)$ |
| | $PostSm1_{t=1}$ | $\sim Uniform(1, 1000)$ |
| | $PostSm2_{t=1}$ | $\sim Uniform(1, 1000)$ |
| | $Sp12_{t=1}$ | $\sim Uniform(1, 100)$ |
| | $Sp21_{t=1}$ | $\sim Uniform(1, 50)$ |
| **Observation process** | | |
| Smolts capture probability | $\mu_{\pi_{Sm}}$ | $\sim Normal(0, 100)$ |
| | $\sigma_{\pi_{Sm}}$ | $\sim Uniform(0, 5)$ |
| Spawners capture probability | $\mu_{\pi_{Sp,1}}$ | $\sim Normal(0, 100)$ |
| | $\sigma_{\pi_{Sp,1}}$ | $\sim Uniform(0, 5)$ |
| Spawners recapture probability | $\mu_{\pi_{Sp,2}}$ | $\sim Normal(0, 100)$ |
| | $\sigma_{\pi_{Sp,2}}$ | $\sim Uniform(0, 5)$ |

**TABLE 11.4**: Prior distribution on parameters.

(see Tables 11.6 and 11.7)) which are retrieved from the returning adults after trapping, either because they die during manipulation or because they are removed for experimental use or hatchery production.

2. *Eggs → 0+ juveniles*

   A density dependent Ricker relationship with unknown parameters $(\alpha, \beta)$ models the freshwater production of juveniles resulting from the reproduction of the spawners returning in year $t$. Environmental variability is introduced via independent and identically distributed LogNormal errors, so that the logarithm of the number of 0+ juveniles can be defined as arising from a Normal probability distribution with variance $\sigma^2$ (see also Chapter 7):

$$0+^*_{t+1} \sim Normal\left(log(W^*_t \times \alpha \times e^{-\beta \times W^*_t}), \sigma^2\right) \qquad (11.24)$$

   where $0+^*_t = 0+_t/h$ and $W^*_t = W_t/h$ are the number of eggs and 0+ juveniles standardized by the surface area of habitat available for juveniles production, $h = 25229m^2$ (see [237] or [258] for more details).

3. *0+ juveniles → Smolts*

   As detailed in Chapter 1, survival and life history choices during the freshwater phase of the life cycle can all be modeled by Binomial distributions to capture the demographic stochasticity. The young-of-the-year $0+_{t+1}$ will survive to the next spring of year $t+2$ as pre-smolts $PSm_{t+2}$, with probability $\gamma_{0+}$ (considered as invariant over time):

$$PSm_{t+2} \sim Binomial(0+_{t+1}, \gamma_{0+}) \qquad (11.25)$$

   A proportion $\theta_{Sm1,t+2}$ of the pre-smolts $PSm_{t+2}$ will migrate as 1+ Smolts, the remaining part will stay one additional year as 1+ Parrs:

$$\begin{cases} Sm1_{t+2} \sim Binomial(PSm_{t+2}, \theta_{Sm1,t+2}) \\ Parr1_{t+2} = PSm_{t+2} - Sm1_{t+2} \end{cases} \qquad (11.26)$$

   Resident 1+ Parrs will survive with probability $\gamma_{Parr1}$ (considered constant) and survivors will migrate as 2+ Smolts with probability 1:

$$Sm2_{t+3} \sim Binomial(Parr1_{t+2}, \gamma_{Parr1}) \qquad (11.27)$$

4. *Smolts → Returning spawners*

   After reaching the sea in spring, 1+ Smolts ($Sm1_{t+2}$) will survive as post-smolts (denoted $PostSm1_{t+3}$) up to the end of the first winter at sea with probability $\gamma_{Sm1,t+2}$:

$$PostSm1_{t+3} \sim Binomial(Sm1_{t+2}, \gamma_{Sm1,t+2}) \qquad (11.28)$$

Hypothesizing a strict homing of adults to their native stream ([238]), a proportion $\theta_{m1}$ of the post-smolts $PostSm1_{t+3}$ will mature and return as one sea-winter fish $Sp11_{t+3}$ in the following summer:

$$\begin{cases} Sp11_{t+3} \sim Binomial(PostSm1_{t+2}, \theta_{m1}) \\ Res1_{t+3} = PostSm1_{t+3} - Sp11_{t+3} \end{cases} \tag{11.29}$$

Nonmaturing fish (denoted $Res1_{t+3}$) will stay one additional year at sea, will survive with probability $\gamma_{Res}$ (considered constant over time) and return as two sea-winter spawners ($Sp12_{t+4}$) in the following spring:

$$Sp12_{t+4} \sim Binomial(Res1_{t+3}, \gamma_{Res}) \tag{11.30}$$

2+ Smolts have a symmetric life history. $Sm2_{t+3}$ survive as post-smolts (denoted $PostSm2_{t+4}$) up to the end of the first winter at sea with probability $\gamma_{Sm2,t+3}$:

$$PostSm2_{t+4} \sim Binomial(Sm2_{t+3}, \gamma_{Sm2,t+3}) \tag{11.31}$$

Post-smolts $PostSm2_{t+4}$ will mature with the same probability $\theta_{m1}$ and return as 1SW spawners $Sp21_{t+4}$ in the following summer:

$$\begin{cases} Sp21_{t+4} \sim Binomial(PostSm2_{t+3}, \theta_{m1}) \\ Res2_{t+4} = PostSm2_{t+4} - Sp21_{t+4} \end{cases} \tag{11.32}$$

Nonmaturing fish ($Res2_{t+4}$) will survive with the same probability $\gamma_{Res}$ and return as 2SW spawners ($Sp22_{t+5}$) in the following spring:

$$Sp22_{t+5} \sim Binomial(Res2_{t+4}, \gamma_{Res}) \tag{11.33}$$

Finally, spawners reproducing in any year $t$ (and used to calculate the number of eggs in Eq. (11.23)) are obtained by the mass balance equation $Sp_t = Sp11_t + Sp12_t + Sp21_t + Sp22_t$. This generates complex dynamics as returns in year $t$ involve three different cohorts. Indeed, $Sp11_t$ are from eggs spawned in year $t-3$, $Sp12_t$ and $Sp21_t$ are from eggs spawned in year $t-4$, and $Sp22_t$ are from eggs spawned in year $t-5$.

In Eq. (11.3) describing the general process equation for a state-space model, the state vector $Z_t$ is the state of the system at time step $t$. Here, $Z_t$ is a multidimensional vector with each component equal to the number of fish in each development stage at time step $t$, *i.e.*, demographic characteristics which are unknown and variable in time. Because of the cohorts overlapping due to the complex life history of A. salmon, $Z_t$ involves 6 different cohorts. In Eq. (11.3), each Markovian transition $[Z_{t+1}|Z_t, \theta_1]$, is fully defined by factorizing Eqs. (11.23)-(11.33) after re-indexing accordingly.

Because of the time dependence between the state variables, initial prior distributions must be specified on most variables composing the state vector for the first year $Z_1$. Diffuse priors are assigned on $0+_{t=1}$, $Psm_{t=1}$, $Sm2_{t=1}$, $PostSm1_{t=1}$, $PostSm2_{t=1}$, $Sp12_{t=1}$ and $Sp11_{t=1}$ (Table 11.4), all other state variables for the first year being defined conditionally on those ones and parameters.

$\theta_1$ is the vector of all parameters involved in the dynamics, formally all the quantities which are unknown and constant over time. It contains vital rates (such as the proportion of fish maturing as one sea-winter fish $\theta_{m1}$) and for which prior distributions have to be specified (see Table 11.4), but also parameters of the hierarchical structures used to represent the between-year variability of some vital rates (such as the proportion of juveniles migrating as 1+ Smolts $\theta_{Sm1,t}$).

### 11.3.2 Modeling the priors on vital rates through an exchangeable hierarchical structures

The model encompasses both demographic and environmental stochasticity by allowing an interannual random variability of the proportion of smolts migrating as 1+ Smolts and the marine survival rate of post-smolts. Interannual random variations are modeled through an exchangeable hierarchical structure ([117]), by assuming that the parameters of all years $t$ are randomly drawn from the same probability distribution conditioned by unknown parameters common to all years. Hierarchical structures allow to probabilistically share information among the different years and can improve the estimation of key parameters ([194]; [204]; [255]). Here, an exchangeable hierarchical structure is used to mimic a between-year variability that is assumed to be random, without any particular time trends or covariates (*i.e.*, climate) to explain the variations. More advanced hierarchical structures such as an autocorrelated random walk can also be used to capture fluctuations in vital rates caused by gradual changes in environmental conditions (see [262] for an example).

An exchangeable hierarchical structure is used to capture the between-year variability of the probability for a pre-smolt to smoltify as 1+ smolt. Instead of considering the $\theta_{Sm1,t}$'s as a priori independently sampled from conditional beta distribution with parameters common to all years (as detailed in Section 9.2 of Chapter 9), the *logit*-transform of the $\theta_{Sm1,t}$'s are a priori drawn in a Normal distribution with mean and variance parameters $\phi_1 = (\mu_{\theta_{Sm1}}, \sigma_{\theta_{Sm1}})$ shared between years. The $\theta_{Sm1,t}$'s are then calculated by the inverse transformation $logit^{-1}$:

$$logit(\theta_{Sm1,t}) \sim Normal(\mu_{\theta_{Sm1}}, \sigma_{\theta_{Sm1}}) \tag{11.34}$$

The same procedure was applied for the survival rate of one year river-aged post-smolts, $\gamma_{Sm1,t}$, with mean and variance parameters shared between years (Eq. (11.35)). The survival rates of two year river-aged post-smolts, $\gamma_{Sm2,t}$, are defined based on the reasonable a priori constraint that the survival rate of two year river-aged post-smolts is always greater than for one year river-aged post-smolts, but with the same time variations. This constraint is specified in the *logit* scale (Eq. (11.35)).

$$\begin{cases} logit(\gamma_{Sm1,t}) \sim Normal(\mu_{\gamma_{Sm1}}, \sigma_{\gamma_{Sm1}}) \\ logit(\gamma_{Sm2,t}) = logit(\gamma_{Sm1,t}) + \delta_\gamma \\ \delta_\gamma \geq 0 \end{cases} \tag{11.35}$$

Finally, the *logit*-transform of the proportions of females in 1SW and 2SW fish, $pf1_t$ and $pf2_t$ are a priori independently drawn in Normal distributions with mean and variance parameters shared between years:

$$\begin{cases} logit(pf1_t) \sim Normal(\mu_{pf1}, \sigma_{pf1}) \\ logit(pf2_t) \sim Normal(\mu_{pf2}, \sigma_{pf2}) \end{cases} \tag{11.36}$$

Informative beta distributions based on prior ecological expertise are set for $\gamma_{0+}$ (survival rates of 0+ Juveniles) and $\gamma_{Parr1}$ (survival rate of 1+ Parrs), with mean 0.5 and 0.66, respectively. All the remaining parameters of the population dynamics are considered constant over time with rather diffuse prior distribution (Table 11.4).

### 11.3.3 Data and observation equations

Yearly noisy observations of both smolt and spawner runs were recorded between 1984 and 2001. However, no observations for eggs ($W$), 0+ juveniles (0+), pre-smolts ($PSm$), resident 1+ Parrs ($Parr1$), 1+ and 2+ post-smolts at sea ($PostSmolts1$, $PostSmolts2$), and nonmaturing fish ($Res1$, $Res2$) are available for these 18 years. The DAG of the model with a simplified representation of the observation models is given in Fig. 11.10. Below we describe the probabilistic models used to sketch the stochastic observation process. Each likelihood term $[y_t|Z_t, \theta_2]$ in the general observation equation as written in Eq. (11.4) is obtained by

| Years | $y_1^{Sm}$ | $y_2^{Sm}$ | $y_3^{Sm}$ | $y_4^{Sm}$ | $y_5^{Sm}$ |
|---|---|---|---|---|---|
| 1984 | NA | NA | NA | NA | NA |
| 1985 | 439 | NA | NA | 439 | 232 |
| 1986 | 887 | 135 | 91 | 887 | 848 |
| 1987 | 283 | 31 | 24 | 283 | 146 |
| 1988 | 307 | 59 | 43 | 307 | 282 |
| ... | | | | | |
| 1997 | 205 | 63 | 31 | 205 | 186 |
| 1998 | 511 | 91 | 44 | 511 | 438 |
| 1999 | 195 | 59 | 45 | 195 | 43 |
| 2000 | 1849 | 300 | 232 | 1849 | 1835 |
| 2001 | 688 | 264 | 123 | 688 | 636 |

**TABLE 11.5**: Capture-mark-recapture and sampling data for smolts at the downstream trap by migration year. $y_{1,t}^{Sm}$: number of smolts caught in the downstream trapping facility during the migration time; $y_{2,t}^{Sm}$: tagged and released smolts; $y_{3,t}^{Sm}$: number of smolts recaptured among $y_{2,t}^{Sm}$; $y_{4,t}^{Sm}$: number of smolts examined for river-age; $y_5^{Sm}$: number of 1+ Smolts in $y_{4,t}^{Sm}$.

multiplying together the probabilistic observation Eqs. (11.37)-(11.42) given below.

The observations (1984 - 2001) have been gathered under an homogeneous experimental design and few data are missing. Each annual survey provides two complementary sources of information, the number of smolts and spawners in each annual run, and the demographic structure (age classes and sex-ratio) in smolts and adult runs.

#### 11.3.3.1 Updating the population size at the various life stages

Available information to estimate the size of smolt and spawner runs does not distinguish between the age classes. For both the smolts and the spawners, a proportion of the migrating population is captured at a partial counting fence and marked for further recapture (see Figure 4.3 in Chapter 4; the interested reader can also refer to Rivot and Prévost [255] for more details). CMR experiments provide data to update the total number of smolts migrating in year $t$, $Sm_t = Sm1_t + Sm2_t$, and the total number of spawners returning in year t, $Sp_t = Sp11_t + Sp12_t + Sp21_t + Sp22_t$.

| Years | $y_1^{Sp}$ | $y_2^{Sp}$ | $y_3^{Sp}$ | $y_4^{Sp}$ | $y_5^{Sp}$ | $y_6^{Sp}$ |
|---|---|---|---|---|---|---|
| 1984 | 167 | 10 | 3 | 154 | 12 | 10 |
| 1985 | 264 | 37 | 11 | 216 | 21 | 4 |
| 1986 | 130 | 28 | 9 | 93 | 5 | 4 |
| 1987 | 16 | 3 | 1 | 12 | 2 | 22 |
| 1988 | 226 | 35 | 8 | 183 | 12 | 0 |
| ... | | | | | | |
| 1997 | 56 | 19 | 3 | 34 | 12 | 3 |
| 1998 | 34 | 3 | 1 | 30 | 6 | 30 |
| 1999 | 154 | 5 | 1 | 148 | 13 | 22 |
| 2000 | 53 | 0 | 0 | 53 | 4 | 33 |
| 2001 | 160 | 1 | 0 | 159 | 31 | 4 |

**TABLE 11.6**: Capture-mark-recapture and sampling data for spawners at the upstream trap by migration year. $y_1^{Sp}$: number of fish trapped at the counting fence during the upstream migration time; $y_2^{Sp}$, $y_3^{Sp}$: one sea-winter (resp. two sea-winters) fish removed from the population; $y_4^{Sp}$: tagged and released fish; $y_5^{Sp}$, $y_6^{Sp}$: number of marked (resp. unmarked) recaptured fish.

### Smolt run

CMR data used to update the unknown number of smolts $Sm_t$ each year $t$ between 1984 and 2001 are given in Table 11.5 (note that there are missing data for the first two years). CMR experiment modeling principles have already been presented in Section 4.1 of Chapter 4, page 84. Here we only recall the main outlines of the model.

CMR experiments for smolts are analogous to the two-stage Petersen experiment ([278]). For each year $t$ between 1986 and 2001, let us denote $y_{1,t}^{Sm}$ the number of smolts caught in the downstream trapping facility during the migration time. Among the $y_{1,t}^{Sm}$ smolts captured, a number $y_{2,t}^{Sm}$ have been tagged and released upstream from the trapping facility used for capture (see Fig. 4.3). Some of these tagged and released smolts, denoted $y_{3,t}^{Sm}$ ($y_{3,t}^{Sm} < y_{2,t}^{Sm}$), will be recaptured at the same downstream trap. Under the hypotheses $H_1$-$H_5$ detailed in Section 4.1, the capture and recapture data for each year $t$ can be modeled by two successive Binomial distributions with the same capture probabilities $\pi_t^{Sm}$ (interpreted as the trapping efficiency that may vary between years):

$$\begin{cases} y_{1,t}^{Sm} \sim Binomial(Sm_t, \pi_t^{Sm}) \\ y_{3,t}^{Sm} \sim Binomial(y_{2,t}^{Sm}, \pi_t^{Sm}) \end{cases} \tag{11.37}$$

Intuitively, the second Binomial equation will contribute to specify

| Years | $y_7^{Sp}$ | $y_8^{Sp}$ | $y_9^{Sp}$ | $y_{10}^{Sp}$ | $y_{11}^{Sp}$ | $y_{12}^{Sp}$ | $y_{13}^{Sp}$ | $y_{14}^{Sp}$ | $y_{15}^{Sp}$ |
|---|---|---|---|---|---|---|---|---|---|
| 1984 | 159 | 113 | 20 | 23 | 3 | 141 | 40 | 26 | 21 |
| 1985 | 211 | 116 | 50 | 43 | 2 | 203 | 69 | 61 | 42 |
| 1986 | 111 | 61 | 19 | 24 | 7 | 93 | 31 | 37 | 26 |
| 1987 | 16 | 13 | 2 | 1 | 0 | 15 | 1 | 1 | 1 |
| 1988 | 197 | 85 | 74 | 36 | 2 | 182 | 63 | 44 | 31 |
| ... | | | | | | | | | |
| 1997 | 52 | 47 | 4 | 1 | 0 | 55 | 21 | 1 | 1 |
| 1998 | 28 | 22 | 5 | 1 | 0 | 33 | 10 | 1 | 1 |
| 1999 | 140 | 105 | 18 | 12 | 5 | 136 | 43 | 18 | 13 |
| 2000 | 51 | 45 | 2 | 2 | 2 | 49 | 26 | 4 | 2 |
| 2001 | 140 | 120 | 14 | 5 | 1 | 151 | 51 | 9 | 3 |

**TABLE 11.7**: Capture-mark-recapture and sampling data for spawners at the upstream trap by migration year (continued). $y_7^{Sp}$: number of fish examined for sea and river-age; $y_8^{Sp}$, $y_9^{Sp}$: number of one sea-winter fish issued from 1+ and 2+ Smolts respectively among $y_7^{Sp}$; $y_{10}^{Sp}$, $y_{11}^{Sp}$: number of two sea-winter fish issued from 1+ and 2+ Smolts respectively among $y_7^{Sp}$. $y_{12}^{Sp}$, $y_{14}^{Sp}$: one sea-winter and two sea-winter fish examined for sex identification; $y_{13}^{Sp}$, $y_{15}^{Sp}$: number of fish identified as females among $y_{12}^{Sp}$ and $y_{14}^{Sp}$.

the trapping efficiency $\pi_{Sm,t}$, while the first one brings information to update the total number of downstream migrating smolts $Sm_t$.

### Spawner run

The experimental design, the trapping methodology and the CMR model used for spawner runs are described in detail in Chapter 9 (see also Rivot and Prévost [255] for comments on the experimental procedure).

The CMR data for the spawners are shown in Tables 11.6 and 11.7. Conversely to smolts, CMR experiments for spawners cannot be assimilated to a simple Petersen experiment. For each year $t$ from 1984 to 2001, $y_{1,t}^{Sp}$ denotes the number of fish trapped at the counting fence (see Fig. 4.3 of Chapter 4 for more details about the trapping device). Among these $y_{1,t}^{Sp}$ fish, $y_{2,t}^{Sp}$ one sea-winter and $y_{3,t}^{Sp}$ two sea-winter fish are removed from the population (as detailed in Section 11.3.1). A number $y_{4,t}^{Sp} = y_{1,t}^{Sp} - (y_{2,t}^{Sp} + y_{3,t}^{Sp})$ of fish are tagged and released upstream from the trap. The recapture sample is gathered during and after spawning by three methods (electrofishing on the spawning grounds, collection of dead fish after spawning, and trapping of spent fish at the downstream trap). Let us denote as $y_{5,t}^{Sp}$ and $y_{6,t}^{Sp}$ the number of marked and unmarked fish among recaptured fish, respectively.

CMR experiments can also be modeled via Binomial distributions. Assuming classical hypotheses (all spawners behave independently and are equally catchable by the upstream trap, with a probability $\pi_t^{Sp,1}$ considered constant over the migration season), $y_{1,t}^{Sp}$ can be modeled as the observed result of a Binomial experiment with the total number of spawners $Sp_t$ as the number of trials and a probability $\pi_t^{Sp,1}$ (Eq. (11.38)). Assuming that *(i)* no spawner runs downstream after getting over the trap; *(ii)* there is no tag shedding between mark and recapture; *(iii)* the recapture probability is the same for all the fish whether or not marked, the recapture of marked ($y_5^{Sp}$) and unmarked fish ($y_6^{Sp}$) can be modeled as the result of Binomial experiments with number of trials $y_{4,t}^{Sp}$ (marked and released fish) and $Sp_t - y_1^{Sp}$ (unmarked fish in the population) respectively, but with probabilities $\pi_t^{Sp,2}$ different from $\pi_t^{Sp,1}$ as recapture experiments do not involve directly the trapping facility with efficiency $\pi_t^{Sp,1}$.

$$\begin{cases} y_{1,t}^{Sp} \sim Binomial(Sp_t, \pi_t^{Sp,1}) \\ y_{5,t}^{Sp} \sim Binomial(y_{4,t}^{Sp}, \pi_t^{Sp,2}) \\ y_{6,t}^{Sp} \sim Binomial(Sp_t - y_{1,t}^{Sp}, \pi_t^{Sp,2}) \end{cases} \tag{11.38}$$

The reasoning underlying how the information contained in the data is used in the three lines of Eq. (11.38) is as follows: the second line essentially conveys information to estimate the recapture probability $\pi_t^{Sp,2}$. Both the third and the first line carry information on the total number of spawners $Sp_t$.

### 11.3.3.2 Hierarchical prior on the trapping efficiencies

The between-year variability of the probabilities of capture $\pi_t^{Sm}$, $\pi_t^{Sp,1}$ and $\pi_t^{Sp,2}$ is modeled through exchangeable hierarchical structures with hyper-parameters denoted $\phi_{Sm}$, $\phi_{Sp,1}$ and $\phi_{Sp,2}$ respectively. As already illustrated in Chapters 9 and 10 (and also detailed in many publications such as [204], [239], [255] or [262]), hierarchical structures significantly improve the inferences as they organize the transfer of information between years to estimate the total population size in years for which CMR data are rather poor (*i.e.*, because of small recapture sample size such as in 1994 for the spawners for which $y_5^{Sp} = 1$ and $y_6^{Sp} = 4$).

In Section 9.2, the probabilities of capture and recapture for spawners for each year $t$ were considered as independently sampled from conditional beta distribution with parameters of the beta distributions common to all years. Here, slightly different hierarchical constructions are proposed. Instead of drawing the $\pi$'s in beta distributions, the *logit*-transform of the $\pi$'s are drawn from Normal distributions with mean and variance parameters shared between years. The $\pi$'s are then calculated by the inverse transformation $logit^{-1}$.

$$\begin{cases} logit(\pi_t^{Sm}) \sim Normal(\mu_{\pi_{Sm}}, \sigma_{\pi_{Sm}}) \\ logit(\pi_t^{Sp,1}) \sim Normal(\mu_{\pi_{Sp,1}}, \sigma_{\pi_{Sp,1}}) \\ logit(\pi_t^{Sp,2}) \sim Normal(\mu_{\pi_{Sp,2}}, \sigma_{\pi_{Sp,2}}) \end{cases} \tag{11.39}$$

$\theta_2 = (\mu_{\pi_{Sm}}, \sigma_{\pi_{Sm}}, \mu_{\pi_{Sp,1}}, \sigma_{\pi_{Sp,1}}, \mu_{\pi_{Sp,2}}, \sigma_{\pi_{Sp,2}})$ is the vector of parameters for the observation process. Diffuse prior distributions were set on parameters in $\theta_2$ (Table 11.4).

#### 11.3.3.3 Updating the demographic structure

In addition to the CMR data, sampling among the fish caught at the trap enables us to update the yearly proportions of the age classes in the smolt and spawner runs $Sm_t$ and $Sp_t$, and the mean sex ratio in the spawner run. Aging of the smolts (river age) and the spawners (river and sea-age) is done by scale reading ([14]; [237]). The sex of the spawners is identified visually based on the sexual dimorphism of salmon at the time they are caught at the trap. We suppose that age and sex determinations are performed without error. The data are given in Tables 11.5, 11.6 and 11.7. For each demographic feature, sampling is modeled by independent Binomial or multinomial processes, detailed in Eqs. (11.40)-(11.42).

#### 11.3.3.4 River age in smolt run

$y_{4,t}^{Sm}$ denotes the sample size of smolts examined for river-age in the run of year $t$ and $y_5^{Sm}$ is the number of 1+ Smolts among them. The proportion of 1+ Smolts in year $t$, denoted $\rho_{Sm1,t}$ is updated by modeling $y_5^{Sm}$ as the result of a Binomial sampling with a sample size $y_{4,t}^{Sm}$ and a probability $\rho_{Sm1,t}$:

$$\begin{cases} \rho_{Sm1,t} = \dfrac{Sm1_t}{Sm1_t + Sm2_t} \\ y_{5,t}^{Sm} \sim Binomial(y_{4,t}^{Sm}, \rho_{Sm1,t}) \end{cases} \tag{11.40}$$

#### 11.3.3.5 Sea and River-age in spawner run

$y_{7,t}^{Sp}$ denotes the sample size of spawners examined for sea and river-age in the run of year $t$. $y_{8,t}^{Sp}$ and $y_{9,t}^{Sp}$ are the number of 1SW fish resulting from 1+ and 2+ Smolts, respectively among $y^{Sp}_{7,t}$, and $y^{Sp}_{10,t}$ and $y^{Sp}_{11,t}$ are the number of 2SW fish issued from 1+ and 2+ Smolts, respectively among $y^{Sp}_{7,t}$. The proportions of the four different age classes in each spawning run (summing to 1) are updated through a multinomial model:

$$\begin{cases} \rho_{Sp,t} = (\frac{Sp11_t}{Sp_t}, \frac{Sp21_t}{Sp_t}, \frac{Sp12_t}{Sp_t}, \frac{Sp22_t}{Sp_t}) \\ (y^{Sp}_{8,t}, y^{Sp}_{9,t}, y^{Sp}_{10,t}, y^{Sp}_{11,t}) \sim Multinomial(y^{Sp}_{7,t}, \rho_{Sp,t}) \end{cases} \quad (11.41)$$

#### 11.3.3.6 Sex ratio in spawner run

Sex-ratio in spawner runs are updated each year $t$ through Binomial sampling experiments considered independent for one and two sea-winter fish. Each year $t$, $y^{Sp}_{12,t}$ 1SW fish (resp. $y^{Sp}_{14,t}$ 2SW fish) are examined for sex identification and $y^{Sp}_{13,t}$ (resp. $y^{Sp}_{15,t}$) are identified as females among them. Proportion of females in the two sea-age classes $pf_{1,t}$ and $pf_{2,t}$ are updated through Binomial sampling:

$$\begin{cases} y^{Sp}_{13,t} \sim Binomial(y^{Sp}_{12,t}, pf_{1,t}) \\ y^{Sp}_{14,t} \sim Binomial(y^{Sp}_{15,t}, pf_{2,t}) \end{cases} \quad (11.42)$$

### 11.3.4 Results

#### 11.3.4.1 MCMC simulations

Three MCMC independent chains were used. For each chain, the first 50,000 iterations were discarded. After this "burn-in" period only one in 10 (thinning) iterations was kept to reduce the MCMC sampling autocorrelation. Inferences were then derived from a sample of 30,000 iterations processed from three chains of 10,000 iterations.

#### 11.3.4.2 Spawner returns and smolt production

Posterior estimates of the total number of spawners returning each year $t$ and of smolts migrating downstream each year $t$ are characterized by a high between-year variability (Fig. 11.11). Posterior medians of the returns $Sp_t$'s range between 46 (year 1991) and 330 (year 1987). Posterior medians of the smolts $Sm_t$'s range between 202 (year 1991) and 2380 (year 2000). The uncertainty around the estimates of the $Sp_t$'s and $Sm_t$'s varies between years and partly stems from CMR experiments. Imprecise inferences about $Sp_t$ in 1987 are due to sparse mark-recapture data (low recapture sample size, see Tables 11.6 and 11.7). No data were available for smolts in 1984, which leads to the large imprecision about the estimation of $Sm_t$'s that year.

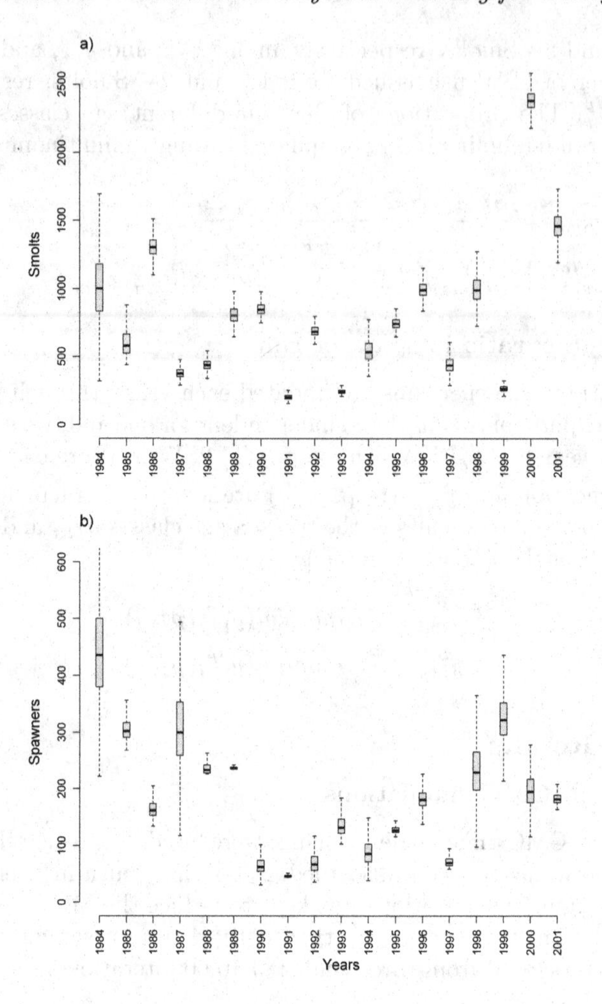


**FIGURE 11.11**: Marginal posterior distributions of (a) the number of migrating smolts ($Sm$) by year of downstream migration and (b) the number of returning spawners ($Sp$) by year of upstream migration. The boxes indicate the interquartile range and the median.

Estimates of the downstream (smolts) and upstream (spawners) trapping efficiencies are given in Fig. 11.12. Trapping efficiencies for spawners exhibit a huge between-year variability (overall mean = 0.67), whereas those for smolts are more constant in time (overall mean = 0.73). Detailed interpretation of the CMR experiments and further results concerning trapping efficiencies are given in Chapter 9 and in Rivot and Prévost [255].

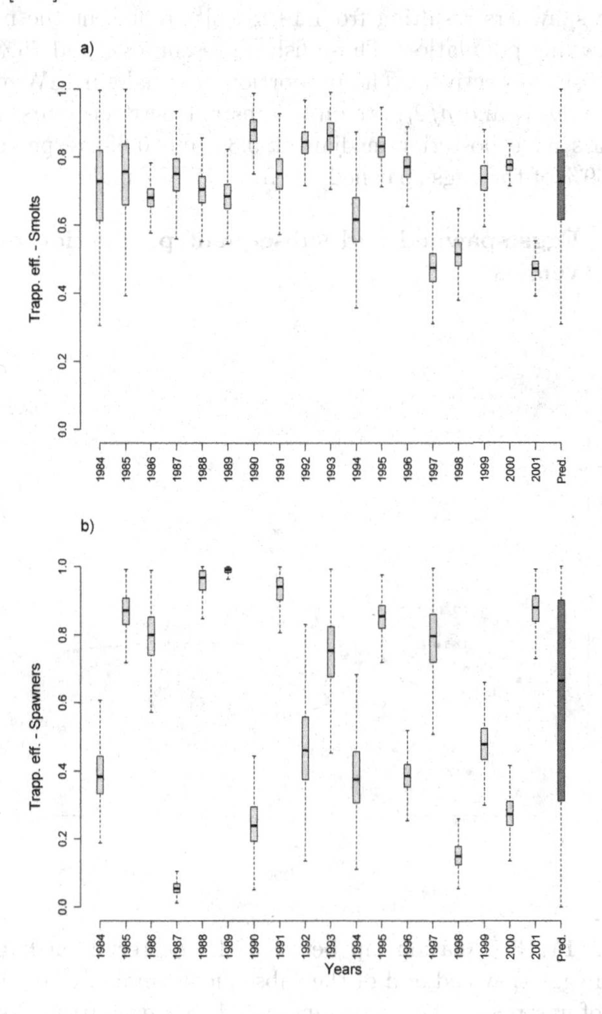


**FIGURE 11.12**: Marginal posterior distributions of trapping efficiencies for (a) downstream migrating smolts by year of migration and (b) upstream migrating spawners by year of migration. The boxes indicate the interquartile range and the median. Boxplots in dark gray fonts at the end of the time series show the posterior predictive of the trapping efficiency.

The posterior pdfs of the main demographic features of the returning spawners exhibit moderate within- and between-year variability. Thus, we only comment on the weighted average calculated across the years. Most of the returning spawners are 1SW fish (posterior median = 87%). Returning spawners resulting from 1+ Smolts represent the main part of the spawning population. Those fish represent 78% and 75% of 1SW and 2SW fish, respectively. The proportion of female in 1SW and 2SW, respectively, $pf1_t$ and $pf2_t$, are quite constant between years. Averages across years have posterior median at 0.34 and 0.68, respectively. On average, 69% of the eggs spawned are from 1SW females.

#### 11.3.4.3 Eggs spawned and subsequent production of 0+ juveniles

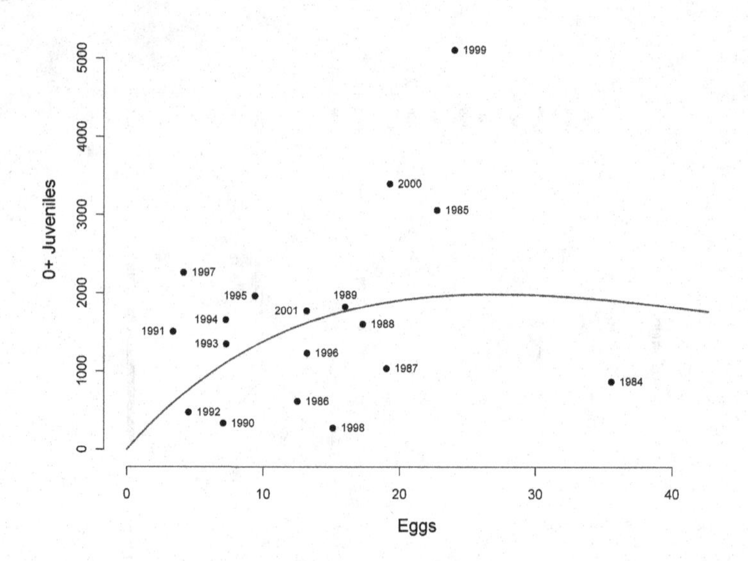


**FIGURE 11.13**: Relationship between the posterior medians of the number of eggs spawned and of the subsequent number of 0+ juveniles. The year of upstream migration is indicated near each point. Years 2000 and 2001 are not represented because the number of smolts cannot be updated due to incomplete data. The Ricker curve (solid line) corresponds to parameters $\alpha$ and $\beta$ set to their posterior medians, 0.01 and 0.047, respectively (Table 11.8).

The relationship between the number of eggs spawned $W_t$ and the subsequent number of 0+ juveniles $0+_{t+1}$ is characterized by a great variability around the mean deterministic Ricker relationship (Fig. 11.13). As $W_t$ is calculated from the number of spawners, the time series of the

$W_t$'s exhibits within- and between-year patterns similar to that of the $Sp_t$'s. The $0+_t$ exhibit the same pattern of between-year variations than the $Sm_{+1}$'s because survival rate between 0+ juveniles and pre-smolts has been considered constant and the pre-recruitment essentially consists of 1+ smolts. The number of 0+ juveniles for the last two years (2000 and 2001) are predictions because the number of smolts in the corresponding cohort are not updated by any data.

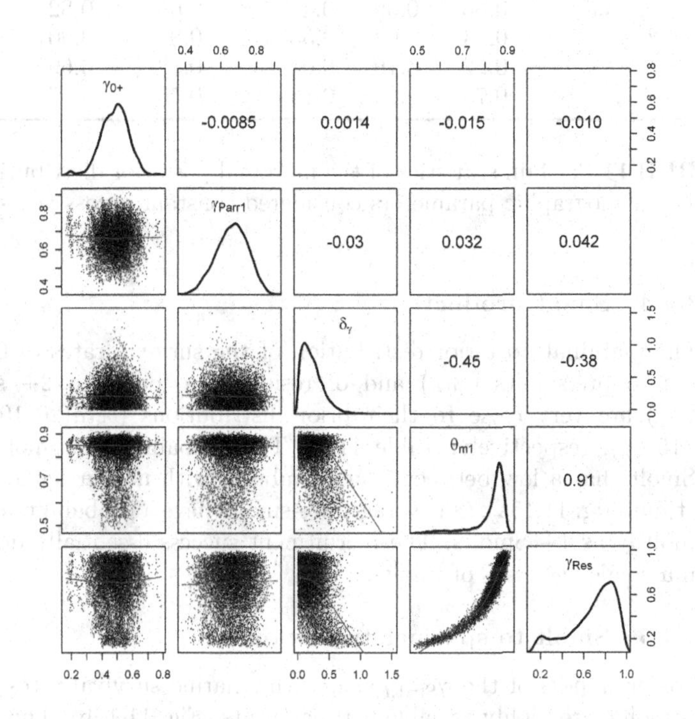


**FIGURE 11.14**: Posterior distributions of the key demographic parameters considered as constant between years. The marginal distributions are shown in the diagonal. Pairwise MCMC plots showing the correlation structure between parameters are shown in the lower part. The upper part shows the linear correlation between the MCMC draws.

The main statistics of the posterior pdfs of the recruitment process parameters ($\alpha$, $\beta$ and $\sigma^2$) are given in Table 11.8. The large standard deviation (in log scale) $\sigma$ reflects the wide dispersion around the Ricker curve (Fig. 11.13). The parameters of the Ricker curve ($\alpha, \beta$) are highly correlated a posteriori (Fig. 11.14). The posterior pdf of $\beta$ seems to be non-null in the neighborhood of 0. The average egg-to-juveniles survival rate calculated across year is 0.006 with a 95% credible interval ranging from 0.004 to 0.01.

| Parameters | Mean | Sd | 2.5% pct. | Median | 97.5% pct. |
|---|---|---|---|---|---|
| $\alpha$ | 0.010 | 0.005 | 0.003 | 0.009 | 0.024 |
| $\beta$ | 0.047 | 0.029 | −0.011 | 0.046 | 0.105 |
| $\sigma^2$ | 0.71 | 0.33 | 0.31 | 0.64 | 1.54 |
| $\gamma_{0+}$ | 0.50 | 0.09 | 0.32 | 0.50 | 0.67 |
| $\gamma_{Parr1}$ | 0.66 | 0.09 | 0.48 | 0.66 | 0.82 |
| $\theta_{m1}$ | 0.84 | 0.08 | 0.55 | 0.86 | 0.90 |
| $\delta_\gamma$ | 0.22 | 0.19 | 0.01 | 0.18 | 0.69 |
| $\gamma_{Res}$ | 0.70 | 0.22 | 0.16 | 0.76 | 0.97 |

**TABLE 11.8**: Main statistics of the marginal posterior distributions of the key demographic parameters considered constant across years.

#### 11.3.4.4 Smolt production

The marginal posterior distribution of the survival rates of 0+ juveniles to pre-smolts ($\gamma_{0+}$) and of resident 1+ parrs to 2+ smolts ($\gamma_{Parr1}$) are very close to their prior distributions ($beta(20, 10)$ and $beta(15, 15)$, respectively; Table 11.8). The probability to smoltify as 1+ Smolts has a low between-year variability with no particular trend over time (Fig. 11.15a). On average, pre-smolts have a probability of 0.84 to smoltify as 1+ Smolts. The recruitment success essentially depends upon a single age class of smolts.

#### 11.3.4.5 Smolt to spawner transitions

Posterior pdfs of the $\gamma_{Sm1,t}$'s, *i.e.*, the marine survival rates of 1+ post-smolts, are highly variable between years (Fig. 11.15b). The posterior medians of the $\gamma_{Sm1,t}$'s vary without any time trend. The estimated posterior medians range between 0.05 (year 1989) and 0.61 (year 1997).

The marine survival rate of 2+ Smolts has the same time variation than for 1+ Smolts, but with a differential measured by $\delta_\gamma$ in the *logit* scale (which translates to Table 11.8). The posterior median of $\delta_\gamma$ is about 0.18 (Table 11.8), which represents an average survival rate of 0.33 for 2+ Smolts versus 0.28 for 1+ Smolts.

On average, post-smolt survivors have a probability of 0.86 to mature after their first winter at sea and return as 1SW fish. Survival rates of fish that do not mature as 1SW is high (post. median = 0.76). However Fig. 11.14 also highlights the high positive correlation between $\theta_{m1}$ and $\gamma_{res}$, thus pointing out the difficulty to disentangle maturation and sur-

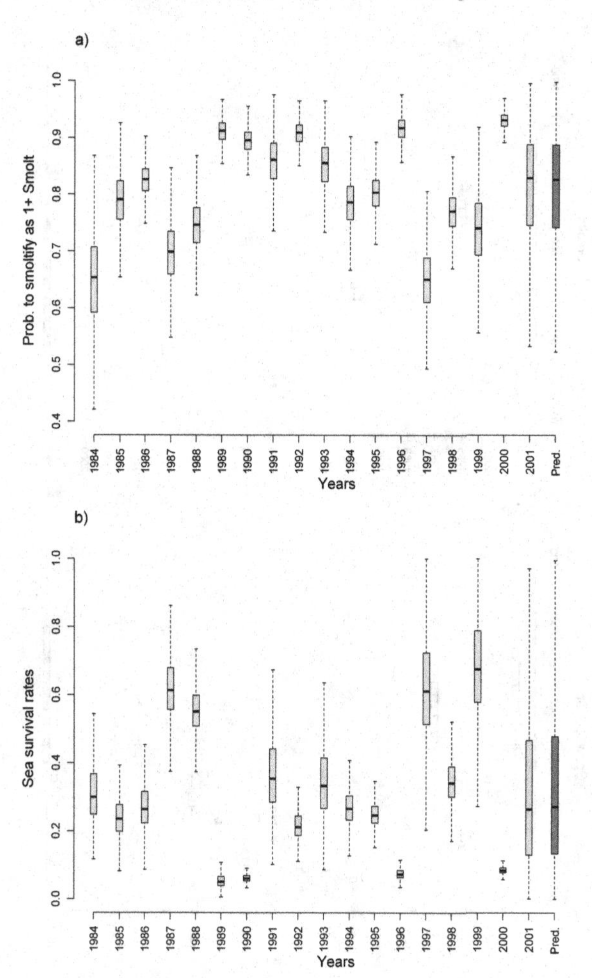


**FIGURE 11.15**: Marginal posterior distributions of (a) the probability to smoltify as 1+ Smolt ($\theta_{Sm1,t}$) (versus delaying the smoltification as 2+ Smolt) by year of smoltification; (b) 1+ Post-smolts marine survival rate $\gamma_{Sm1,t}$ by year of smolts migration. The boxes indicate the interquartile range and the median. Boxplots in dark gray fonts show the posterior predictive of $\theta_{Sm1}$.

vival in the marine phase of salmon, since no intermediate observation exists between smolts and returns.

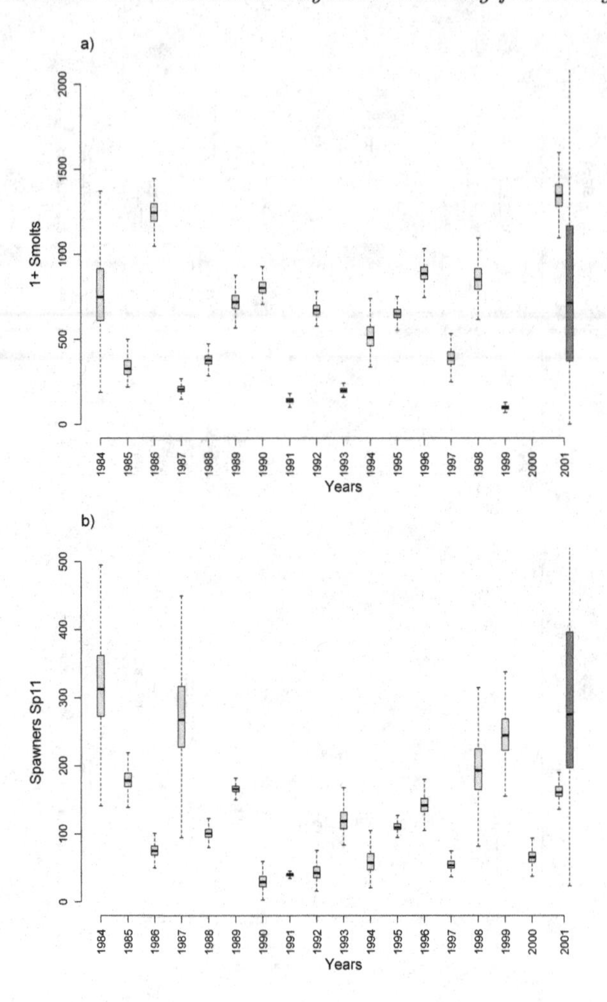


**FIGURE 11.16**: Ability for one step ahead forecasting. Marginal posterior distributions of (a) the number of migrating 1+ Smolts by migration year ($Sm1_t$) and (b) the number of returning spawners $Sp11_t$ by migration year. Light gray font: posterior distributions obtained with the complete data set; dark gray font (only for year 2001): posterior distribution obtained after deleting all the CMR and other sampling data for the last year 2001 (forecasting).

#### 11.3.4.6 Forecasting ability

To assess the model ability for one step ahead forecasting, we performed statistical tests analogous to cross-validation tests. We assess the ability in forecasting the smolt production and the number of returning spawners by comparing the estimates of $Sm1_{2001}$ and $Sp11_{2001}$ obtained after deleting all the CMR and other sampling data for the last year 2001. When no CMR and sampling data for smolts are available in 2001, $Sm1_{2001}$ is directly derived from the forecasts of 0+ juveniles and pre-smolts from Eqs. (11.24)-(11.26) with parameters drawn from their posterior or posterior predictive distributions (such as for $\theta_{Sm1}$) derived from the dataset with data of year 2001 deleted. Similarly, when no CMR and sampling data for smolts are available in 2001, $Sp11_{2001}$ is predicted from Eqs. (11.28)-(11.29) with $\gamma_{Sm1}$ drawn from its posterior predictive distribution and $\theta_{m1}$ from its posterior distribution.

Figure 11.16 emphasizes the poor forecasting ability of the model. The posterior distribution of $Sm1_{2001}$ and $Sp11_{2001}$ obtained after deleting the data for year 2001 are more dispersed and highly biased compared to the posterior updated by data of year 2001.

### 11.3.5 Discussion

#### 11.3.5.1 A powerful data synthesis method

The state-space modeling approach relies on a rigorous and coherent probabilistic framework to fit a realistic life cycle model for A. salmon. Embedding the model within a statistical approach enables us to synthesize the 18-year time series of field data from the Oir River with regards to the structure and the functioning of its salmon population. The parameters of the various between-stage transitions are estimated, and in some cases their inter-annual variation is assessed. Given the Southern European location of the Oir River, the results are consistent with the ecological knowledge on A. salmon life history ([42]; [97]; [143] ).

#### 11.3.5.2 Balancing ecological realism with available data

The model tries to balance biological realism and parsimony to ensure that all parameters are statistically identifiable. Some intermediate development stages (*e.g.*, 0+ juveniles or resident 1+ Parrs) have been represented to improve biological realism, although no direct observations are available to update the number of fish in these stages and parameters of the associated transitions. Informative priors on the associated parameters have been used to compensate for the lack of information in the data. Other parameters have been considered constant over time so as to ensure all parameters are identifiable. For instance, in some two-step transitions, such as the smolt-to-spawner transition that depends upon both the post-smolt marine survival rate $\gamma_{Sm1,t}$ and the probability to mature as one sea-winter fish $\theta_{m1}$; the latter parameter has been considered constant.

Our analysis also stresses the limited ability of our model to reproduce the evolution of the population abundance over time. The cross-validation tests show that the predictive power of our model concerning the smolt outputs and the spawner returns is rather poor. The cumulative effects of two types of misspecification explain this lack of fit. First, the influence of nonmodeled environmental factors blurs the demographic relation between some stages. Second, some critical processes may have been forgotten. The results are useful in setting priorities for model enhancement because they point out the portions of the model which particularly failed and should be improved.

In particular, we restrict our analysis to the dome-shaped Ricker curve with constant parameters and an i.i.d. LogNormal error structure. The variance around the Ricker function is very large (Table 11.8 and Fig. 11.13). This dispersion points out a departure from the specified LogNormal Ricker model. One needs to better understand and improve the modeling of the density-dependent regulation mechanisms during the juvenile phase. The posterior pdf of the $\beta$ parameter does not provide strong evidence for over-compensatory mortality of the eggs. Alternative shapes could be tested, such as the Beverton-Holt, the Deriso-Schnute or the Shepherd curve (see Chapter 7 but also [275] or [207]). But whatever the deterministic function used, the variation around it will remain large. Hence, it is clearly another major challenge to unravel the density-independent mechanisms which influence the egg-to-smolt survival rate ([150]; [207]). Using the available data on the abundance of 0+ Juveniles as developed in Chapter 10 could provide insight on the dynamics of the juvenile phase of the life cycle.

The estimates of the $\gamma_{Sm1,t}$'s are characterized by a large between-year variability. This variability is considered purely random in our model, and it plays a major role in its poor prediction power as revealed by the cross-validation tests (Fig. 11.16). Some key demographic processes explaining these yearly variations in the smolt-to-spawner transition may have been missed. The $\gamma_{Sm1,t}$'s, result from at least three combined processes: *(i)* the natural mortality of smolts during their stay at sea; *(ii)* the fishery mortality; *(iii)* the emigration/immigration processes when the adults return to freshwater. Unfortunately, the data available do not enable us to unambiguously disentangle these three components. But our results stress that the emigration/immigration processes are likely to be non-negligible at the spatial scale of our study. In particular, return rates exceeding 30% (Fig. 11.15) are extremely unlikely, and a significant net positive immigration of fish issued from other rivers than the Oir River might be accounted for to explain these high values ([259]).

## 11.4 Epilogue: A powerful reasoning logic for the Ecological Detective

Beyond the limitations of the illustrative case studies discussed in Sections 11.2 and 11.3, these two examples illustrate how the Bayesian state-space modeling framework can make a major contribution regarding the following two important challenges for the Ecological Detective.

### 11.4.1 Handling the link between population dynamics models and field data

State-space modeling is a highly flexible framework for analyzing sequential data which can handle both process and observation errors within a single consistent probabilistic rationale. Multiple examples can be found in the statistical ecology literature and illustrate the capacity of Bayesian state space modeling (BSSM) for embedding population dynamics within a statistical inferences framework ([40];[41]; [121]; [164]; [219]; [259]; [293]). BSSM has been widely applied for age- or stage-structured fish population dynamics models ([103]; [176]; [196]; [206]; [279]; [290]), but also for modeling population dynamics in a context of conservation ([53]), harvest regulation ([298]; [318]) or animal invasions ([141]). BSSM also proved successful to analyze multiple mark-recapture data ([42]), animal movements ([149]; [210]; [227]) or even trajectory of fishing boats ([307]).

Thanks to the flexibility offered by Bayesian modeling methods coupled with MCMC inferential techniques, complex system dynamics including many latent variables can be combined with sophisticated descriptive statistical models for various sources of information. Hence, it becomes workable to analyze a wide range of state-space models with nonlinear relationships in the dynamic and observation equations, and non-Gaussian error structure as well. The Ecological Detective can treat their data with the model they actually want to use rather than with a model that has been imposed by some mathematical convenience. Because BSSMs free the Ecological Detective from many modeling restrictions, they encourage the investigation of several competing models. It is possible to improve, extend or even replace the dynamic model while keeping the data assimilation scheme unchanged. Supplementing the data does not cause any further complication as it requires only the specification of their relations to the state variables by means of stochastic observation equations. Incomplete series of data can be used as well.

The two examples in this chapter show a progression in the dimension and complexity of both the process and observation equations. The biomass production model example describes the dynamics of a highly aggregated state variable, namely the A. salmon population dynamics model is a stage-structured model of higher dimension including multiple life histories and complex stage-structured interactions. The specific observation model developed in the A. salmon example avoids the simplifying assumptions which are made in traditional filtering methods to ensure the observation errors and the process stochasticity can be disentangled. The form of the distribution of observation errors is often assumed known and constant across the years, a LogNormal structure being a classical choice such as in the Biomass production model example. The ratio of the process versus the observation error variances is often fixed arbitrarily (the ratio $\sigma_p/\sigma_o = 1$ is fixed in the Biomass production model example). These hypotheses are hardly justified; the measurement errors may vary over time, particularly if sample sizes or the data gathering methods change during the observation period. In the detailed observation model developed in the A. salmon example, observation errors are specified on a yearly basis from a realistic stochastic observation model, based on a random sampling process and CMR experiments.

### 11.4.2 Setting diagnostics and management decisions in a consistent probabilistic framework

In recent years, BSSM has been increasingly applied as a quantitative tool to synthesize information and quantify uncertainty in decision analysis ([94]). Applications in fisheries sciences have been numerous. This is most likely related to the prevalence of uncertainties stemming from stochasticity in the dynamics of the system and measurement and sampling errors in the data, and to the potentially disastrous biological and socio-economical consequences of overlooking major uncertainties ([133]; [184]).

The framework offers many advantages for dealing with uncertainty and is naturally connected to the assessment of population status and of management actions within a formal decision theoretic approach. Posterior pdfs provide readily interpretable probability statements of any quantities derived from the model. The Bayesian approach is well suited for simulation based studies that aim to predict possible results of alternative management actions in a probabilistic framework that accounts for all sources of uncertainty. MCMC samples from the joint posterior pdfs of parameters (such as in Figs. 11.3 and 11.14) can be used as direct inputs for Monte Carlo forward projections, thus allowing for an honest incorporation of our remaining partial ignorance about the parameters.

The BSSM framework is also well suited for the introduction of control variables within the dynamic model and for finding the optimal control conditionally on the data. Chapter 12 reconsiders the Atlantic salmon life cycle model in a formal decision analysis to answer questions such as: (*i*) How many returning spawners can be harvested without significantly impacting the population dynamics? or (*ii*) How can the relative performances of competing harvest strategies be assessed?

# Chapter 12

# Decision and planning


## Summary

In this closing chapter, we illustrate how models can be used in a deductive way as a tool for prediction, by using simulations designed to explore the response of the system under different scenarios. Statistical decision theory is used as a natural extension of the Bayesian paradigm that we relied on throughout the book.

As a case study, we reconsider the decision analysis from the case study of Rivot and Prévost [256]. We use the dynamics of the A. salmon population in the estuary of two coastal rivers forming the Sée-Sélune River network located in Mount Saint Michel Bay (France). Angling takes place all along both river banks during spawning migration between March and October. Population dynamics modeling allows to estimate key management reference points such as the sustainable level of exploitation, and to predict the dynamics of the resource under alternative scenarios for future exploitation. We develop a model to simulate the response of the system under 10 alternative management rules (limited fishing periods, quotas, etc.) and try to evaluate and compare their performances. The alternative management strategies are compared through performance criteria designed to balance objectives of preserving natural heritage in the long range and of optimizing the captures in the short term.


## 12.1 Introduction

The quantification of impacts of human activities on ecosystems and their consequences on related resources and services has generated increasing scientific and social interests ([72]; [314]). A better understanding of the mechanisms driving the response of wildlife populations to various natural and human-induced stresses remains both a fundamental ecological question and a requirement for sound management. This requires appropriate tools to unravel the underlying mechanisms and accurately predict the response of populations under plausible future scenarios with a fair appraisal of all sources of uncertainties ([66]; [94]; [133];[184]). Considerable attention has been paid to alterations of marine ecosystems. Human pressures like fishing mortality or the degradation of marine and coastal habitat are responsible for dramatic drops of fish populations ([58]; [79], [315]), alteration of biodiversity ([228]; [314]) or even regime shifts in marine ecosystems ([142]; [214]). Various examples of deeply adversely affected species resulting from inappropriate and ineffective regulation strategies raise many questions among scientists. Keeping a very modest scale with Salmon as our species of preference, this chapter exemplifies some mathematical contributions to ecosystem-based management of fish stocks.

In Chapter 11, dynamic models for fish stocks appear under the general formalism of a Markov chain:

$$Z_{t+1} = f(Z_t, \theta_1, \varepsilon_t) \tag{12.1}$$

with transition function $f(\cdot)$ driving the dynamics of the system. $\theta_1$ denotes the parameters of the dynamics with $\varepsilon_t$ the process noise (in order to simplify the mathematical content of this book, models and methods are restricted to discrete time case). Equation 12.1 is also often written in the equivalent compact form of the conditional pdf $[Z_{t+1}|Z_t, \theta_1]$, sometimes called the stochastic transition kernel.

As a rather sophisticated example, the components of the state vector $Z_t$ for the salmon life cycle model in Section 11.3 are essentially the number of individuals at year $t$ belonging to the different life stages:

- 0+ juveniles born in year $t$ or older resident juveniles;
- Smolts of one and two years migrating to the sea year $t$;
- Nonmature or mature returning adults at year $t$ with the four different life histories (one or two years spent in freshwater before the smolt migration combined with one or two winters spent at sea).

The *transition* function $f(\cdot)$ stands for the structural form of the transition between time steps with the strong stationary assumption that the underlying ecological mechanism of population dynamics does not change with time. The multidimensional parameter $\theta_1$ essentially contains the coefficients of the *eggs* $\rightarrow$ *0+ juveniles* Ricker stock-recruitment relationship, the variance of the logNormal environmental noise, several

**FIGURE 12.1**: Recreational fly fishing on a salmon river (photo credit to Gabriel Gerzabek).

survival rates as well as transition probabilities governing the life history choices modeled as Binomial trials. The process noise $\varepsilon_t$ arises from the random nature of the stock-recruitment relationship as well as the randomness of the life history traits.

Equivalently, Eq. (12.1) says that, were the parameter $\theta_1$ and present state $Z_t$ known, the analyst would be able to describe the future state $Z_{t+1}$ by making draws from the conditional pdf $[Z_{t+1}|Z_t, \theta_1]$. Starting from initial condition $Z_{t=0} = Z_0$, successive draws $Z_{0:T} = (Z_0, Z_1, ..., Z_T)$ obtained by iterating Eq. (12.1) can be understood as a *scenario* describing some possible trajectory of the uncertain future states hypothesizing perfect knowledge of $\theta_1$ and conditioned upon initial state $Z_0$.

Most often, the homogeneous Markov chain given by iterating Eq. (12.1) along successive periods will be only partially observed, with a noisy observation $Y_t|Z_t$ or, equivalently an observation equation at time $t$:

$$Y_t = g(Z_t, \theta_2, \omega_t) \tag{12.2}$$

In the salmon cycle example given in Section 11.3, the observation vector $Y_t$ consists mainly of captures of adult returns and smolt runs. The function $g(\cdot)$ corresponds to the experimental protocol for capture-mark-recapture (for migrating smolts or spawners) or successive removals (for juveniles) experiments with observation errors $\omega_t$ conveying randomness into its experimental results. Unknown parameters from Eq. (12.2) such as the trap or the electrofishing efficiencies must also be added under the banner of the vector $\theta_2$. In the joint formulation for the process and observation model given by Eqs. 12.1 and 12.2, the state variable $Z_t$ becomes *latent*, (*i.e.*, hidden since not observed directly) and when a prior $[\theta]$ is provided for the parameters $\theta = (\theta_1, \theta_2)$ and initial state $Z_0$, Bayesian inference allows to infer the hidden states and the parameters through the joint posterior distribution $[Z_{0:T}, \theta | Y_{0:T}]$.

Yet, the intervention of man, although often dramatically influential, seems to be absent from this abstract mathematical formulation. Eq. (12.1) must be complexified by introducing covariates $D_t$ that depict the influence of some forcing variables to be considered as deterministic, such as known harvests or known catches from fisheries:

$$Z_{t+1} = f(Z_t, D_t, \theta_1, \varepsilon_t) \tag{12.3}$$

The conditional pdf for the Markovian transition is now written $[Z_{t+1}|Z_t, \theta_1; D_t]$, and conditioned upon initial state $Z_0$, parameters $\theta_1$ and the series of decisions $D_{1:T-1} = D_1, ..., D_{T-1}$, the joint distribution of all hidden states is then written $[Z_{0:T}|\theta_1; D_{1:T-1}]$. Given the observations $Y_{0:T}$, the joint posterior distribution is written $[Z_{0:T}, \theta | Y_{0:T}; D_{1:T-1}]$.

In this chapter, we focus on $D_t$ in Eq. (12.3) to be considered as a set of *control variables* (*e.g.*, harvesting; Fig. 12.1). From a decision analysis perspective, the way the conditional pdf $[Z_{t+1}|\theta_1, Z_t, D_t]$ responds to the possible variations of the control variable $D_t$ is of special interest. However, considering a predetermined *open loop* control through the mapping $t \mapsto D_t$ is rarely efficient or even realistic. Rules of management are indeed *feedback loops* in reaction to the state of the system $Z_t$. Therefore, a more general framework is to consider a *policy* as a management rule $\delta$ that allows for a certain decision $D_t$ (*e.g.*, an upper limit for the number of fish that could be harvested) as a function of the system state $Z_t$ (*e.g.*, the abundance of fish at time $t$):

$$D_t = \delta(Z_t) \tag{12.4}$$

In practice, however, decisions can rarely be made depending upon the true state of the system: instead of $Z_t$, mostly unknown, only some *clues* - *i.e.*, $Y_t$ - can be observed from the system via the observation Eq. (12.2). In addition, there may be implementation error $\chi_t$ when trying to control a natural system (*e.g.*, some departure between fishing quotas and harvest that actually takes place, etc.):

$$D_t = \delta(Y_t, \chi_t) \tag{12.5}$$

Consequently, the general formulation of a controlled state-space system is rather given by assembling Eqs. 12.3, 12.2 and 12.5 to yield the conditional pdf $[Z_{t+1}|Z_t, \delta, \theta]$ under the form:

$$Z_{t+1} = f(Z_t, \theta_1, \varepsilon_t, \delta(g(Z_t, \theta_2, \omega_t, \chi_t))) \qquad (12.6)$$

Here, we reconsider the decision analysis from the case study of [256], and illustrate three issues:

1. Given a management rule $\delta$ that impacts the life cycle, what are its consequences on the returns?
2. What can be said about the mid- and long-term behavior of the system under strategy $\delta$?
3. Considering $K$ competing strategies $(\delta_1, \delta_2, ..., \delta_K)$, how to recommend the "best" one?

## 12.2 Case study: Managing salmon harvest in the Sée-Sélune River network

The estuary of two coastal rivers forming the Sée-Sélune River network is located in Mount Saint Michel Bay (Fig. 12.2), where both rivers flow into the eastern Channel. Their watersheds cover 1010 $km^2$ and 459 $km^2$ for the Sélune and Sée Rivers, respectively. These two small rivers are colonized by Atlantic salmon (the whole stream for Sélune but only the first 16 kilometers for Sée due to an impassable dam).

Angling takes place all along both river banks during spawning migration between March and October. Different management strategies (limited fishing periods, quotas, etc.) have been tested during the last 20 years, with frequent changes, making it difficult if not impossible to assess their consequences on the system in the past. Comparing the relative performance of alternative management actions is also useful for future management plans. We hereafter develop a model to simulate the response of the system under various management rules and try to evaluate and compare their performances.

As in many cases, the management of salmon exploitation casts light on conflicting objectives. Roughly speaking, from the point of view of resource conservation, leaving the fish alone is a matter of preserving natural heritage in the long term (several decades). From the point of view of the exploitation, it is often rather a question of optimizing the

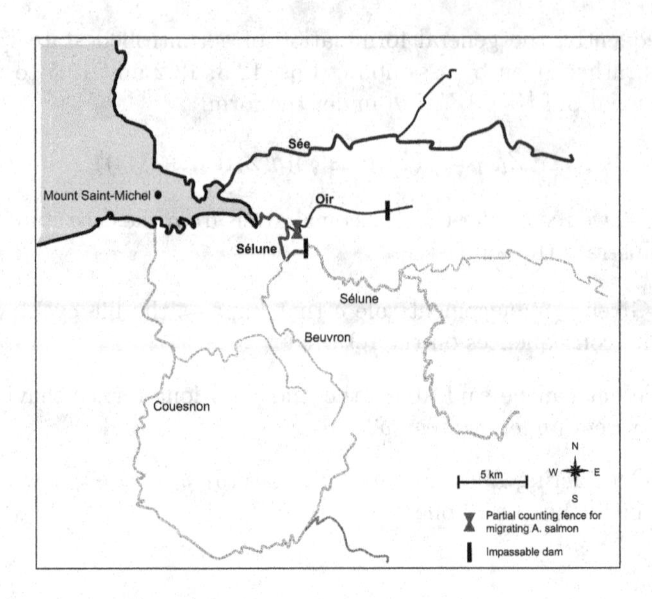


**FIGURE 12.2**: The Sée-Sélune River network. The river sections colonized by A. salmon appear in a bold line.

captures over a much shorter term (a few years). This conflict between conservation and exploitation is exacerbated by the rarefaction of the biggest individuals, the two sea winter salmon (2SW) that are the most sought-after fish but also the most precious spawners for the renewal of generations.

## 12.3 Salmon life cycle to model the dynamics of the resource

Figure 12.3 recalls the main stages of the A. salmon life cycle. The dynamics are quite similar to the ones depicted in Section 11.3, although most of the transitions were simplified to remain focused on the four main life histories and most of the parameters have been considered constant through time. Male and female are not distinguished and the date of each phase is to be read in years starting from year $t$ (*i.e.*, the egg stage). $D_{1,t}$

and $D_{2,t}$ are the control variables standing for the harvest of one and two sea-winter fish, respectively (detailed hereafter).

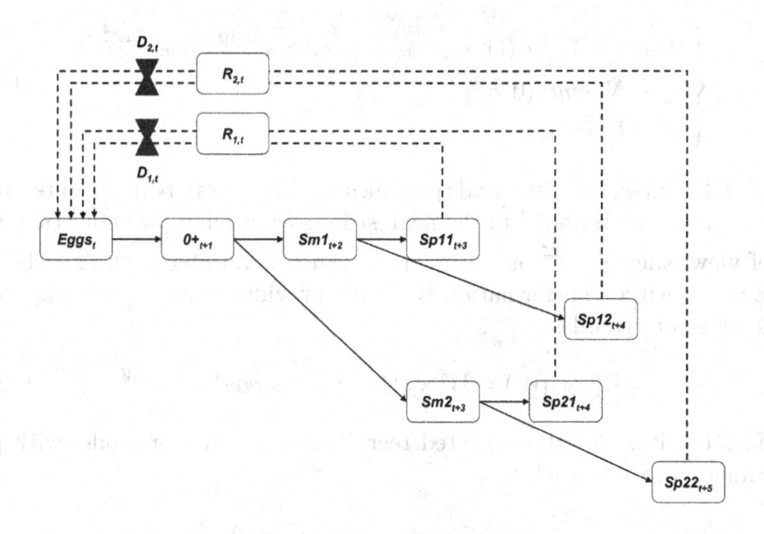


**FIGURE 12.3**: Atlantic Salmon cycle as a controlled Markov chain. $D_{1,t}$ and $D_{2,t}$ are the control variables standing for the harvest of one and two sea-winter fish, respectively.

### 12.3.1 Production of 0+ juveniles

The transition between eggs $W_t$ and juveniles $0+_{t+1}$ , *i.e.*, the recruitment process, is modeled as a mixture between two different regimes, a high recruitment success regime and a low recruitment success regime. These two regimes are captured by two Ricker models (see Eq. (7.4), page 151) with slope at the origin $\alpha_1$ (low) and $\alpha_2$ (high; $\alpha_2 > \alpha_1$) respectively, but with the same parameter ruling the density dependence $\beta_1 = \beta_2 = \beta$. We denote $p$ the probability of being in the lower recruitment regime (such as a flood washing out most of the eggs, or very low flows in the tributaries that prevent the adults from reaching the upstream zones of spawning). As in Chapter 7, an environmental noise blurs the expected value provided by the Ricker model. Recourse is made to a logNormal variable identically and independently distributed over the years.

To sum it up, the number of juveniles $0+_{t+1}$ appears under the form of the mixture of two Ricker models with $\tau_t$ defining a Bernoulli indicator variable taking the value 0 (low recruitment regime) with probability $p$ and 1 with probability $1-p$ (high recruitment regime):

$$\begin{cases} 0+_{t+1} = W_t \times ((1-\tau_t)\alpha_1 + \tau_t\alpha_2) \times e^{-\beta W_t} \times e^{\varepsilon_t - \frac{\sigma^2}{2}} \\ \varepsilon_t \sim Normal(0, \sigma^2) \\ \tau_t \sim Bernoulli(p) \end{cases} \tag{12.7}$$

Of course, as eggs and juveniles are integers, real outputs from Eq. (12.7) are rounded to the nearest integer. From a probabilistic point of view, since $e^{\varepsilon_t - \frac{\sigma^2}{2}}$ is a logNormal random variable with expectation $\mathbb{E} = 1$, such a transformation $W \to 0+$ provides in average an expected number of juveniles:

$$\mathbb{E}(0+|W) = W \times ((1-p)\alpha_1 + p\alpha_2) \times e^{-\beta W} \tag{12.8}$$

This equation for the expected recruitment is a Ricker model with parameters $(\alpha', \beta')$ such that

$$\begin{cases} \alpha' = ((1-p)\alpha_1 + p\alpha_2) \\ \beta' = \beta \end{cases} \tag{12.9}$$

### 12.3.2 Smolts

After spawning at the end of year $t$, we assume all the adults die. Juveniles $0+_{t+1}$ emerge in the spring of year $t+1$ from the spawning sites where $W_t$ eggs were deposited. The description of intermediate stages (*e.g.*, pre-smolts) is skipped here for the sake of simplicity, and survival from 0+ to smolts is assumed to be 100%. Some juveniles will migrate as 1+ Smolts in the spring of year $t+2$. The remaining juveniles will spend one additional year in the river and migrate downstream as 2+ Smolts in year $t+3$. Transition from 0+ Juveniles to Smolts is modeled via a Binomial probability distribution with parameter $\gamma_{Sm1}$ , to be interpreted as the probability that a 0+ Juvenile survive and takes part in first smolt run. Assuming a 100% survival rate during the additional year in the river, the number of 2+ Smolts is obtained via a simple balance equation:

$$\begin{cases} Sm1_{t+2} \sim Binomial(\gamma_{Sm1}, 0+_{t+1}) \\ Sm2_{t+3} = 0+_{t+1} - Sm1_{t+2} \end{cases} \tag{12.10}$$

### 12.3.3 Marine phase

As described in Section 11.3, salmon spend one or two winters at sea before getting back to their native river to spawn. Survival at sea and life history choices are simply modeled as Binomial distributions with parameters $(\gamma_{11}, \gamma_{12}, \gamma_{21}, \gamma_{22})$. These parameters are interpreted as return rates for each life history that incorporate both survival at sea and life history choices:

$$\begin{cases} Sp11_{t+3} \sim Binomial(\gamma_{11}, Sm1_{t+2}) \\ Sp12_{t+4} \sim Binomial(\gamma_{12}, Sm1_{t+2}) \\ Sp21_{t+4} \sim Binomial(\gamma_{21}, Sm2_{t+3}) \\ Sp22_{t+5} \sim Binomial(\gamma_{22}, Sm2_{t+3}) \end{cases} \quad (12.11)$$

### 12.3.4 A bivariate decision for angling

In Eq. (12.11) above, $Sp11$, $Sp12$, $Sp21$, and $Sp22$ denote the number of adults homing back to their native river before they are exploited by a professional fleet in the estuary or by anglers along the river. Here we consider harvesting through catches of one or two sea-winter fishes. Catches are to be understood as a bivariate vector $D_t = \begin{pmatrix} D_{1,t} \\ D_{2,t} \end{pmatrix}$. No matter the smolt age class from which they arose, fish that have spent only one winter at sea (1SW fish or *grilse*), denoted $R_1$ (see Fig. 12.3) return mainly from June to August of year $t$:

$$R_{1,t} = Sp11_t + Sp21_t$$

Out of these one sea-winter returning adults $R_{1,t}$, anglers will take $D_{1,t}$ and we denote by $h_1$ the exploitation rate for 1SW fish:

$$h_{1,t} = \frac{D_{1,t}}{R_{1,t}}$$

Bigger fish (2SW fish or *spring salmon*) that have spent two winters at sea, denoted by $R_{2,t}$ (see Fig. 12.3) return earlier in spring of year $t$:

$$R_{2,t} = Sp12_t + Sp22_t$$

We denote $D_{2,t}$ the removals from the home water recreational fishery and $h_{2,t}$ the corresponding exploitation rate:

$$h_{2,t} = \frac{D_{2,t}}{R_{2,t}}$$

As detailed in Section 11.3, as the reproduction occurs after the fishery, the number of eggs $W_t$ spawned by the females of both age classes in year $t$ is calculated as:

$$\begin{aligned} W(t) =& (R_{1,t} - D_{1,t}) \times pf_1 \times fec_1 \\ &+ (R_{2,t} - D_{2,t}) \times pf_2 \times fec_2 \end{aligned} \tag{12.12}$$

with $pf_i$ and $fec_i$ denoting the proportion of females and their fecundity in returns of sea-age $i$ ($i = 1$ for grilse and $i = 2$ for spring salmon). Overall, with values of $pf_1$, $fec_1$, $pf_2$ and $fec_2$ given in Table 12.1, the average fecundity of a one spring salmon is worth a little more than three times the fecundity of a grilse.

### 12.3.5 Assessing parameter values

The parameter values considered in this application are given in Table 12.1. Parameters have been estimated from the retrospective analysis of the data available for this river network ([256]). The full inference of the state-space model follows the ideas developed in Section 11.3. It is obtained by adding an observation layer 12.2 to link the dynamic submodel 12.3 to the available data in this system (20 years of observations). This provides the posterior pdf of parameters, but here we only consider point estimates (posterior means), *i.e.*, simulations for the controlled Markov chain given in Eq. (12.6) will be run as if perfect knowledge (no uncertainty) were attainable for these parameters. However, it is worth noting that most of the uncertainty in the dynamics stems from the stochasticity of the recruitment process in Eq. (12.7) (random mixture of two recruitment regimes and LogNormal distribution with variance $\sigma^2$). The 90% confidence interval for a logNormal distribution (with standard deviation $\sigma = 58\%$ as given in Table 12.1) spans from 10% to 300% of its mean! This gives the order of magnitude of the natural variability of the natural cycle (given a hypothesized value of $\theta$). Furthermore, Eq. (12.8) indicates that its mean is ruled by $p$, the probability of a catastrophic event. $p$ is itself poorly known: with only twenty years of data, the posterior distribution for $p$ should remain fairly diffuse. That is why $p$ happens to be the component of parameter $\theta$ that is subject to the highest uncertainty, and we only focus our attention on the uncertainty attached to this component. In what follows, we perform a sensitivity analysis on $p$, the most uncertain parameter of the system 12.6 considered in this application. Up to now, we pick $p = 0.155$ as a representative value in Table 12.1. We will further assess the sensitivity of the results to alternative values of $p$ in $\{0.01, 0.09, 0.13, 0.155, 0.19, 0.24\}$.

| Parameters | Value | Unit |
|---|---|---|
| Adults → Eggs | | |
| $pf_1$ | 0.34 | Female ratio in 1SW |
| $pf_2$ | 0.68 | Female ratio in 2SW |
| $fec_1$ | 4635 | Eggs per female 1SW |
| $fec_2$ | 7965 | Eggs per female 2SW |
| Eggs → 0+ | | |
| $\alpha_1$ | 0.0043 | Ricker coeff. (good year) |
| $\alpha_2$ | 0.0012 | Ricker coeff. (bad year) |
| $\beta$ | $7.75 \times 10^{-08}$ | Ricker coeff. |
| $p$ | 0.155 | Prob. of a bad year |
| $\sigma$ | 0.58 | LogNormal std. deviation |
| 0+ → Smolts | | |
| $\gamma_{Sm1}$ | 0.89 | 0+ → 1+ Smolts |
| Smolts → Adults | | |
| $\gamma_{11}$ | 0.13 | 1+ Smolts → 1SW |
| $\gamma_{12}$ | 0.03 | 1+ Smolts → 2SW |
| $\gamma_{21}$ | 0.22 | 2+ Smolts → 1SW |
| $\gamma_{22}$ | 0.04 | 2+ Smolts → 2SW |

**TABLE 12.1**: Point estimates for parameters used for the Salmon cycle in the Sée-Sélune system.

## 12.4 Long-term behavior: Collapse or equilibrium?

For a given operating rule $\delta$ (*e.g.*, fixed exploitation rates $h_1$ and $h_2$ over years) and a known parameter vector $\theta$, Eq. (12.6) defines the transition for the *homogeneous* Markov chain $[Z_{t+1}|Z_t, \delta, \theta]$.

We know from statistical theory that the long-term behavior of most dynamic stochastic systems is to quickly forget their initial conditions and to visit states $Z_t$ according to some limiting distribution known as the *ergodic* distribution ([253],[297]). If there were such a stable distribution $l(.)$ in the present case study, then starting from $Z_t$, one (marginal) draw at random from $l(\cdot)$, $Z_{t+1}$, obtained via the (conditional) random transition $[Z_{t+1}|Z_t, \delta, \theta]$ could also be considered as a (marginal) draw at random from $l(\cdot)$. Conversely to most deterministic cases, the steady state will not be represented here by an equilibrium at a single point,

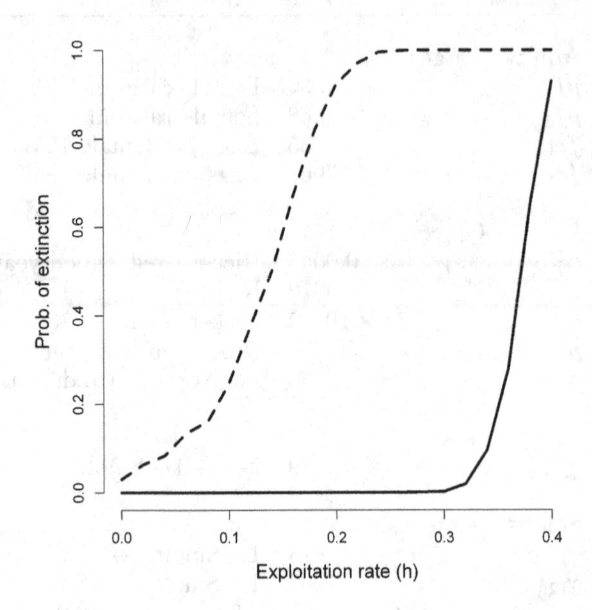


**FIGURE 12.4**: Probability of extinction on a 1000 year period for the Salmon Markovian cycle as a function of the exploitation rate. Solid line: With overlapping generations; Dotted line: Without overlapping generations (only 1+ smolts giving returns of 1SW).

but rather by this ergodic probability distribution $l(\cdot)$, to be interpreted as the long-run frequency of visits of the possible state values of the population.

Unfortunately in the present case study defined by Eqs. 12.7-12.12, the states are countable (the stocks in each age class are integer quantities) and there exists a trapping state which precisely corresponds to no individual left in any of the possible stages of the system. Indeed, even with $h = 0$, the extinction of the population can be reached from any state with a very small (but theoretically not null) probability: Imagine that all Binomial stochastic links depicted in Fig. 12.3 provide zero as draws! The mathematical fate of such system is therefore known to collapse to that absorbing null state, and it can be mathematically proven that this extinction asymptotic behavior is the only ergodic distribution $l(\cdot)$.

From a given starting state, assuming a given rule $\delta$ of stock exploitation, one can estimate the probability of collapse over some period of the future by simulation, and this probability can obviously be high for some strategies not leaving enough fish in the system to guarantee its natural renewal. Figure 12.4 estimates this probability per millennium as the ratio of collapsing trajectories under contrasted exploitation rates (assuming $h = h_1 = h_2$). A thousand initial stocks have been launched and followed over 1000 successive years for exploitation rates $h$ varying between 0 and 40%. But in the present case study, the influence of initial conditions is rather low, and Fig. 12.4 has been drawn with initial conditions equal to the abundance at the reference points given in the next section (see Table 12.2).

The very strong robustness observed for low and medium exploitation rates, stems from the various life strategies of the Atlantic Salmon: every year, three generations of animals (born 3, 4 and 5 years before) contribute to the egg deposition, dampening spawner variability, since catastrophic events occurring in the early stages of the cycle would then reduce only one component of the four life histories that form the yearly spawner runs. For the sake of comparison, we can assess the same collapse probability by running simulations with a life cycle having only a single life strategy (1+ Smolts giving one sea-winter fish) but keeping all other parameters as in Table 12.1. As shown in Fig. 12.4, only by assuming reproduction without any mixing of generations, the extinction probability increases by several orders of magnitude. This first simulation exercise points out that keeping the variability in the life histories portfolio is critical for long term conservation of the resource.

But rather than following this avenue of thought that questions the concept of a sustainable population (see Reed [250] for instance), in what follows, we will simply undertake stochastic simulations of the system given by Eq. (12.6) to study the impact of various policies for the near future, for instance the next century.

## 12.5 Management reference points

Parameter values from Table 12.1 are meaningful quantities for the population dynamics, but do not speak much to the decisionmaker that could be more interested in management reference points such as the optimal spawning escapement or the optimal harvest rate as defined in Section 7.3.2 of Chapter 7.

The approach described below is similar to the one detailed in Section 7.3.2 except that the Ricker stock-recruitment model is included within a stage-structured life cycle model. In order to suggest management reference points for the model in Fig. 12.3, let's imagine a simplified deterministic version of the stochastic system given by Eqs. 12.7-12.12. When ignoring stochasticity and considering the expectation of all transitions, the equilibrium behavior of the system is driven by the shape of the expectation of the mixture Ricker model in Eq. (12.8) with parameters $\alpha'$ and $\beta'$ defined in Eq. (12.9). Then, on average, $W$ eggs spawned will produce $\Re(W, \alpha', \beta')$ 0+ juveniles, with

$$\Re(W, \alpha', \beta') = W \times \alpha' \times e^{-\beta' W} \tag{12.13}$$

Now how many eggs will these 0+ juveniles spawn once they have completed their life cycle? A number $N$ of juveniles will produce on average $N \times \Psi$ eggs in the future, where $\Psi$ is a coefficient representing the average number of eggs that will be spawned when averaging over all possible life histories from juveniles to spawners (see [50] for more detailed considerations on life cycle). For a harvested population with exploitation rates $h_1$ and $h_2$ of 1SW and 2SW fish, respectively, $\Psi$ depends on $h_1, h_2$. Given the life cycle in Fig. 12.3 and parameters in Table 12.1:

$$\begin{cases} \Psi(h_1, h_2) = (1 - h_1) \times \Psi_1 + (1 - h_2) \times \Psi_2 \\ \Psi_1 = (\gamma_{Sm1} \times \gamma_{11} + (1 - \gamma_{Sm1}) \times \gamma_{21}) \times pf_1 \times fec_1 \\ \Psi_2 = (\gamma_{Sm1} \times \gamma_{12} + (1 - \gamma_{Sm1}) \times \gamma_{22}) \times pf_2 \times fec_2 \end{cases}$$

Because our model does not consider any inheritance of life histories, *an egg is worth an egg* whatever the life history of the spawners they are coming from. Thus, the significant harvest rate in terms of population dynamics is the weighted average of $h_1$ and $h_2$:

$$h = \frac{h_1 \times \Psi_1 + h_2 \times \Psi_2}{\Psi_1 + \Psi_2}$$

and

$$\Psi(h_1, h_2) = (1 - h) \times (\Psi_1 + \Psi_2) \tag{12.14}$$

Now, given the *eggs* $\rightarrow$ 0+ *juveniles* Ricker relationship in Eq. (12.13) and the average 0+ $\rightarrow$ *eggs* coefficient in Eq. (12.14), the average number of eggs that will be produced when $W$ eggs are spawned is

$$\Re(W, \alpha', \beta') \times (1 - h) \times (\Psi_1 + \Psi_2).$$

This defines an *eggs* $\rightarrow$ *eggs* Ricker average relationship with parameters $\alpha_{eggs}$ and $\beta_{eggs}$ that now depend upon the exploitation rate $h$:

$$\begin{cases} \alpha_{eggs}(h) = \alpha' \times (1 - h) \times (\Psi_1 + \Psi_2) \\ \beta_{eggs} = \beta' \end{cases} \tag{12.15}$$

The equilibrium condition of such an *eggs* → *eggs* relationship corresponds to a quantity of eggs such that:

$$W(h) = \alpha_{eggs}(h) \times W(h) \times e^{-\beta_{eggs} W(h)} \tag{12.16}$$

Reasoning as in Section 7.3.2, special reference points $(W^*, h^*, R^*)$ can be obtained (see Table 12.2). For instance, the optimum exploitation rate $h^*$ yielding the maximum sustainable yield (see Eq. (7.7), page 155), is the solution of:

$$h^* - log(1 - h^*) = log(\alpha_{eggs}(h^*))$$

The corresponding egg stock at maximum sustainable yield is such that $W^* = W(h^*)$ calculated from Eq. (12.16). The corresponding expected returns for adults are respectively $R_1^*$ for grilse and $R_2^*$ for spring salmon:

$$\begin{cases} R_1^* = \dfrac{W^*}{\Psi_1 + \Psi_2} \times (\gamma_{Sm1} \times \gamma_{11} + (1 - \gamma_{Sm1}) \times \gamma_{21}) \\ R_2^* = \dfrac{W^*}{\Psi_1 + \Psi_2} \times (\gamma_{Sm1} \times \gamma_{12} + (1 - \gamma_{Sm1}) \times \gamma_{22}) \end{cases}$$

| Reference Point | Value |
|---|---|
| $\Psi_1 + \Psi_2$ | 389 eggs per 0+ juvenile |
| $\alpha'$ | 0.00456 |
| $\beta'$ | $7.78 \cdot 10^{-08}$ |
| $W*$ | $3.4 \cdot 10^6$ eggs |
| $h*$ | 0.265 |
| $N*$ | 8 759 0+ juveniles |
| $R_1^*$ | 1 225 1SW spawners |
| $R_2^*$ | 272 2SW spawners |
| $R^* = R_1^* + R_2^*$ | 1 497 spawners |
| $S^*$ | 1 100 spawners |

**TABLE 12.2**: Reference points for the Salmon cycle in the Sée-Sélune system.

As shown in Appendix C, these reference points are meaningful quantities in a scale understandable by the system manager, but do not correspond to remarkable features of the stochastic behavior of the system dynamics, such as the mean or the median of some equilibrium distribution.

## 12.6 Management rules and implementation error

In Chapter 5 of Prévost *et al.* [238], Hilborn and Walters suggested a general linear framework for formulating harvest strategies:

$$D_t = \alpha + \beta \times R_t$$

Of course, the catch $D_t$ must be positive and cannot be greater than the total returns $R_t$. In what follows, we focus on three classical types of management rules:

1. Fixed harvest rate $h^*$: $\beta = h^*$; $\alpha = 0$,
2. Constant catch $C^*$: $\beta = 0$, $\alpha = C^*$,
3. Constant escapement $S^*$ : $\beta = 1$, $\alpha = -S^*$.

We know from optimal control theory (see simple models in Appendix C) that constant escapement policies allow for the maximum cumulated returns under a wide family of cost functions. In real life, however, overlapping generations and impossibility to reliably or fully implement those control rules in practice (implementation uncertainty) may invalidate these theoretical results. Indeed, in practice, management authorities have to set their management rules under unknown returns of adults of many age classes and noninstantaneous records of catches. Also, one of the main tools the manager has in hand for this recreative fishery is the opening and closing dates of the fishery. In the case of migratory fishes like salmon, opening and closing dates have to be considered in interaction with the migration phenology of 1SW and 2SW spawners.

Let's first assume that the weekly rates of salmon returns $(r_{1,w}, r_{2,w})$ are invariant between years (using indices $w$ for week, $i = 1$ for grilse and $i = 2$ for large salmon). The cumulative return rates shown in Fig. 12.5 have been obtained from a long series of observations collected on all the nearby sites of the region. A multinomial trial can mimic the weekly return variability between years. Each cohort of returning adults $(R_{1,t}, R_{2,t})$ will be split within the weeks as:

$$\begin{aligned}(R_{i,t,1}, ..., R_{i,t,w}, ..., R_{i,t,52}) \\ \sim Multinomial(R_{i,t}, (r_{i,1}, ..., r_{i,w}, ..., r_{i,52}))\end{aligned} \tag{12.17}$$

In the remainder of this section, we take for granted that, under the present fishing pressure, the probability that a one-sea winter (respectively two-sea winter) returning spawner caught is $p_1 = 0.05$ (respectively $p_2 = 0.051$) every week of the fishing period. As a consequence, fixing the opening and closing dates of the fishery controls the harvest rate on the returning stock. We rely on this hypothesis to illustrate how Monte Carlo simulations can help evaluate 10 alternative management policies (in Table 12.3) by accounting for some specificities of the case study.

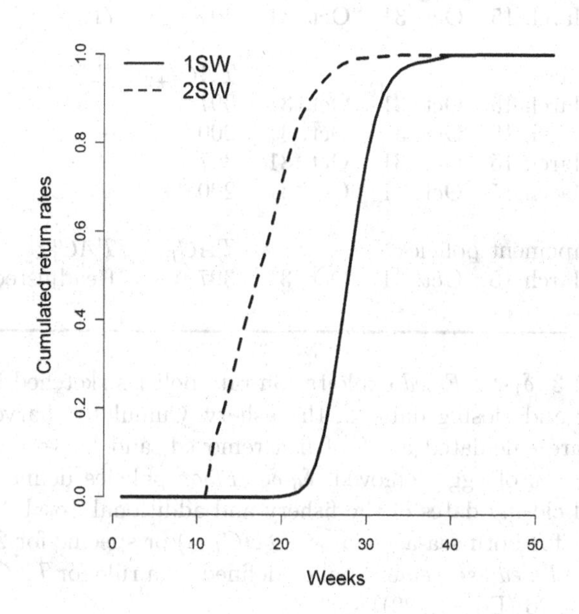


**FIGURE 12.5**: Cumulative return rates for one- and two-sea winter salmon.

- *Fixed exploitation rate policies*

  Fixed exploitation rate policies (but with implementation noise) can be controlled by the opening and closing dates of the fishery, $w_{i,open}$ and $w_{i,close}$, that may differ for grilse $i = 1$ and large salmon $i = 2$. For any week $w$ in between the opening and closing dates, captures for grilse $C_{1,t,w}$ and for large salmon $C_{2,t,w}$ can be modeled as Binomial trials in the escapements ($E_{1,t,w-1}$ and $E_{2,t,w-1}$) from the previous weeks augmented by the incoming adults at current week $w$ :

$$\begin{cases} C_{i,t,w} \sim Binomial(E_{i,t,w-1} + R_{i,t,w}, p_i) \\ E_{i,t,w} = E_{i,t,w-1} - C_{i,t,w} + R_{i,t,w} \end{cases} \tag{12.18}$$

| Policy | $w_{open}$ | $w_{1,close}$ | $w_{2,close}$ | | | |
|---|---|---|---|---|---|---|
| Constant harvest rate policies | | | | $\mathbb{E}(h_1)$ | $\mathbb{E}(h_2)$ | $\mathbb{E}(h)$ |
| $\delta_1$ | - | - | - | 0% | 0% | 0% |
| $\delta_2$ | March 15 | June 15 | June 15 | 1% | 28% | 22% |
| $\delta_3$ | March 15 | July 31 | July 31 | 12% | 46% | 38% |
| $\delta_4$ | March 15 | Aug. 24 | Aug. 24 | 26% | 55% | 49% |
| $\delta_5$ | March 15 | Oct. 31 | Oct. 31 | 50% | 71% | 66% |
| Floor policies | | | | $TAC_{1+2}$ | $TAC_2$ | |
| $\delta_6$ | March 15 | Oct. 31 | Oct. 31 | 397 | – | |
| $\delta_7$ | March 15 | Oct. 31 | Oct. 31 | 200 | – | |
| $\delta_8$ | March 15 | Oct. 31 | Oct. 31 | 397 | 72 | |
| $\delta_9$ | March 15 | Oct. 31 | Oct. 31 | 200 | 40 | |
| Fixed escapement policies | | | | $TAC_{1+2}$ | $TAC_{1+2}^{new}$ | |
| $\delta_{10}$ | March 15 | Oct. 31 | Oct. 31 | 397 | Readjusted | |

**TABLE 12.3**: $\delta_1$-$\delta_5$: *Fixed* exploitation rate policies sketched by fixing the opening and closing dates of the fishery. Cumulated harvest rates $h_1$ and $h_2$ are calculated as % of fish removed, and harvest rate $h$ is calculated as % of eggs removed. $\delta_6$-$\delta_9$: *Floor* policies defined by the opening and closing dates of the fishery and additional Total Allowable Catch $TAC$, for both sea-age classes ($TAC_{1+2}$) or specific for 2SW fish ($TAC_2$). $\delta_{10}$: *Fixed escapement* policy defined by a rule for $TAC$ adjustment at week 26 (Eq. (12.20)).

In the past, management authorities often applied the same policy to the two types of spawners, setting $w_{1,open} = w_{2,open}$ and $w_{1,close} = w_{2,close}$. Table 12.3 gives the opening and closing dates of the fishing season under various possible strategies that have already been tested in the past. They range by increasing exploitation rates (computed by % of eggs removed) between rule $\delta_1$ corresponding to the absence of exploitation (indeed we allow only one week of harvest to mimic poaching) and rule $\delta_5$ corresponding to the largest fishing period, between week 12 (March 15) and week 43 (October 31). Making here the difference between grilse ($i = 1$) and larger spring salmon ($i = 2$), one can evaluate the effective exploitation rates at the end of the fishing season as:

$$h_{i,t} = \frac{\sum_{w=w_{i,open}}^{w_{i,close}} C_{i,t,w}}{R_{i,t}}$$

Although the corresponding rules are called *fixed* exploitation rate policies, implementation randomness stemming from randomness in return rhythms (Eq. (12.17)) and in catch rates (Eq. (12.18)) renders catches and harvest rates $h_{i,t}$ random.

- *Floor policies*

  Floor policies correspond to variable harvest rates as their catches are limited by total allowable catches ($TAC$). In addition to the opening and closing dates $w_{i,open}$ and $w_{i,close}$ of the fishery, one introduces a total allowable catch $TAC_{1+2}$ for spawners. Once this amount is reached, the corresponding fishing season is closed. Implementation of floor policies assumes that anglers declare their catches without cheating to the management authority, as required by fishing regulations concerning wild Salmon on the Sée-Sélune rivers.

  In what follows, we consider policy $\delta_6$ as trying to comply to the Ricker sustainable stock by controlling catches at the level of the maximum sustainable yield

$$\begin{aligned} TAC_{1+2} &= (R_1^* + R_2^*) \times h^* \\ &= 397 \text{ salmon} \end{aligned}$$

  and policy $\delta_7$ that considers a less optimistic total $TAC_{1+2} = 200$ salmon.

  In the past, fishing authorities often apply variant policies by considering an additional specific $TAC_2$ for the most sought after spring salmon. $\delta_8$ and $\delta_9$ are similar to $\delta_6$ and $\delta_7$, respectively, except that an additional $TAC_2$ for spring salmon has been set at 72 ($= R_2^* \times h^*$) and 40 fish, respectively.

- *Fixed escapement policy*

  Fixed escapement policies imply that the catches are adapted each year to ensure that the spawning escapement at the end of the fishing season matches as close as possible the spawning target $S^*$. In theory (*e.g.*, without any implementation bias and uncertainty), they are defined by the following rule:

$$\begin{cases} R_{1,t} + R_{2,t} < S^* \Rightarrow C_{1,t} = C_{2,t} = 0 \\ R_{1,t} + R_{2,t} > S^* \Rightarrow C_{1,t} + C_{2,t} = (R_{1,t} + R_{2,t}) - S^* \end{cases} \tag{12.19}$$

However, given that the fishery can only be managed through fixing the opening and closing dates and that the total returns $R_{1,t}+R_{2,t}$ remain largely unknown at the time of the decision (even if the catches have been declared), fixed escapement strategies remain largely inapplicable as such in the present application.

The rules that were enforced after 2003 on the Sée-Sélune system consist of adapting the catches to the interannual variations of returns by gaining information about the total return during the first part of the fishing season. Early in the season, a tentative $TAC^*_{1+2}$ is set (mostly equal to the maximum sustainable yield $(R^*_1 + R^*_2) \times h^*$). Then the cumulated catches at week 26 $C_{1,t,1:26} + C_{2,t,1:26}$ are compared to $TAC^*_{1+2}$. If they are greater, the fishing season is immediately closed. Otherwise, a readjustment of the $TAC$ is eventually made and the closing date will be decided later when the readjusted $TAC$ is reached. The readjustment rule is based on the comparison between the estimated spawning escapement at week 26 ($\hat{E}_{26}$) and the optimal spawning escapement $S^* = (R^*_1+R^*_2)\times(1-h^*)$. Given the cumulated return rhythm in Fig. 12.5 and the weekly exploitation rate ($p_1 = 0.05$ and $p_2 = 0.051$), it is easy to show that cumulated catches at week 26 should represent on average 20% of the cumulated returns at the same date. One can therefore estimate the total escapement at week 26 from the catches:

$$\hat{E}_{26} = \frac{1-0.20}{0.20} \cdot (C_{1,t,1:26} + C_{2,t,1:26})$$

Management rule $\delta_{10}$ is based on the principle that the readjusted $TAC$, denoted $TAC^{new}_{1+2}$ should be reduced if $\hat{E}_{26}$ is low and could eventually be increased if $\hat{E}_{26}$ is high with regards to the reference $S^*$:

$$\begin{cases} \hat{E}_{26} < 0.5 \times S^* \Rightarrow TAC^{new}_{1+2} = 0.66 \times TAC^*_{1+2} \\ 0.5 \times S^* < \hat{E}_{26} < S^* \Rightarrow TAC^{new}_{1+2} = TAC^*_{1+2} \\ S^* < \hat{E}_{26} < 1.5 \times S^* \Rightarrow TAC^{new}_{1+2} = 1.5 \times TAC^*_{1+2} \\ 1.5 \times S^* < \hat{E}_{26} \Rightarrow TAC^{new}_{1+2} = 2 \times TAC^*_{1+2} \end{cases} \tag{12.20}$$

## 12.7 Economic model

### 12.7.1 Cumulative discounted revenue

In this chapter, it is convenient to adopt the fiction of a sole owner, who may be imagined as the management authority, searching for a satisfying management rule. This implies that the single stakeholder adopts the same objective as a public social manager attempting to maximize social welfare, the stock being an open access fishery resource. Few present day naturally renewable resources satisfy these idealized conditions completely. Additionally, we assume the objective $J(\delta, \theta)$ can be quantified as the expectation of the sum (defined in a long period of time $t = 1, ..., T$) of discounted revenues $L_t(D_t, Z_t)$ depending both on the unknown natural resource $Z_t$ and the harvest $C_t$ governed by the decision $D_t$ under the management rule $\delta$ from Eq. (12.5):

$$J(\delta, \theta) = \mathbb{E}_Z \left[ \left( \sum_{t=1}^{T} e^{-\lambda \cdot (t-1)} \times L_t(D_t, Z_t) \right) | \theta \right] \tag{12.21}$$

Note that Eq. (12.21) is written using conditioning by $\theta$ explicit to keep in mind that parameters $\theta$ are to be known to simulate the dynamics of the resource over the period $t = 1, ..., T$. The parameter $\lambda \in \theta$ in the objective function $J$ is the *discount factor* detailed below. In what follows, values of the objective function were calculated by summing over 100 years ($T = 100$).

### 12.7.2 Revenue per time period

The discount factor $\lambda$ allows for intertemporal trade-offs. We will also use it here to weigh the conflicting objectives between maximizing harvest in the short term and ensuring conservation of the natural resource in the long term. A large value of $\lambda$ in Eq. (12.21) will favor short-term income whereas a low value of $\lambda$ will give more weight to long-term income and thus promotes more conservative harvest rules that preserve escapement to ensure the renewal of the resource. For numerical applications, we take $\lambda = 0.03$, which is commonly considered in civil engineering, yielding an equivalent discounted return period of $\sum_{t=1}^{\infty} e^{-\lambda(t-1)} \approx 34$ years. The sensitivity of the results can also be assessed by considering lower ($\lambda = 0.01$) and higher ($\lambda = 0.07$) discount rates.

One can find in the economic literature many possible expressions of the benefit from the catch. For the case study, we will simply assume an expression under the form

$$\begin{aligned} L_t(D_t, Z_t) = & q_1 \times (h_{1,t} \times R_{1,t}) + q_2 \times (h_{2,t} \times R_{2,t}) \\ & - q_3 \times (h_{1,t} + h_{2,t}) \end{aligned} \tag{12.22}$$

as if $q_1$ and $q_2$ were the *prices* (independent from the resource) obtained when selling grilse and large Spring salmon on the market and $q_3$ was a constant marginal investment cost for a unit fishing effort. Of course, such assumptions of a hypothetical market and constant prices are highly specialized but allow for a simple expression. Much more sophisticated expressions making recourse to *utility theory* can be found in [7], [94], [133], [165], [180], [289]. For the numerical applications of the next section, to take into account that Spring salmon are the most sought after, both for conservation and sport, we try $q_3 = 0$, $q_1 = 1$, $q_2 = 10$ (Spring salmon catches highly valued) and $q_3 = 0, q_1 = 1, q_2 = 2$ (Spring salmon catches only moderately rewarded).

### 12.7.3 Uncertainties

The expectation in Eq. (12.21) is to be taken on all future revenues that are random quantities due to the stochastic nature of the dynamic system (Eq. (12.6)). For the application in this chapter, the main source of stochasticity in the salmon cycle stems from the stock-recruitment submodel given by Eq. (12.7). We used the standard deviation $\sigma = 0.58$ for the logNormal randomness in the recruitment. Up to now, we have worked within a perfect information situation, *i.e.*, the estimations of Table 12.1 are adopted as true values for the components of the parameter $\theta$. But after a Bayesian analysis of the data collection, we assume that judgmental uncertainty about $\theta$ is encoded by a pdf, noted as usual $[\theta]$. Therefore the optimal policy should result from the integration of Eq. (12.21) over the pdf $[\theta]$ :

$$\begin{aligned} \bar{J}(\delta) &= \mathbb{E}_\theta \left[ J(\delta, \theta) \right] = \int_\theta J(\delta, \theta, \lambda) \times [\theta] \times d\theta \\ &= \mathbb{E}_\theta \left[ \mathbb{E}_{Z|\theta} \left( \sum_{t=1}^{T} e^{-\lambda \cdot (t-1)} \times L_t(D_t, Z_t) \right) \right] \end{aligned} \tag{12.23}$$

In this application, however, instead of considering a pdf $[\theta]$ and evaluating $\underset{\delta}{Max} \bar{J}(\delta)$, we assess the sensitivity of $\delta^*(\theta)$, the argument of $\underset{\delta}{Max} J(\delta, \theta)$, to alternative values of $p$, the most uncertain component of the vector parameter $\theta$, as explained in Section 12.3.5.

### 12.7.4 Other indicators of performance

In addition to the criterion $\bar{J}(\delta)$ to be optimized, indicators of performance are quantities designed to capture the key concerns about yield, variability of the stock, average number of spawners and sustainability of the population. Simply by iterating Eq. (12.6), it is easy to obtain the distributions for:

- Number of eggs and juveniles;
- Number of spring salmon and grilse returning for spawning;
- Catches of spring salmon and grilse.

We extracted meaningful statistics (expected values, variance, quantiles, etc.) from each quantity of interest. We also focused on the average harvest of spawners (particularly spring salmon) and their variability, the average stock size and its variance, as well as on probability of extinction, all calculated over $T = 100$ years.

---

## 12.8 Results

Simulations over 100 years were performed given the management policies $\delta_1 - \delta_{10}$ and the previous hypotheses. Below are summarized the main results.

### 12.8.1 Extinction probabilities

Each line of Table 12.4 gives the estimated probability of extinction resulting from adopting each of the 10 policies.Clearly, for all strategies, extinction becomes more likely with the various values for the probability $p$ of a catastrophic recruitment (from left to right). Because of the early closure of the fishing season, some advantage is given to constant harvest rate policies $\delta_2$-$\delta_3$. Policies $\delta_4$ and $\delta_5$ must be discarded because they authorize harvest rates that are too high and could lead to the mid-term extinction of the population. Floor policies also give quite high extinction probabilities, although some advantage can be given to $\delta_7$ and $\delta_9$ that both have more precautionary $TAC$. The constant escapement policy $\delta_{10}$ generates only the very small probability of extinction even with high values of $p$.

| Policy | $p = 0.01$ | $p = 0.09$ | $p = 0.13$ | $p = 0.155$ | $p = 0.19$ | $p = 0.24$ |
|---|---|---|---|---|---|---|
| Constant harvest rate policies | | | | | | |
| $\delta_1$ | 0 | 0 | 0 | 0 | 0 | 0 |
| $\delta_2$ | 0 | 0 | 0 | 0 | 0 | 0 |
| $\delta_3$ | 0 | 0 | 0 | 0 | 0 | 0 |
| $\delta_4$ | 0 | 0 | 0 | 0 | 0 | 0.01 |
| $\delta_5$ | 0 | 0.03 | 0.03 | 0.12 | 0.26 | 0.61 |
| Floor policies | | | | | | |
| $\delta_6$ | 0 | 0.97 | 1 | 1 | 1 | 1 |
| $\delta_7$ | 0 | 0.3 | 0.61 | 0.77 | 0.91 | 0.96 |
| $\delta_8$ | 0 | 0.72 | 0.8 | 0.95 | 0.97 | 1 |
| $\delta_9$ | 0 | 0.05 | 0.16 | 0.26 | 0.45 | 0.79 |
| Fixed escapement policies | | | | | | |
| $\delta_{10}$ | 0 | 0 | 0 | 0 | 0.01 | 0.07 |

**TABLE 12.4**: Extinction probabilities estimated under various policies and for a range of values for the probability $p$ (occurrence of a catastrophic recruitment).

### 12.8.2 Indicators of performance

The detailed features in the performance indices have been computed for the value of $p = 0.155$ and are illustrated in Figs. 12.6 and 12.7. Policy $\delta_1$ (no exploitation except a little poaching) provides the most important average adult run, with a large variability for both grilses and spring salmon (see Fig. 12.6). Increasing the harvest rate ($\delta_1 \rightarrow \delta_5$) yields a decrease in the mean escapement together with an increase in the between-years variability of the returns. Strategy $\delta_5$ strongly exploits both grilses and Spring salmon (Fig. 12.7), leaves the smallest escapement for both sea-age classes and triggers a strong variability in returns. Mean catches decrease progressively when increasing the harvest rate (Fig. 12.7) but mean catches seem to reach a maximum for $\delta_2$-$\delta_3$ (grilses) or $\delta_3$ (spring salmon) and then decrease to 0 for $\delta_5$, which is a typical response for population dynamics governed by a density dependent production function like the Ricker stock-recruitment relationship. Fixed quota strategies have weak performances, except $\delta_9$ which allows for a better escapement of Spring salmon ($TAC_2$=40 fish). The fixed escapement strategy $\delta_{10}$ produces good results in terms of catches and returns and low extinction probability (Fig. 12.7).

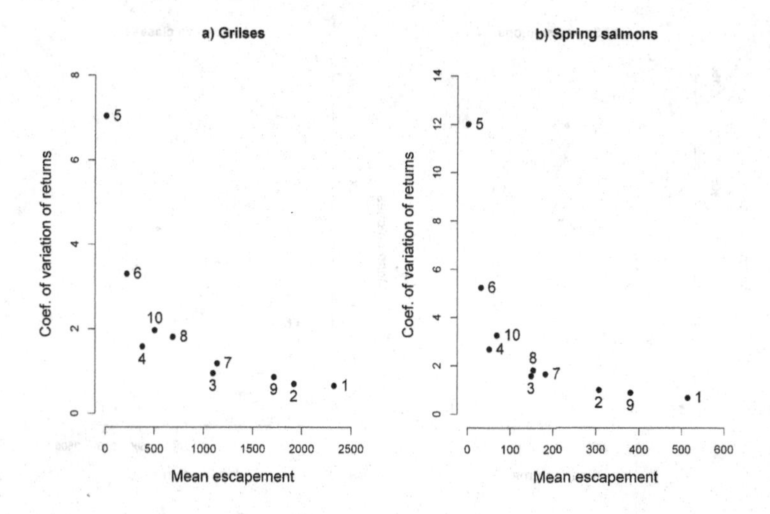


**FIGURE 12.6**: Coefficient of variation of returns as a function of the mean escapement (calculated with $p = 0.155$) for (a) grilse, and (b) Spring salmon.

### 12.8.3 Cumulated benefits

Table 12.5 shows the cumulated benefits calculated with discount rate $\lambda = 0.03$ for each of these strategies under increasing values for the main unknown $p$ and high reward of spring salmon catches. We note that increasing frequencies of catastrophic years for the stock-recruitment yield degrade the overall performance but does not significantly change the relative ranks of the competing strategies. Therefore, averaging over $p$ to compute the expected integrated benefit would not modify the results obtained with $p = 0.155$. Strategies $\delta_3$ and $\delta_4$ (constant exploitation rate with $31st$ of July and $24th$ of August as a closing dates for the fishing season) and $\delta_{10}$ (escapement policy with cautious TAC and reevaluation) seem to overcome the other competitors when the reward for Spring salmon is high ($q_1 = 1$, $q_2 = 10$). These rankings remain robust when considering the other discounting factors $\lambda = 0.01$ and $\lambda = 0.07$ (figures not shown here).

As a conclusion of this decision analysis, fixed harvest rate policies $\delta_3$ or $\delta_4$ would be in practice the best choice because of their good performance and their ease of implementation with only the closing date to be enforced.

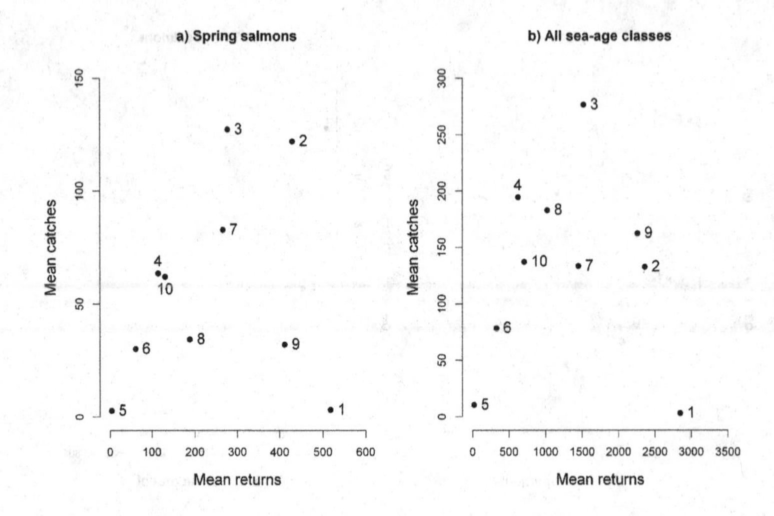


**FIGURE 12.7**: Exploitation rate (Captures versus Returns; calculated with $p = 0.155$) for (a) Spring salmon, and (b) all sea-age classes in the spawning run.

---

## 12.9 Discussion

The aim of this chapter was to illustrate how stochastic simulations can be used to compare harvesting management strategies through the eyes of the specific case of A. salmon recreational fisheries. We hope that the statistical models and methods exposed in this book could provide useful tools for people diving in the stream of research needed to improve the population ecology and management. Obviously, modeling of probabilistic tools as exposed here is only one of the issues in the much wider problematic natural resources and ecosystems management. Admittedly, our coverage of the topic of HBM has been far from complete. To go many steps further, the interested reader should now work and strengthen his skills in various important domains: (i) There are computational issues to develop MCMC algorithms of his own. There comes a time when relying on WinBUGS for inference is no longer satisfactory. Among others, books like [2], [164] and [261] can help writing R programs so as to understand MCMC techniques in depth by practice. However R may be too slow, and when requiring minimal run-time support for large dataset modeling, the experienced analyst will turn to faster pro-

| Policy | $p = 0.01$ | $p = 0.09$ | $p = 0.13$ | $p = 0.155$ | $p = 0.19$ | $p = 0.24$ |
|---|---|---|---|---|---|---|
| Constant harvest rate policies | | | | | | |
| $\delta_1$ | 0.2 | 0.2 | 0.2 | 0.2 | 0.2 | 0.2 |
| $\delta_2$ | 10.8 | 9.2 | 8.5 | 7.9 | 7.2 | 6.1 |
| $\delta_3$ | 20.5 | 16 | 14.1 | 12.3 | 10.7 | 8.5 |
| $\delta_4$ | 19.1 | 12.4 | 10.5 | 8.9 | 7.2 | 5.0 |
| $\delta_5$ | 2.3 | 1.3 | 1.1 | 1 | 0.7 | 0.5 |
| Floor policies | | | | | | |
| $\delta_6$ | 12.6 | 8.2 | 6.4 | 5.1 | 4.2 | 2.9 |
| $\delta_7$ | 10.5 | 9.3 | 8.3 | 7.7 | 6.6 | 5.3 |
| $\delta_8$ | 13.2 | 10.8 | 9.6 | 8.2 | 6.9 | 5.2 |
| $\delta_9$ | 6.8 | 6.8 | 6.7 | 6.5 | 6.3 | 5.4 |
| Fixed escapement policies | | | | | | |
| $\delta_{10}$ | 16.6 | 11.3 | 9.1 | 7.6 | 6.2 | 4.1 |

**TABLE 12.5**: Cumulated benefits under sustainable policies and for a range of values for the probability $p$ (occurrence of a catastrophic recruitment) with a discount rate $\lambda = 0.03$ and a strong reward for spring salmon ($q_1 = 1$, $q_2 = 10$).

gramming languages, like C. (ii) Ecological data are nowadays mostly georeferenced and we see an explosion of spatial statistical research in environmental studies with considerable computational demands. Such statistical models try to express the idea that nearby observations tend to be more alike. Studying [91], [13], [18] and [114] provides excellent entry points in the corresponding literature. (iii) Time also plays a major role in ecological processes and getting background knowledge in multivariate random time series and non linear system dynamics reveals most fruitful when elaborating efficient models in ecology. Books that we find particularly helpful in understanding such a perspective are [65], [67] and [76].

Below we conclude with some bibliographic notes to enlarge the concepts of decision analysis illustrated in this chapter.

### 12.9.1 Optimum solution?

We have used stochastic simulation to compare competing policies and pick the best one. These realistic policies are submitted to implementation randomness but consist of rules easily enforced (opening and closing dates for the fishing season, total allowable catches, etc.). Of course, a better solution can certainly be found outside this limited prefixed set of potential strategies. There are advanced mathematical techniques ([108], [291]) to obtain the optimal refined solution of Eq. (12.21): dynamic programming ([23]) is the most popular procedure to derive a closed loop feedback control (*i.e.*, a management rule) for discrete time models ([24]). Appendix C makes recourse to this backward recursion algorithm to show that escapement policies are optimal in the case of a stochastic Ricker model with nonoverlapping generations. Walters [308] applied this procedure to obtain optimal harvest curves for the Skeena River Sockeye and suggested to approximate them by simplified strategies not requiring close monitoring of escapement during the fishing season which are difficult to apply in the field and are error prone. Optimal control application in fisheries include [22], [59], [61], [62]. But an exponential computational burden occurs as soon as system states become multidimensional (the so-called *curse of dimensionality* in dynamic programming) and various *ad hoc* approximation methods must be invented (see for instance [146]). This computational task brings trouble when searching for the optimal solution as soon as one considers age-structured populations like the Atlantic salmon's cycle of this chapter, or multispecies models mimicking more complex representations of ecosystems (for instance a marine food web with prey, predators, competitors and planktonic organisms) not addressed in this book.

### 12.9.2 Expected utility

#### 12.9.2.1 Decisionmakers' behavior

The key feature of simulation is to consider all possible futures and evaluate the expected benefit of a policy by weighting the return of each scenario by its probability of occurrence. Furthermore, it makes no difference whether these probabilities are representing judgmental uncertainty about possible values of parameters or natural stochastic variations of unknowns. The principle that decision makers behave in risky situations as optimizers of Eq. (12.21) and that their state of knowledge can be described by the means of a random variable is grounded on the five mathematical axioms of Pratt *et al.* [235]. During the last 50 years, experiments on behavior under risk have exhibited a series of "paradoxes"([4]; [5]) mostly linked to discrepancies between observable rationality of a decisionmaker under risk ([231]) and the expected utility optimization principle. New models of behavior have been developed ([186], [243]; [301]), trying to take into account these other types of rationality under risk ([212]).

However, many reasons advocate to keep the rationale associated with Eq. (12.23), since:

1. It agrees with common actuarial practice for public works;
2. It guarantees an always positive value of information ( *i.e.*, a better state of information can never worsen the decision in a sequential setting);
3. It works coherently within the Bayesian statistical framework ([26]) for ecological modeling developed in this book.

#### 12.9.2.2 Robustness

In the Salmon case addressed in this chapter, the ranks of the competing strategies do not change significantly when $p$ (the probability of bad recruitment) varies. Therefore the recommended policies are quite robust with regards to the choice of the prior pdf encoding the uncertainty of this parameter. In general, the influence of priors should be carefully checked. The techniques to correctly "elicit" experts' prior beliefs, *i.e.*, to encode their knowledge as probabilistic judgements (see [155]) are becoming a mature field of research at the interface between psychology and probability (see [223]).

Symmetrically, the influence of the criterion to be optimized should also be examined with great care. In this case study, changing the discount factor $\lambda$ within reasonable ranges does not change the decision for the family of cost functions from Eq. (12.21) but other utility-based criteria – including a deformation of costs and benefits to take into account explicitly risk aversion– such as in [158] or [256] could also have been tried. Theoretical work have shown some robustness of optimal decisions with regard to a relatively large class of loss functions ([1]). All this study suggests that the main results from this risk analysis can be used with reasonable confidence for monitoring the Atlantic salmon cycle, even if important efforts remain to be done to understand the stock recruitment variations and to improve our knowledge about environmental and implementation noises.

### 12.9.3 Viability theory

Viability models do not lead to optimizing a time-related criterion as in Eq. (12.23), but instead try to encompass all viable evolutions of the system in Eq. (12.6). As detailed in Delara and Doyen [83], *viable* here means being able to satisfy at each future time step specified constraints, such as keeping the spawning stock biomass above some reference point or restricting the fishing effort so that the fish mortality remains below a threshold. Viability theory is based on the formal mathematical treatment given in Aubin [10] and [11]. The set fulfilling constraints includes the *viability kernel,* a key notion of the theory. The viability kernel contains all initial states from which at least a sequence of decisions can drive a trajectory that never violates the constraints. Stakeholders have to adjust their requirements so as to check that the viability kernel is a nonempty subset, allowing for a workable multi-criteria decision aid. This mathematical notion provides an interesting quantitative and formal link between sustainability and stewardship issues, decision and precaution problems in the management of natural resources. Its application concerning fisheries, is advocated ([80]) and developed ([25], [84], [191]). One promising outcome of viability theory is that sustainable yields need not be defined species by species (contrarily to current practice) but may jointly depend on the whole ecosystem dynamics, in the spirit of the Ecosystemic Approach to Fisheries ([78]; [85]; [110]; [232]).

### 12.9.4 More than one player in the field

Real life no longer holds the assumption of a sole owner, working for the common welfare by making trade-offs between long-term conservation and myopic catches. Many actors interplay at various scales but overexploitation of the resource is more often the rule than the exception.

#### 12.9.4.1 The tragedy of commons

Diamond [90] points out five factors triggering societal collapse: damage to the environment, climate change, hostile neighbors, decay of trade partners and cultural beliefs. Fisheries provide a worldwide damage to the environment from present societies. It can be understood as a classic example of the tragedy of the commons ([131]). The access to the resource is open, it is difficult to establish and to enforce rights to fish in the sea. Therefore the selfish rule of capture prevails. Why should a fishing boat not try to maximize its harvest today, knowing that other competitors might benefit from its efforts to maintain abundance tomorrow?

#### 12.9.4.2 Cooperation

In addition to playing against nature as in optimal control theory, the many players involved in the search for fish play one against another. Recourse to game theory ([217]) can help to better understand such behaviors. Some economists argue that the waste associated with this problem could be alleviated by privatizing the commons, that is, creating individual private property rights for common-pool resources. Optimum regulation is then equivalent to the search for a standard general economic equilibrium. Other game scientists think that fish exploitation rather belong to a family of problems known as the *iterated prisoner's dilemma.* Axelrod [12] formalized this problem by the means of a tournament of computer algorithms simulations. He discovered that when encounters of many players are repeated over a long time period, selfish strategies tend to do very poorly in the long run while more altruistic strategies work better. Practical experiments tend to prove that such a cooperation emerges when the group is small enough so that everyone can known many other members of the community. In the fisheries context, such evolution toward fruitful cooperation behavior may occur by learning from experience and by self-regulation of the group of fishers.

# Appendix A

## The Normal and Linear Normal model

The two parts of this appendix can help understand the detailed calculations for Chapters 3, 6 and 7:

- The first part describes the Student distributions (centered and noncentered). They have also received the name of T-distributions. This distribution appears in many problems related to Normal models such as the salmon fish farm example (Chapter 3), the Thiof abundance ANCOVA of Chapter 6 and the regression approach in the stock/recruitment model in Chapter 7.
- The second part recalls the essentials of the Bayesian approach for the linear Gaussian model and its convenient priors.

---

## A.1 Centered and Noncentered Student distributions

### A.1.1 Student distributions as a ratio

Let $U$ be a Gaussian random variate $N(0,1)$ and $Y$ a Gamma random variate with a unit scale parameter $Y \sim Gamma(a,1)$. We take $U$ and $Y$ independently, so that the joint distribution reads:

$$[u, y] \propto y^{a-1} e^{-y-\frac{u^2}{2}}$$

We consider the random variable $T$, a function of $U$ and $Y$ such that:

$$t = \sqrt{a}\frac{u}{\sqrt{y}}$$

Changing variables, from $(u, y)$ to $(t, y)$ with Jacobian $\frac{D(u,y)}{D(t,y)}$ =

$\sqrt{y/a}$, leads to the joint distribution of $T$ and $Y$:

$$[t, y] \propto y^{a-1/2} e^{-y(1+\frac{t^2}{2a})}$$

The marginal distribution for $T$ is obtained by integrating $y$ out:

$$[t] = const \times \frac{1}{[1 + \frac{t^2}{2a}]^{\frac{2a+1}{2}}}$$

The constant is such that $\int [t]dt = 1$, so that:

$$[t] = \frac{\Gamma(\frac{2a+1}{2})}{\Gamma(a)\sqrt{2a\pi}} \frac{1}{[1 + \frac{t^2}{2a}]^{\frac{2a+1}{2}}} \tag{A.1}$$

Equation (A.1) for the pdf of $T$ is referred to as $dStudent(t, \nu)$, the Student distribution with $\nu = 2a$ degrees of freedom. It exhibits a symmetric distribution with variance defined only for $a > 1$

$$\mathbb{V}(T) = \frac{a}{a-1}.$$

This distribution can be extended to the case when the numerator of the ratio defining $T$ is a Normal variate centered on $\delta$, *i.e.*, $U \sim N(\delta, 1)$ (with $U$ and $Y$ remaining independent). In this case, $[t]$ is said to follow a decentered Student distribution with $\delta$ as a decentered parameter. The analytical expression for the decentered Student pdf does not write nicely but it can be computed by all statistical packages and, of course, generating a decentered Student random variable is easily programmed in any scientific software language.

### A.1.2 Student distributions as a mixture

There is another way to generate Student distributions ; the most interesting results are obtained by conditioning. Consider $Y$ again Gamma distributed with a unit scale parameter

$$Y \; Gamma(a, 1)$$

but the random variable $U$ is now conditionally Normal with precision (inverse of the variance) $y$:

$$U \; Normal(0, y^{-\frac{1}{2}})$$

If we look for the marginal pdf of $U$, we write for $[U, Y]$ formula very similarly to the previous section, and it is straightforward to show that the rescaled version of $U$, $T = \sqrt{a}U$ is marginally Student distributed with $\nu = 2a$ degrees of freedom. Consequently, one can obtain the marginal pdf of $U$:

$$
\begin{aligned}
[u] &= [t]\frac{dt}{du} \\
&= dStudent(\sqrt{a}(u-\delta), 2a) \times \sqrt{a}
\end{aligned}
$$

### A.1.3 Multivariate Student distributions

The multivariate Student-distribution for a random vector $t$ with dimension $p$, depends on the following parameters:

- $\nu$: degrees of freedom
- $\Sigma$: variance-covariance matrix

This pdf is the multivariate extension of the same Gamma-Normal mixture following the very same conditional construction in two steps:

- First consider Gamma distributed with a unit scale parameter $Y \sim Gamma(a, 1)$
- Given $Y = y$, take the random variable $U$ as a multinormal distribution with precision matrix $y \times \Sigma^{-1}$:

$$U\ simMultiNormal(\delta, \frac{\Sigma}{y})$$

Marginally, the rescaled version of $U$, $T = \sqrt{a}(U - \delta)$ is a multivariate Student distribution with $\nu = 2a$ degrees of freedom. The probability density function $[t|\nu, \Sigma]$ $= dmStudent(t, \Sigma, \nu)$ is obtained as the multivariate generalization of Equation (A.1):

$$[t|\nu, \Sigma] = (\nu\pi)^{-\frac{p}{2}} \frac{\Gamma(\frac{\nu+p}{2})}{\Gamma(\nu/2)\sqrt{|\Sigma|}}[1 + \frac{t'\Sigma^{-1}t}{\nu}]^{-\frac{\nu+p}{2}} \tag{A.2}$$

and

$$[u|\delta, \Sigma, a] = dmStudent(\sqrt{a}(u-\delta), \Sigma, 2a) \times a^{\frac{p}{2}}$$

The variance-covariance matrix $\frac{\nu}{\nu-2}\Sigma$ of the multivariate Student random vector is only defined for $\nu > 2$.

## A.2 The linear Normal regression model

### A.2.1 Likelihood

Consider $n$ independent Normal random variables $Y_i$ with the same variance parameter $\sigma^2$ but different expectation depending linearly on $p$ covariates $X_{ij}$ (with $1 \leq j \leq p$). We define $Y$ as the vector whose components are the $n$ quantities $Y_i$ (with $1 \leq i \leq n$) and $\mathbf{X}$ the model matrix with $n$ row and $p$ columns. The first column is commonly made of $1's$ (constant effect) and the other columns for $j = 2$ to $p$ are obtained by taking the $n$ coordinates of the remaining explanatory variables. With theses notations, the expectation vector is expressed as:

$$\mu = E(Y) = \mathbf{X}\beta$$

with $\beta = \begin{pmatrix} \beta_0 \\ \beta_1 \\ \cdots \\ \beta_p \end{pmatrix}$, a $p$ -dimension vector. This model belongs to the exponential family and is parameterized by $p+1$ unknowns, $\theta = (\beta, \sigma^2)$. The likelihood is thus obtained, using matrix operations:

$$\begin{aligned} [Y|\mathbf{X},\theta] &= \frac{1}{(\sqrt{2\pi}\sigma)^n}\exp\left(-\frac{(Y-\mathbf{X}\beta)'(Y-\mathbf{X}\beta)}{\sigma^2}\right) \\ &= \frac{1}{(\sqrt{2\pi}\sigma)^n}\exp\left(-\frac{\sum_{i=1}^{n}(Y_i - \sum_j \beta_j X_i^{(j)})^2}{\sigma^2}\right) \end{aligned}$$

We consider the max-likelihood statistics $\hat{\beta}$ and assume that $\mathbf{X}$ is of full rank:

$$\hat{\beta} = (\mathbf{X}'\mathbf{X})^{-1}\mathbf{X}'Y$$

Every component of $\hat{\beta}$ is a linear combination of the elements $\mathbf{X}'Y$ whose coefficients are the rows of the inverse matrix $(\mathbf{X}'\mathbf{X})^{-1}$.

One can prove that:

$$(Y-\mathbf{X}\beta)'(Y-\mathbf{X}\beta) = (Y-\mathbf{X}\hat{\beta})'(Y-\mathbf{X}\hat{\beta}) + (\beta-\hat{\beta})'\mathbf{X}'\mathbf{X}(\beta-\hat{\beta})$$

Working with the precision parameter $h = \frac{1}{\sigma^2}$, the likelihood for the Normal linear model relies on the two sufficient statistics $\hat{\beta}$ and $(Y-\mathbf{X}\hat{\beta})'(Y-\mathbf{X}\hat{\beta})$:

$$[Y|\mathbf{X},\beta,h] = \left(\sqrt{\frac{h}{2\pi}}\right)^n \exp\left(-\frac{h(Y-\mathbf{X}\hat{\beta})'(Y-\mathbf{X}\hat{\beta})}{2} - \frac{h(\beta-\hat{\beta})'\mathbf{X}'\mathbf{X}(\beta-\hat{\beta})}{2}\right)$$

In the frequentist setting, an unbiased estimator of $\sigma^2$ is provided by $\hat{\sigma}^2 = \frac{(Y-\mathbf{X}\hat{\beta})'(Y-\mathbf{X}\hat{\beta})}{n-p}$ and $\hat{\sigma}^2(\mathbf{X}'\mathbf{X})^{-1}$is the common estimation for the covariance matrix of $\hat{\beta}$.

### A.2.2 Conjugate prior

The exponential form of the likelihood shows that natural conjugate priors can be found for this model:

1. A gamma distribution for the precision $h = \frac{1}{\sigma^2}$,

$$h \sim Gamma(\frac{n_0}{2}, \frac{S_0}{2})$$

2. Given the precision, a conditional $p-$multivariate Normal distribution for $\beta$, with prior expectation $\beta_0$ and $p \times p$ covariance (invertible) matrix $\sigma^2 V_0$,

$$\beta \sim= MultiNormal\left(\beta, \beta_0, \frac{V_0}{h}\right)$$

With such a prior on $(\beta, h)$, the *a posteriori* distribution $[\beta, h|Y, \mathbf{X}]$ is derived:

- Given $h$, $Y$, and $X$, the posterior conditional distribution for $\beta$ is a MultiNormal

$$\beta|h, Y, \mathbf{X} \sim MultiNormal\left(\beta_y, \frac{V_y}{h}\right)$$

with updated parameters:

$$\begin{cases} (V_y)^{-1} = \mathbf{X}'\mathbf{X} + V_0^{-1} \\ (V_y)^{-1}\beta_y = \mathbf{X}'\mathbf{X}\hat{\beta} + V_0^{-1}\beta_0 \end{cases}$$

This is interpreted in terms of a gain in precision and a barycentric position for the posterior mean because of the equality $0 = X'X(\beta_y - \hat{\beta}) + V_0^{-1}(\beta_y - \beta_0)$.

- Given $\beta$, $Y$, and $X$, the posterior for the precision $h$ is

$$h|\beta, Y, \mathbf{X} \sim Gamma(\frac{n_{y,\beta}}{2}, \frac{S_{y,\beta}}{2})$$

such that:

$$\begin{aligned} n_{y,\beta} &= n_0 + n + p \\ S_{y,\beta} &= S_0 + (Y - \mathbf{X}\hat{\beta})'(Y - \mathbf{X}\hat{\beta}) \\ &\quad + (\beta - \hat{\beta})'\mathbf{X}'\mathbf{X}(\beta - \hat{\beta}) + (\beta - \beta_0)V_0^{-1}(\beta - \beta_0) \end{aligned}$$

- Given $Y$ and $X$, but unconditionally to $\beta$, the posterior for the precision $h$ is

$$h|Y, \mathbf{X} \sim Gamma(\frac{n_y}{2}, \frac{S_y}{2})$$

such that:

$$\begin{aligned} n_y &= n_0 + n \\ S_y &= S_0 + (Y - \mathbf{X}\hat{\beta})'(Y - \mathbf{X}\hat{\beta}) + E_y \end{aligned}$$

In other words, $\sigma^2 = \frac{1}{h}$ is *a posteriori* distributed according to an inverse gamma pdf, with

$$\begin{aligned} E_y &= \hat{\beta}'\mathbf{X}'\mathbf{X}\hat{\beta} + \beta_0' V_0^{-1}\beta_0 - \beta_y'(V_y)^{-1}\beta_y \\ &= \hat{\beta}'\mathbf{X}'\mathbf{X}\hat{\beta} + \beta_0' V_0^{-1}\beta_0 - \beta_y'(V_y)^{-1}(V_y)(V_y)^{-1}\beta_y \\ &= \hat{\beta}'\mathbf{X}'\mathbf{X}\hat{\beta} + \beta_0' V_0^{-1}\beta_0 - \left(\mathbf{X}'\mathbf{X}\hat{\beta} + V_0^{-1}\beta_0\right)' V_y(\mathbf{X}'\mathbf{X}\hat{\beta} + V_0^{-1}\beta_0) \\ &= \hat{\beta}'\mathbf{X}'\mathbf{X}\hat{\beta} + \beta_0' V_0^{-1}\beta_0 + \\ &\left(-\mathbf{X}'\mathbf{X}\hat{\beta} + V_0^{-1}(\hat{\beta} - \beta_0) - V_0^{-1}\hat{\beta}\right)' V_y(\mathbf{X}'\mathbf{X}\left(\hat{\beta} - \beta_0\right) + V_0^{-1}\beta_0 + \mathbf{X}'\mathbf{X}\beta_0) \\ &= \hat{\beta}'\mathbf{X}'\mathbf{X}\hat{\beta} + \beta_0' V_0^{-1}\beta_0 + \\ &\left(V_0^{-1}(\hat{\beta} - \beta_0) - V_y^{-1}\hat{\beta}\right)' V_y(\mathbf{X}'\mathbf{X}\left(\hat{\beta} - \beta_0\right) + V_y^{-1}\beta_0) \\ &= \hat{\beta}'\mathbf{X}'\mathbf{X}\hat{\beta} + \beta_0' V_0^{-1}\beta_0 + \\ &\left(\hat{\beta} - \beta_0\right)' V_0^{-1}V_y\mathbf{X}'\mathbf{X}\left(\hat{\beta} - \beta_0\right) + \left(\hat{\beta} - \beta_0\right)' V_0^{-1}\beta_0 - \hat{\beta}'\mathbf{X}'\mathbf{X}\hat{\beta} - \hat{\beta}'V_0^{-1}\beta_0 \\ &= \left(\hat{\beta} - \beta_0\right)' V_0^{-1}V_y\mathbf{X}'\mathbf{X}\left(\hat{\beta} - \beta_0\right) \\ &= \left(\hat{\beta} - \beta_0\right)' V_0^{-1}(V_0^{-1} + \mathbf{X}'\mathbf{X})^{-1}\mathbf{X}'\mathbf{X}\left(\hat{\beta} - \beta_0\right) \end{aligned}$$

This expression makes it clear that the term $E_y$ is positive. Providing $\mathbf{X}'\mathbf{X}$ is invertible, some algebra leads to the expression given in Marin and Robert [189] page 54:

$$E_y = (\hat{\beta} - \beta_0)'(V_0 + (\mathbf{X}'\mathbf{X})^{-1})^{-1}(\hat{\beta} - \beta_0)$$

Consequently, since given $h$, $\beta_j$ the $jth$ component of $\beta$ is *a posteriori* distributed according to the $Normal(\beta_y(j), \frac{V_y(j,j)}{h})$ . Therefore $\frac{\beta_j - \beta_y(j)}{\sqrt{\frac{V_y(j,j)}{S_y}}}$ is $Normal(0, \frac{1}{S_y h})$, with $S_y \times h$ Gamma distributed. Taking into account the results from the previous section, we deduce that $\sqrt{n_0 + n} \times (\beta_j - \beta_y(j)) \times \sqrt{\frac{S_y}{V_y(j,j)}}$ is marginally distributed as a Student pdf with $2n_y$ degrees of freedom.

Because of the conjugate property, the prior predictive, useful for model choice, is explicitly obtained:

$$\begin{aligned}
[Y|\mathbf{X}] &= \frac{[Y|\mathbf{X}, \beta, h][\beta, h|\mathbf{X}]}{[\beta, h|\mathbf{X}, Y]} \\
&= \frac{dmnorm(y, \mathbf{X}\beta, I/h) \times dmnorm\left(\beta, \beta_0, \frac{V_0}{h}\right) \times dgamma(h, \frac{n_0}{2}, \frac{S_0}{2})}{dmnorm\left(\beta, \beta_y, \frac{V_y}{h}\right) \times dgamma(h, \frac{n_y}{2}, \frac{S_y}{2})} \\
&= (\sqrt{\frac{h}{2\pi}})^n \exp(-\frac{h(Y - \mathbf{X}\hat{\beta})'(Y - \mathbf{X}\hat{\beta})}{2} - \frac{h(\beta - \hat{\beta})'\mathbf{X}'\mathbf{X}(\beta - \hat{\beta})}{2}) \\
&\times (\sqrt{\frac{h}{2\pi}})^p \sqrt{|V_0^{-1}|} \exp(-\frac{h(\beta - \beta_0)V_0^{-1}(\beta - \beta_0)}{2})) \\
&\times \frac{h^{\frac{n_0}{2}-1} \exp(-h\frac{S_0}{2})}{\Gamma(\frac{n_0}{2})} \left(\frac{S_0}{2}\right)^{\frac{n_0}{2}} \left(\frac{S_y}{2}\right)^{-\frac{n_y}{2}} \frac{\Gamma(\frac{n_y}{2})}{h^{\frac{n_y}{2}-1} \exp(-h\frac{S_y}{2})} \\
&\times \frac{1}{(\sqrt{\frac{h}{2\pi}})^p \sqrt{|V_y^{-1}|} \exp(-\frac{h(\beta - \beta_y)V_y^{-1}(\beta - \beta_y)}{2}))} \\
&= \left(\sqrt{\frac{1}{\pi}}\right)^n \frac{\sqrt{|V_y|}}{\sqrt{|V_0|}} \times \frac{\Gamma(\frac{n_y}{2})}{\Gamma(\frac{n_0}{2})} \times \frac{(S_0)^{\frac{n_0}{2}}}{(S_y)^{\frac{n_y}{2}}} \qquad \text{(A.3)}
\end{aligned}$$

It is not obvious to see that Eq. (A.2) reveals that $[Y|\mathbf{X}]$ is indeed a $n-$multivariate Student with $n_0$ degrees of freedom, a mean vector $\mathbf{X}\beta_0$ and a variance covariance matrix $\Sigma = \frac{S_0}{2}(I_n + \mathbf{X}V_0\mathbf{X}')$ . Given $\sigma^2$, one can consider that $Y = \mathbf{X}\beta + \varepsilon, \beta = \beta_0 + \eta$ with $\varepsilon \sim Normal(0, \sigma^2 I)$ and $\eta \sim N(0, \sigma^2 V_0)$. Marginalizing over $\beta$, $Y = \mathbf{X}\beta_0 + \varepsilon + \mathbf{X}\eta$ *i.e.*, $Y|\sigma^{-2} \sim Normal(\mathbf{X}\beta_0, \sigma^2(I + \mathbf{X}V_0\mathbf{X}'))$. Setting $h = \frac{S_0}{2}\sigma^{-2}$, $Y$ is obtained following the Gamma-multivariate Normal mixture construction given page 335 with $a = \frac{n_0}{2}$

$$\begin{cases} Y|h \sim Normal(\mathbf{X}\beta_0, \frac{1}{h}\frac{S_0}{2}(I + \mathbf{X}V_0\mathbf{X}')) \\ h \sim Gamma(\frac{n_0}{2}, 1) \end{cases}$$

Therefore, $T = \sqrt{\frac{n_0}{2}}(Y - \mathbf{X}\beta_0)$ is a multivariate Student distribution with $\nu = 2 \times \frac{n_0}{2} = n_0$ degrees of freedom and $\Sigma = \frac{S_0}{2}(I + \mathbf{X}V_0'\mathbf{X}')$ and

$$\begin{aligned}
&[y|\mathbf{X}, \beta_0, \Sigma, n_0, S_0] = \\
&dmStudent(\sqrt{\frac{n_0}{2}}(y - \mathbf{X}\beta_0), \Sigma, n_0) \times \left(\frac{n_0}{2}\right)^{\frac{n}{2}} \\
&= \left(\frac{n_0}{2}\right)^{\frac{n}{2}} (n_0\pi)^{-\frac{n}{2}} \frac{\Gamma(\frac{n_0+n}{2})}{\Gamma(n_0/2)\sqrt{|\Sigma|}} \\
&\times [1 + \frac{\sqrt{\frac{n_0}{2}}(Y - \mathbf{X}\beta_0)'\Sigma^{-1}\sqrt{\frac{n_0}{2}}(Y - \mathbf{X}\beta_0)}{n_0}]^{-\frac{n_0+n}{2}} \\
&= \left(\frac{2}{2\pi S_0}\right)^{\frac{n}{2}} \frac{\Gamma(\frac{n_0+n}{2})}{\Gamma(n_0/2)\sqrt{|I + \mathbf{X}V_0'\mathbf{X}'|}} \\
&\times [1 + \frac{(Y - \mathbf{X}\beta_0)'(I + \mathbf{X}V_0\mathbf{X}')^{-1}(Y - \mathbf{X}\beta_0)}{S_0}]^{-\frac{n_0+n}{2}} \qquad \text{(A.4)}
\end{aligned}$$

Getting back to notations $n_y = n_0 + n$, $V_0V_y^{-1} = V_0\left(\mathbf{X}'\mathbf{X} + V_0^{-1}\right)$ so that $|I + \mathbf{X}V_0\mathbf{X}'| = \frac{|V_0|}{|V_y|}$, one get

$$\begin{aligned}
[y|\mathbf{X}, \beta_0, \Sigma, n_0, S_0] = & \frac{S_0^{-\frac{n_0}{2}}\pi^{-\frac{n}{2}}\Gamma(\frac{n_y}{2})}{\Gamma(n_0/2)}\sqrt{\frac{|V_0|}{|V_y|}} \\
& \times \left(S_0 + (Y - \mathbf{X}\beta_0)'(I + \mathbf{X}V_0\mathbf{X}')^{-1}(Y - \mathbf{X}\beta_0)\right)^{-\frac{n_y}{2}}
\end{aligned}$$

One could check that $S_y = S_0 + (Y - \mathbf{X}\beta_0)'(I + \mathbf{X}V_0\mathbf{X}')^{-1}(Y - \mathbf{X}\beta_0)$ so that Eqs. (A.4) and (A.3) are identical but (A.3) is computationally the most efficient form (only a $p \times p$ matrix inversion instead of a $n \times n$ one).

### A.2.3 Zellner's G-prior

Zellner's G-prior relies on a conditional Normal prior for $\beta$, with $V_0 = c \times (\mathbf{X}'\mathbf{X})^{-1}$. As $c$ increases, the prior becomes more and more diffuse. Making the prior variance depends on the explainatory variables

$\mathbf{X}$ has sometimes been argued as cheating because the prior should not be data dependent, but anyway the whole model is conditioned on $\mathbf{X}$.

$$[\beta|h] \propto h^{\frac{p}{2}} \exp(\frac{h(\beta-\beta_0)(\mathbf{X}'\mathbf{X})(\beta-\beta_0)}{2c})$$

and an improper (Jeffreys) prior for $h$, i-e a limiting form of the Gamma pdf with $n_0 \longrightarrow 0$ and $S_0 \longrightarrow 0$

With this particular prior, the posterior simplifies into

- $h = \frac{1}{\sigma^2}$ is *a posteriori* distributed according to an inverse Gamma pdf $dgamma(h, \frac{n}{2}, \frac{(Y-\mathbf{X}\hat{\beta})'(Y-\mathbf{X}\hat{\beta})+\frac{c}{c+1}(\hat{\beta}'(\mathbf{X}'\mathbf{X})\hat{\beta})}{2})$
- Given $h$, the posterior conditional distribution for $\beta$ is a multivariate $Normal(\beta_y, \frac{V_y}{h})$ with updated parameters:

$$V_y = \frac{c}{1+c}\left(\mathbf{X}'\mathbf{X}\right)^{-1} \tag{A.5}$$

and

$$\begin{aligned}\beta_y &= \frac{c}{1+c}\left(\mathbf{X}'\mathbf{X}\right)^{-1}\left(\mathbf{X}'\mathbf{X}\hat{\beta} + \frac{\mathbf{X}'\mathbf{X}}{c}\beta_0\right)\\ &= \frac{\beta_0 + c\hat{\beta}}{1+c}\end{aligned} \tag{A.6}$$

These equations highlight the role of $c$ which expresses the strength of the prior information: setting $c = 1$ is equivalent to putting the same weight on the prior information and on the sample, setting $c = n$ means that the prior information is worth only one data record.

---

## A.3 The Zellner Student as a prior

Let $h = \frac{c}{\sigma^2}$ and consider the following distribution

$$\begin{cases} \beta|h, X \sim N_p\left(\beta_0, (h \times (X'X))^{-1}\right) \\ h \sim Gamma(a, 1) \end{cases}$$

As shown in Eq. (A.1), the marginal (predictive prior) distribution for $\sqrt{a}(\beta - \beta_0)$ is a multivariate Student distribution, which we may call here a Zellner Student prior:

$$[\beta|X] = \left(\frac{a}{2\pi a}\right)^{\frac{p}{2}} \sqrt{|X'X|}\frac{\Gamma(a+\frac{p}{2})}{\Gamma(a)}\left(1+\frac{(\beta-\beta_0)'(X'X)^{-1}(\beta-\beta_0)}{2a}\right)^{-a-\frac{p}{2}}$$

and

$$\begin{cases} \mathbb{E}(\beta) = \beta_0 \\ \mathbb{V}(\beta) = \dfrac{a}{a-1}(X'X)^{-1} \end{cases}$$

# Appendix B

# Computing marginal likelihoods and DIC

## B.1 Computing predictive distribution

In any case, one requires a numerical evaluation of the predictive distribution of the data, denoted $[y]$, which is obtained by the integration of the likelihood over the whole parameters space (one also speak about the *marginal likelihood*):

$$[y] = \oint [y, \theta] d\theta.$$

Such an integration can be approximated by Monte Carlo methods as shown below.

### B.1.1 Method of Marin and Robert

Importance sampling methods can be used to compute a Monte Carlo estimation of $[y]$. Let $\pi(\theta)$ be the importance distribution from which a sample of size $G$ of parameters *theta*, $\left\{\theta^{(g)}\right\}_{g=1:G}$, can be generated. Using the importance sampling distribution $\pi(\theta)$, the marginal likelihood $[y]$ can be written:

$$[y] = \oint \left( \frac{[y, \theta]}{\pi(\theta)} \right) \pi(\theta) d\theta \tag{B.1}$$

The integral (B.1) can be estimated from the Monte Carlo sample in $\pi(\theta)$ as:

$$\begin{cases} \theta^{(g)} \sim \pi(\cdot) \\ [y] \approx \dfrac{1}{G} \displaystyle\sum_{g=1}^{G} \dfrac{[y, \theta^{(g)}]}{\pi(\theta^{(g)})} \end{cases} \tag{B.2}$$

Since $[y] = \oint [y, \theta] d\theta = \oint [\theta|y][y] d\theta$, the best candidate for the sampling distribution $\pi$ would of course be the posterior distribution $[\theta|y]$.

Unfortunately, $[\theta|y]$ is analytically unavailable (otherwise its normalizing constant $[y]$ would be known!). But a good approximation of $[\theta|y]$ is available from the posterior MCMC runs.

A classical way to proceed consists in fitting a multi-Normal pdf on the posterior sample $\left\{\tilde{\theta}^{(k)}\right\}_{k=1:M}$ (*e.g.*, by the method of moments), and using the fitted multi-Normal as the importance distribution in (B.2).

If no MCMC sample in $[\theta|y]$ is available, the asymptotic Normal distribution of the frequentist estimate for $\theta$ is generally a good candidate for $\pi$. In this method, the posterior sample is of no use by itself, but is needed to design some reasonable importance function $\pi$, in particular the number of importance replicates $G$ can be much bigger than the posterior sample size $M$.

### B.1.2 Extension of the method of Raftery

When the size of the available sample $\left\{\tilde{\theta}^{(k)}\right\}_{k=1:M}$ is not big enough to ensure a good Normal approximation of the importance function, one would rather like to rely on the posterior draws directly, *i.e.*, use the posterior pdf $[\theta|y]$ itself as importance distribution.

From the Bayes' theorem, it is easy to obtain

$$[y]^{-1} = \frac{[\theta|y]}{[\theta]}[y|\theta]^{-1} \tag{B.3}$$

This equality (B.3) holds whatever the value of the parameter $\theta$. It holds for all values of $\theta$ taken in the support of any distribution $\pi(\theta)$. Consequently,

$$\begin{aligned}
[y]^{-1} &= \oint [y]^{-1}\pi(\theta)d\theta \\
&= \oint \frac{[\theta|y]}{[\theta]}[y|\theta]^{-1}\pi(\theta)d\theta \\
&= \oint \frac{\pi(\theta)}{[\theta]}[y|\theta]^{-1}[\theta|y]d\theta
\end{aligned}$$

Based on samples of $\theta$ in the posterior $[\theta|y]$, a Monte Carlo approximation of $[y]^{-1}$ can then be computed as follows:

$$\begin{cases}
[y]^{-1} \approx \dfrac{1}{M}\displaystyle\sum_{k=1}^{M}\frac{\pi(\theta^{(k)})}{[\theta^{(k)}]}[y|\theta^{(k)}]^{-1} \\
\theta^{(k)} \sim [\theta|y]
\end{cases} \tag{B.4}$$

Taking $\pi(\theta)$ equal to the prior $[\theta]$, Raftery et al. ([245]) considered that "the predictive is the posterior harmonic mean of the likelihood":

$$\begin{cases} [y] \approx \left( \frac{1}{M} \sum_{k=1}^{M} [y|\beta^{(k)}]^{-1} \right)^{-1} \\ \theta^{(k)} \sim [\theta|y] \end{cases} \tag{B.5}$$

Therefore, provided that one can evaluate the likelihood for each value of the posterior sample (for instance the WinBUGS *Deviance* instruction computes minus two times the loglikelihood), the posterior draws of the MCMC algorithm can straightforwardly be re-used to obtain the predictive $[y]$.

Unfortunately this estimate is known to be unstable since small values of the likelihood (this can sometimes occur even though the likelihood and the posterior are close to one another) would make rare but big jumps of the successive levels of the estimation as $M$ tends to infinity. This may hinder the comparison of two models when their difference in credibility is low. If the contrast between the credibility of the different models is high, the approximation gives results which are accurate enough for identifying the most credible model(s).

A more efficient method of estimation is then to rely on

$$\begin{cases} [y] \approx \left( \frac{1}{M} \sum_{k=1}^{M} \frac{\pi(\theta^{(k)})}{[\theta^{(k)}]} [y|\theta^{(k)}]^{-1} \right)^{-1} \\ \theta^{(k)} \sim [\theta|y] \end{cases} \tag{B.6}$$

with $\pi()$ a probability distribution that behaves like the likelihood in the regions that are not likely (such a Normal approximation of the posterior) or cut off the tails of it (such as a triangular distribution centered on the posterior mean).

In any case, as the marginal likelihood (and then the $BF$) is known to be sensitive to the prior choice, it a good idea to check the robustness of alternative prior configurations when ranking models.

---

## B.2 The Deviance Information Criterion

The Deviance Information Criterion (*DIC*; [283]) is a measure of complexity and fit designed to compare hierarchical models of arbitrary structure. Its rationale and interpretation are different than those of the $BF$ and are more analogue to the frequentist Akaike Information Criterion: the smaller the $DIC$ the more favoured is a model. It combines a measure of the goodness of fit, defined as the posterior mean of the deviance $\overline{Dev(\theta)}$ (where the deviance is $-2$ times the loglikelihood), with a measure of the model complexity $pD$, acting as a penalty term:

$$DIC = \overline{Dev(\theta)} + pD \tag{B.7}$$

$pD$ is defined as $\overline{Dev(\theta)}$ minus the deviance calculated at the posterior mean of the parameter vector $Dev(\widehat{\theta})$. We approximate $pD$ by $\frac{1}{2}\, Var(Dev(\theta))$ as proposed by Gelman *et al.* [117]. The $DIC$ is easily calculated by sampling techniques and its computation avoids the instability that may occur when estimating $BF$. To interpret $DIC$ values, Spiegelhalter *et al.* [283] suggest to adopt the rule of thumb proposed by Burnham and Anderson [43], for Akaike Information Criterion.

An explanation of the success of the $DIC$ criterion among the community of applied statisticians is that it gives a pragmatic solution to the problem of model choice, and is now routinely available in the softwares WinBUGS, OpenBUGS, or JAGS, but many criticisms remain about its interpretation, as shown by the discussion of Spiegelhalter *et al.* [283] or in [233]. Celeux *et al.* ([52]) point out many of its flaws in the context of missing data models, with special emphasis on mixture models.

# Appendix C

## More on Ricker stock-recruitment

This supplementary material is an appendix to Chapters 7 and 12 that can help to understand useful Ricker stock-recruitment concepts.

## C.1 A closer look at the Ricker model

### C.1.1 Origin

The Ricker stock-recruitment relationships (see Fig. 7.3 in Chapter 7) stems from the hypothesis that the *per capita* mortality rate (due to predation, disease, cannibalism, etc.) of larvae during their growing phase ($0 < t < T$) is linearly dependent on the population size of spawners ($S_t$) with slope and intercept function of the fluctuating environment ([244]):

$$\frac{1}{N_t}\frac{dN_t}{dt} = k_1(t) - k_2(t) \times S_t$$

Hypothesizing:

1. no mortality of adults during the larval growing phase,*i.e.*,$S_t = S_0 = S$,
2. initial condition that the eggs are proportional to the number of spawners $N_0 \propto S$,
3. terminal condition yielding the recruitment $R = N_t$,

the previous equation can be solved as

$$\begin{aligned} N_T &= N_0 \times \exp\left\{\int_0^T (k_{1,t} - k_{2,t} \times S)\, dt\right\} \\ &= N_0 \times \exp\left(\int_0^T k_{1,t}dt - S \times \int_0^T k_{1,t}dt\right) \end{aligned}$$

leading to the classical Ricker form already given by Eq. (7.4)

$$R = \alpha S \times \exp(-\beta \times S) \tag{C.1}$$

### C.1.2 A deterministic controlled state equation

Therefore an elementary model with nonoverlapping generations can be made of a simple deterministic Ricker stock-recruitment cycle with recruits $R$, surviving as adults with a natural mortality $\pi$ and becoming spawners $S$ after catch $C$. This is depicted by the following equations:

$$\begin{cases} R_t = \alpha S_{t-1} \times \exp(-\beta S_{t-1}) \\ S_t = \pi R_t - C_t \end{cases}$$

Rearranging the terms it comes:

$$R_{t+1} = \alpha \times \left(R_t - \frac{C_t}{\pi}\right) \times \exp\left(\log(\pi) - \beta\pi\left(R_t - \frac{C_t}{\pi}\right)\right)$$

Thus

$$\begin{aligned} &\frac{\beta\pi}{log(\alpha) + \log(\pi)} R_{t+1} = \\ &\frac{\beta\pi}{log(\alpha) + \log(\pi)} \times \left(R_t - \frac{C_t}{\pi}\right) \\ &\times \alpha \times \exp\left(\log(\pi) \times \left(1 - \frac{\beta\pi}{log(\alpha) + \log(\pi)}\left(R_t - \frac{C_t}{\pi}\right)\right)\right) \end{aligned}$$

and rescaling variables $x = R \times \frac{\beta\pi}{log(\alpha)+\log(\pi)}$, $r = \alpha \times \exp(\log \pi)$ and $u = C \times \frac{\beta}{log(\alpha)+\log(\pi)}$ for the ease of notations, one highlights the special role played by the intrinsic log-survival rate $r$:

$$x_{t+1} = (x_t - u_t) \times e^{-r \times \{1-(x_t - u_t)\}} \tag{C.2}$$

Note that this simplified deterministic salmon cycle with nonoverlapping generations is a particular member of the general class of deterministic models in discrete time ([24]):

$$x_{t+1} = F(x_t - u_t) \tag{C.3}$$

with $F(z) = z \times e^{-r \times (1-z)}$.

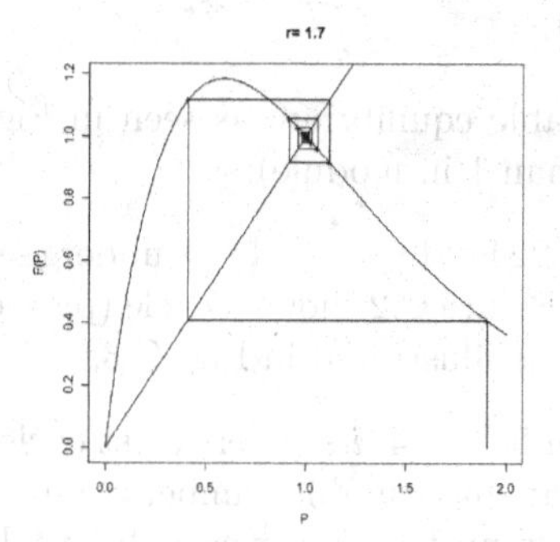


FIGURE C.1: Ricker stock-recruitment behavior with $r = 1.7$ (stable equilibrium point).

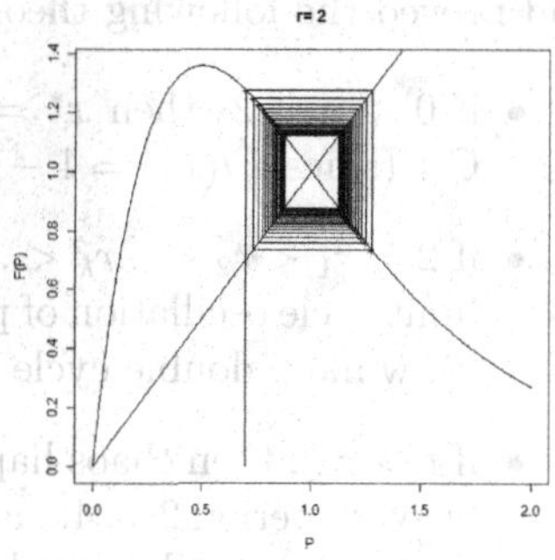

FIGURE C.2: Ricker stock-recruitment behavior with $r = 2$ (limit cycle of period 2).

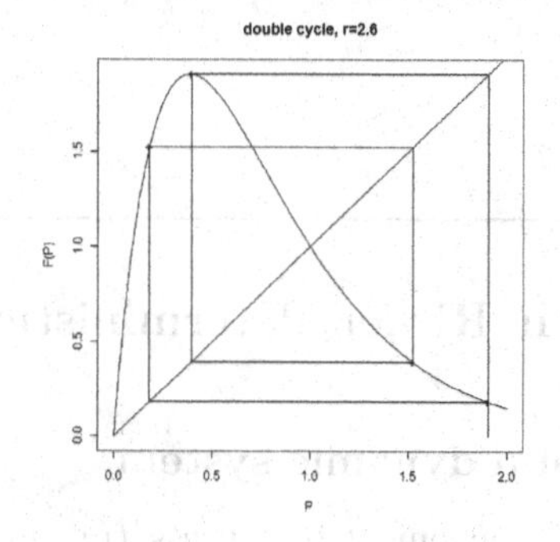


FIGURE C.3: Ricker stock-recruitment behavior with $r = 2.6$ (limit cycle of period 4).

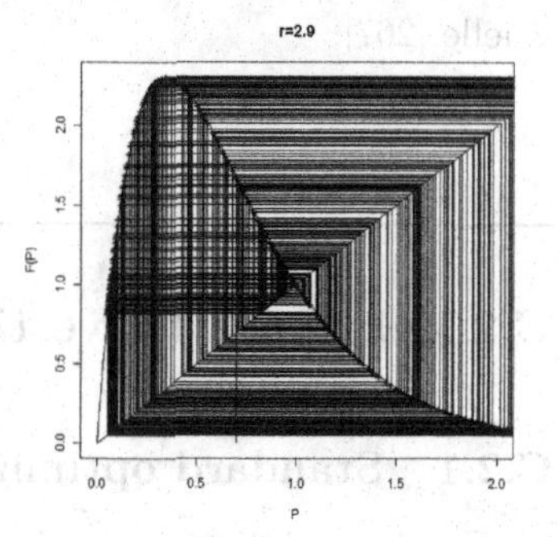


FIGURE C.4: Ricker stock-recruitment iterations with $r = 2.9$ (chaotic behavior).

### C.1.3 Attractors and chaos

Setting $u = 0$ in the previous equation, the idealized representation of a renewable resource system unaffected by human influence starts by studying the recursive equation:

$$x_{n+1} = F(x_n) \tag{C.4}$$

Equilibrium points $x^* = F(x^*)$ are of special interest. Among the many dome-shaped functions $F$, May ([193]) considered simple ones such as the rescaled Ricker recursion having equilibrium point $x^* = 1$,

$$F(x) = x \times e^{r(1-x)}$$

and proved the following theorem ([192]):

- if $0 < r < 2$ then $x^* = 1$ is a stable equilibrium as seen in Fig. C.1 (since $F\prime(x^*) = 1 - r$ is less than 1 in module);
- if $2 = r_1 < r_2 < ...r_k < ...r_\infty = 2.6924$, the series $\{x_n\}$ undergoes limit cycle oscillation of period $2^k$. Figure C.2 shows a cycle (period 2), while a double cycle (period 4) is illustrated in Fig. C.3;
- if $r > r_\infty$, then chaos happens as in Fig. C.4, *i.e.*, there exist cycles of every period $2, 3, 4...$ along with an uncountable number of initial population levels for which the system does not eventually settle into any finite cycle.

Elements about the chaotic behavior theory for dynamics systems following Eq. (C.4) can be found in textbooks such as Iooss [145] or Ruelle [267].

---

## C.2 Optimal harvesting with Ricker deterministic behavior

### C.2.1 Standard optimization of a dynamic system

For dynamic systems, it is sometimes convenient to express $f(x_t, u_t,$ the increase of population state $x_t$ from one time step to the next one under the control $u_t$:

$$f(x, u) = F(x - u) - x$$

Control theory deals with the search of a sequence of successive controls $u_{1:T} = \{u_t\}_{t=1:T}$ so as to optimize cumulated benefits encountered by the trajectory of a dynamic system $x_{t+1} = x_t + f(x_t, u_t)$ (starting at time 0 from initial state $x_0$ ) until a final time $T$.

$$J(u_{1:T}) = \underset{u_{1:T}}{Max} \sum_{t=1}^{T} L_t(x_t, u_t) \tag{C.5}$$

Cumulated benefits are mostly discounted quantities of the same reward so that $L_t(x, u) = \rho^{t-1}(B(x, u))$. In fishery economic literature, benefits often exhibit two components corresponding to selling the harvest $u$ at constant price $p$ and implementing the fishing effort necessary to harvest the stock $s$ from $s = x$ to $s = x - u$.

### C.2.2 Hamiltonian in the discrete case

To find the optimal sequence, the mathematical trick is to work with a dual variable $\lambda_t$, the so-called Lagrange multiplier to be interpretated as shadow price. The Lagrangian $\sum_{t=1}^{T} L_t(x_t, u_t) + \sum_{t=1}^{T} \lambda_t \{x_{t+1} - x_t + f(x_t, u_t)\}$ corresponds to the discrete time optimization given by Eq. (C.5) under constraints given by Eq. (C.3). Its partial derivatives with regards to $u$ and $\lambda$ have to be set to zero. Introducing the Hamiltonian $H_t(x, u, \lambda)$ as :

$$H_t(x, u, \lambda) = \rho^{t-1}(B_t(x, u)) + \lambda_t \times f(x, u)$$

necessary conditions for optimality are derived as follows:

$$\begin{cases} \dfrac{\partial H_t(x,u,\lambda)}{\partial u} = 0 \\ -\dfrac{\partial H_t(x,u,\lambda)}{\partial x} = \lambda_t - \lambda_{t-1} \\ \dfrac{\partial H_t(x,u,\lambda)}{\partial \lambda} = x_{t+1} - x_t \end{cases}$$

As a consequence, the first derivative of the Hamiltonian yields to:

$$\frac{\partial H_t(x,u,\lambda)}{\partial u} = \rho^{t-1}(\frac{\partial B(x_t, u_t)}{\partial u}) - \lambda_t F'(x_t - u_t)$$

$$0 = \rho^{t-1}\frac{\partial B(x_t, u_t)}{\partial u} - \lambda_t F'(x_t - u_t) \quad \text{(C.6)}$$

and the second one yields to:

$$-\frac{\partial H_t(x,u,\lambda)}{\partial x} = -\rho^{t-1}\frac{\partial B(x_t, u_t)}{\partial x} - \lambda_t\left(F'(x_t - u_t) - 1\right)$$

$$\lambda_t - \lambda_{t-1} = -\rho^{t-1}\frac{\partial B(x_t, u_t)}{\partial x} - \lambda_t\left(F'(x_t - u_t) - 1\right)$$

$$-\lambda_{t-1} = -\rho^{t-1}\frac{\partial B(x_t, u_t)}{\partial x} - \lambda_t\left(F'(x_t - u_t)\right)$$

Combining with Eq. (C.6), this can be simplified as:

$$\lambda_{t-1} = \rho^{t-1}\frac{\partial B(x_t, u_t)}{\partial x} + \rho^{t-1}\frac{\partial B(x_t, u_t)}{\partial u} \quad \text{(C.7)}$$

and we can rewrite Eq. (C.6) at time $t-1$ if we later want to get rid of $\lambda_{t-1}$ :

$$\lambda_{t-1} \times F'(x_{t-1} - u_{t-1}) = \rho^{t-2}\frac{\partial B(x_{t-1}, u_{t-1})}{\partial u}$$

### C.2.3 Searching for a stationary decision rule

Equations (C.6) and (C.7) can be solved with a singular solution if we let $T \to \infty$ and admit some equilibrium point $(x^*, u^*)$ of the stock-recruitment model with harvest $u^*$ such that $u^* = cste$, $F(x^* - u^*) = x^*$. This asymptotic behavior obtained by Clark [60], page 237, is such that

$$\rho^{-1} = F'(x-u)\frac{\frac{\partial B(x,u)}{\partial x} + \frac{\partial B(x,u)}{\partial u}}{\frac{\partial B(x,u)}{\partial u}}$$

There are many expressions for the fishing effort. Among other possibilities, we can imagine a (decreasing) marginal cost $c(s)$ so that the last fish are the most costly to extract. In what follows, we assume:

$$\begin{aligned} B(x,u) &= p \times u - \int_{x-u}^{x} c(s)ds \\ &= \int_{x-u}^{x} \{p - c(s)\}\, ds \end{aligned}$$

and

$$\begin{cases} \dfrac{\partial B}{\partial u} = p - c(x-u) \\ \dfrac{\partial B}{\partial x} = -c(x) + c(x-u) \end{cases}$$

With the previous hypotheses, the long-term optimal escapement $x^* - u^*$ is such that:

$$\rho^{-1} = F'(x^* - u^*)\frac{p - c(F(x^* - u^*))}{p - c(x^* - u^*)} \tag{C.8}$$

### C.2.4 Special cases

If there is no exploitation cost ($c(s) = 0$):

$$\begin{cases} x^* = F(x^* - u^*) \\ F'(x^* - u^*) = 1 + \left(\dfrac{1-\rho}{\rho}\right) \end{cases}$$

$F'^{-1}(\rho^{-1})$ gives the optimum escapement, with $\left(\frac{1-\rho}{\rho}\right)$ being the interest rate.

The mean sustainable yield is a very special sub-case obtained for $\rho = 1$, *i.e.*, weighting present and future linear benefits of harvests equivalently. By setting $S^* = x^* - u^*$ and $C^* = u^*$, one retrieves the equations already presented in Chapter 7, page 154.

$$\begin{cases} F'(S^*) - 1 = 0 = \alpha \times \exp(-\beta S^*)\{1 - \beta S^*\} - 1 \\ \alpha \times S^* \exp(-\beta S^*) - S^* = C^* \end{cases}$$

or, equivalently:

$$\begin{cases} \beta = \frac{1}{S^*}\{\frac{C^*}{C^* + S^*}\} \\ log(\alpha) = \log(\frac{C^* + S^*}{S^*}) + \{\frac{C^*}{C^* + S^*}\} \end{cases}$$

---

## C.3 Stochastic Ricker model

### C.3.1 A stochastic controlled state equation

Let's introduce stochasticity on the stock-recruitment phase of the Atlantic Salmon cycle. Equation (7.4) actually belongs to some more general situation proposed by Reed [250] with Eq. (C.9), where $\zeta$ is a stochastic multiplier with mean 1. The random variables $\{\zeta_t\}_{t=1:T}$ are assumed *iid* with probability distribution function $\phi$.

$$\begin{cases} x_{t+1} = \zeta_t \times F(x_t - u_t) \\ \zeta_t \sim \phi() \\ \mathbb{E}(\zeta_t) = 1 \end{cases} \tag{C.9}$$

The optimization program (see Eq. (C.5)) is to be modified by an expectation operator so as to take into account the stochastic terms:

$$J(u_{1:T}) = \underset{u_{1:T}}{Max}\left(\mathbb{E}_{\zeta_1,\zeta_2,\cdots}\left(\sum_{t=1}^{T} L_t(x_t, u_t)\right)\right)$$

### C.3.2 Dynamic stochastic programming

The common approach to solve the previous equation is to make recourse to dynamic programming ([23]) and consider some piece of an optimum trajectory from current state $x_t$ to the terminal period $T$. The Bellman function $V_t(x_t)$ is the cumulated optimum expected benefit along this path when starting from state $x_t$ at time $t$.

$$V_t(x_t) = \underset{u_{t+1:T}}{Max}\left(\mathbb{E}_{\zeta_{t+1},\zeta_{t+2},\cdots}\left(\sum_{k=t+1}^{T} L_k(x_k, u_k)\right)\right)$$

It is straightforward to show that optimum is fulfilled thanks to the sequential backward Bellman equation:

$$V_t(x) = \underset{u}{Max}\{\mathbb{E}_{\zeta_t}(B(x,u) + \rho V_{t+1}\{\zeta_t \times F(x-u)\})\}$$

Assuming regular smooth behavior with regard to expectation and differentiability of the Bellman function, the following recursive differential Bellman equation is of very special interest:

$$\frac{\partial B(x,u)}{\partial u} - \rho \times F'(x-u) \times \int_z V'_{t+1}\{z \times F(x-u)\}\,\phi(z)dz = 0$$

$$\mathbb{E}_\zeta\left(V'_{t+1}\{\zeta \times F(x-u)\}\right) = \rho^{-1} \times \frac{\frac{\partial B(x,u)}{\partial u}}{F'(x-u)} \tag{C.10}$$

### C.3.3 Dynamic stochastic programming gives optimal escapement level

In what follows, we keep on with the hypothesis $B(x,u) = \int_{x-u}^{x}\{p - c_t\}\,dt = K(x) - K(x-u)$. Starting from terminal condition at $t = T$, the optimal control is obviously to harvest all the possible stock, *i.e.*, $u^*(T) = x_t$ since $B(x,u)$ is an increasing function of $u$:

$$V(x_T) = \rho^T\{K(x_t) - K(0)\}$$

At time step $t = T-1$, Eq. (C.10) provides the optimal decision $u$ as a function of some equilibrium escapement $e = x_t - u_t$. Equation (C.11) below is the stochastic counterpart of Eq. (C.8).

$$0 = \{p - c(x-u)\} - \rho \times F'(x-u) \times \left(p - \int_\zeta c\{\zeta \times F(x-u)\}\,\phi(\zeta)d\zeta\right)$$

$$\rho^{-1} = F'(e^*)\frac{(p - \mathbb{E}_\zeta(c\{\zeta \times F(e^*)\}))}{p - c(e^*)} \tag{C.11}$$

$$u^* = \begin{cases} x - e^* & \text{if } x > e^* \\ 0 & \text{otherwise} \end{cases}$$

As in the deterministic case, it can be recursively proven that the optimal control sequences for time steps $t < T-1$ are to enforce, inasmuch as the constraints allow, the same $e^*$ escapement policy given by Eq. (C.11). If there is no cost of exploitation ($c = 0$), the stochastic optimal escapement, if any, takes the same value as in the deterministic case.

### C.3.4 Stochastic case: theoretical and practical issues

There are subtle issues inherent to the stochastic case (see [251]). For instance, the mathematical derivations in Eq. (C.11) and above are grounded on the principle that recruitment is greater than escapement. To achieve such self sustainability, one must assume for any time step $t$:

$$\zeta_t F(x_t - u_t) > (x_t - u_t)$$

which needs *ad hoc* features for the replacement function $F$ and the stochastic perturbation $\zeta$.

In current practice, in addition to the discrete nature of the states, there are many limitations to work with the theoretical optimal policy obtained by Eq. (C.11). First, the current stock is unknown at the time when optimal escapement is specified! Second, such policy is optimal only in error free configurations with no implementation perturbations nor parameter uncertainty. Finally many of the costs and benefits, and even sometimes decisionmakers themselves may remain hidden to the naive mathematical analyst.

---

## C.4 The Ricker management point is not the mean of (quasi) equilibrium distribution $l(\cdot)$

In this section, we point out that the special reference points $(W^*, h^*, R^*)$ (see Table 12.2) calculated from a deterministic vision given by Eq. (12.16), have nothing to do with any limiting equilibrium quantities of the general stochastic model.

Section 12.5 of Chapter 12 showed a simplified model version such that, given the number $z_t$ (of eggs) at stage $t$ and assuming constant average conditions for survival in the soft and marine phases yielding to (the potential deposit of eggs) $z_{t+1}$ at the next stage, the *conditional* expectation for $Z_{t+1}$, is some Ricker function $\Re_{a,b}(z_t)$ of $z_t$, with coefficients $a = a_{eggs}$, $b = b_{eggs}$ in this example where the stock of eggs is the state variable:

$$\mathbb{E}(Z_{t+1}|z_t) = \Re_{a,b}(z_t)$$

It is therefore tempting to interpret the Ricker reference points of $\Re_{a,b}$, for instance the deterministic equilibrium stock $Z^* = \Re_{a,b}(Z^*)$, as some limiting trend of the corresponding stochastic system behavior. Suppose we have some (quasi) equilibrium distribution $l(\cdot)$ for the system state $Z_t$ (hypothesizing we can neglect the probability of extinction),

$$l(z_t) = \int_n [z_t|z_{t-1}]l(z_{t-1})dz_{t-1}$$

Will the pdf $l(\cdot)$ be waving around the previous reference point (for eggs in this example)? Can $Z^*$ be considered as some sort of expected value?

Confusion should not arise between marginal probabilities $[Z_{t+1}]$ and conditional probabilities $[Z_{t+1}|Z_t]$. Denoting $\mu$ the first moment of $l(\cdot)$ if any, one gets as a consequence:

$$\mathbb{E}(Z_{t+1}) = \mathbb{E}_{z_t}\left(\mathbb{E}_{Z_{t+1}|z_t}(Z_{t+1})\right)$$
$$\int_z z \times l(z)dz = \mu = \int_z \Re_{a,b}(z) \times l(z)dz \tag{C.12}$$

$\mu = \Re_{a,b}(\mu)$ would in addition means that the equilibrium point of the mean deterministic Ricker model would be the mean of the equilibrium distribution of the corresponding stochastic model. The rest of this appendix shows that, if $l(\cdot)$ is approximatively Normal, then relation (C.12) involves the variance and, consequently, $\mu \neq \Re_{a,b}(\mu)$!

### C.4.1 Recalling the Normal characteristic function

We know that, $u$ being a Normal($\mu$,$\sigma^2$) random variate

$$\int_{-\infty}^{+\infty} e^{ku}[u]du = e^{k\mu}e^{\sigma^2\frac{k}{2}^2}$$

Therefore

$$\int_{-\infty}^{+\infty} ue^{ku}[u]du = \frac{\partial \mathbb{E}\left(\exp kU\right)}{\partial k} = (\mu + k\sigma^2)e^{k\mu}e^{\sigma 2\frac{k}{2}^2}$$

As a consequence,

$$\mathbb{E}(ue^{-\beta u}) = (\mu - \beta\sigma^2)e^{-\beta\mu}e^{\frac{1}{2}\beta^2\sigma^2}$$

### C.4.2 SR transition

These formulae apply to the stochastic stock/recruitment relationship generalizing Eq. (C.1) with a logNormal noise; given $S$, we assume that log$R$ is a $Normal(0, \sigma^2)$ distribution ,

$$\mathbb{E}(R|S) = \alpha \times Se^{-\beta S}e^{\frac{\sigma^2}{2}} = \Re_{a+\frac{\sigma^2}{2},b}(S)$$

Unconditional to $S$, supposedly Normal with mean $\mu_S$ and variance $\sigma_S^2$

$$\mathbb{E}(R) = \alpha e^{\frac{\sigma^2}{2}}(\mu_S - \beta\sigma_S^2)e^{-\beta\mu_S}e^{\frac{1}{2}\beta^2\sigma_S^2}$$

Therefore, identifying $S$ with $z$ in relation (C.12), the mean of the pdf $l(S)$ is such that:

$$\begin{aligned}\mu_S &= \alpha e^{-\beta\mu_S}(\mu_S - \beta\sigma_S^2)e^{\frac{\sigma^2}{2}}e^{\frac{1}{2}\beta^2\sigma_S^2}\\ &= \Re_{a+\frac{\sigma^2}{2},b}(\mu_S)\times\left(1-\frac{\beta\sigma_S^2}{\mu_S}\right)\times e^{\frac{1}{2}\beta^2\sigma_S^2}\end{aligned}$$

For the deterministic case when $\sigma_S^2 = \sigma^2 = 0$, we do check $\mu_S = \Re_{a,b}(\mu_S)$. But by no means does this equation imply that $\mu_S = \Re_{a+\frac{\sigma^2}{2},b}(\mu_S)$!

# Appendix D

# Examples of predictive and full conditional distributions

## D.1 Predictive for the Logistic model

Let's denote

$$\log \frac{p_i}{1-p_i} = X_i\beta$$

the linear form in the *logit* scale of the probability $p_i$, or equivalently

$$p_i = \frac{\exp(X_i\beta)}{1+\exp(X_i\beta)}$$

Let the data $y$ be distributed following a Binomial distribution

$$[y|\beta] = \prod_i^n dbinom(y_i, n_i, p_i)$$

and the parameters $\beta$ be a priori distributed following a multidimensional Student distribution

$$[\beta] = dmStudent(\sqrt{a}(\beta-\beta_0), (X'X)^{-1}, 2a) \times a^{\frac{p}{2}}$$

Then the joint distribution $[y, \beta]$ writes

$$[y,\beta] = \prod_i^n \frac{\Gamma(1+n_i)p_i^{y_i}(1-p_i)^{n_i-y_i}}{\Gamma(1+y_i)\Gamma(1+n_i-y_i)} \times \left(\frac{1}{2\pi a}\right)^{\frac{p}{2}} \sqrt{|X'X|}$$
$$\times \frac{\Gamma(a+\frac{p}{2})}{\Gamma(a)} \left(1+\frac{(\beta-\beta_0)'(X'X)(\beta-\beta_0)}{2a}\right)^{-a-\frac{p}{2}} \quad \text{(D.1)}$$

## D.2 Predictive for the LogPoisson model

Let's denote the linear form in the *log* scale for the expected abundance $\lambda_i$

$$\lambda_i = Log(X_i \beta)$$

Let the data $y$ be distributed following a Poisson distribution

$$[y|\beta] = \prod_i^n dPois(y_i, \lambda_i)$$

and let the parameters $\beta$ be a priori distributed following a multidimensional Student distribution

$$\beta = dmStudent(\sqrt{a}(\beta - \beta_0), (X'X)^{-1}, 2a) \times a^{\frac{p}{2}}$$

Then the joint distribution $[y, \beta]$ writes

$$[y, \beta] = \left( \prod_i^n \frac{e^{-\lambda_i} \lambda_i^{y_i}}{\Gamma(1 + y_i)} \right) \left( \frac{1}{2\pi c} \right)^{\frac{p}{2}} \sqrt{|X'X|}$$
$$\times \frac{\Gamma(a + \frac{p}{2})}{\Gamma(a)} \left( 1 + \frac{(\beta - \beta_0)'(X'X)(\beta - \beta_0)}{2a} \right)^{-a-\frac{p}{2}} \quad \text{(D.2)}$$

## D.3 Full conditional for the categorial probit model

We detail here how the data augmentation approach of Albert and Chib [3] renders easy the inference of the ordered multinomial probit model presented on page 187 of Chapter 8 for the skate data.

We further assume *prior* independence for $[\delta, \beta]$

$$[\delta, \beta] = [\delta] \times [\beta]$$

More specifically, we pick priors in the Normal family for conjugacy reasons:

$$\begin{cases} [\delta] = dmnorm(\delta, \delta_0, D) 1_{\delta_1 < \ldots < \delta_{J-1} < \delta_{J-1}} \\ [\beta] = dmnorm(\beta, \beta_0, \Sigma_0) \end{cases}$$

The full conditionals are as follows:

1. Since

$$[\mathbf{Z}\,|\delta,\beta,\mathbf{y},\mathbf{x}] \propto \prod_{t=1}^{T}\left\{1_{\delta_{y_{t-1}}\leqslant Z_t<\delta_{y_t}} \times dnorm\left(Z_t,\beta x_t,1\right)\right\} \quad \text{(D.3)}$$

given $\delta$,$\beta$, $\mathbf{y}$, $\mathbf{x}$, the $Z_t$ are truncated independent variables.

2. The vector $\delta$ contains the noninfinite bounds of each category, its full conditional reads as a truncated joint multivariate Normal distribution:

$$[\delta\,|\mathbf{Z},\alpha,\beta,\mathbf{y},\mathbf{x}] \propto dmnorm\left(\delta,\delta_0,D\right) \times I_\delta \times \prod_{t=1}^{T}\left\{1_{\delta_{y_{t-1}}\leqslant Z_t<\delta_{y_t}}\right\}$$

The constraints $I_\delta : \delta_1 < \delta_2 < ... < \delta_{J-1}$ must be respected. Suppose that in addition to that constraint, we took independent priors for each component, *i.e.*, a diagonal prior variance, $D = diag\left(\sigma^2_{\delta_j}\right)$. One can work componentwise. The relationship

$$[\delta\,|\mathbf{Z},\beta,\mathbf{y},\mathbf{x}] \propto dmnorm\left(\delta,\delta_0,diag\left(\sigma^2_{\delta_j}\right)\right) \times \prod_{t=1}^{T}\left\{1_{\delta_{y_t-1}\leqslant Z_t<\delta_{y_t}}\right\}$$

is equivalent to

$$[\delta_j\,|\mathbf{Z},\beta,\delta_{-j},\mathbf{y},\mathbf{x}] \propto dnorm\left(\delta_j,\delta_{j0},\sigma^2_{\delta_j}\right) \times 1_{\delta_j^{\inf}\leq\delta_j\leq\delta_j^{\sup}} \quad \text{(D.4)}$$

for $\delta_j$, $j = 1,...,J-1$ with $\delta_j^{\inf} = \max\left\{\max\left\{Z_t : y_t = j\right\};\delta_{j-1}\right\}$ and $\delta_j^{\sup} = \min\left\{\min\left\{Z_t : y_t = j+1\right\};\delta_{j-1}\right\}$. If needed, the normalizing constant $\left(\int_{\delta_j^{\inf}}^{\delta_j^{\sup}} \mathbf{N}\left(\delta_j\left|\delta_{j0},\sigma^2_{\delta_j}\right.\right)d\delta_j\right)^{-1}$ can be evaluated via a simple routine based on the univariate Normal cumulative function.

3. The full conditional for $\beta$ stems from the Normal conjugate property. From

$$[\beta\,|Z_0,\mathbf{Z},\delta,\mathbf{y}] \propto dmnorm\left(\beta,\beta_0,\Sigma_0\right) \times \prod_{t=1}^{T}\left\{\mathbf{N}\left(Z_t\,|\beta x_t,1\right)\right\}$$

one can derive:

$$[\beta\,|\mathbf{Z},Z_0,\delta,\mathbf{y}] = dmnorm\left(\beta,\widehat{\beta}_Z,\Sigma^{-1}\right) \quad \text{(D.5)}$$

with

$$\begin{cases} \Sigma = X'X + \Sigma_0 \\ \widehat{\beta}_Z = \Sigma^{-1}\left(X'Z + \Sigma_0\beta_0\right) \end{cases}$$

Equations (D.3)+(D.4)+(D.5) can be handled and a Gibbs algorithm can be designed for the Bayesian inference of the probit model. The Gibbs algorithm iterates between the steps:

$$\begin{cases} \textit{sample } [Z_t \,|Z_{\neq t}\,, \delta, \beta, \mathbf{y}], \\ \textit{sample } [\delta_j \,|\mathbf{Z}, \delta_{\neq j}, \beta, \mathbf{y}], \\ \textit{sample } [\beta \,|\mathbf{Z}, \delta, \mathbf{y}] \end{cases} \tag{D.6}$$

# Appendix E

# *The baseball players' historical example: A simple introduction to hierarchical modeling*

## E.1 The baseball players' example

Efron and Morris [98] brilliantly exemplified via an interesting sport dataset that the James-Stein estimator ([286]) based on each player's first 45 at bats does perform better at predicting subsequent performance than their observed averages. In this appendix[1], we revisit this example as a simple introduction to hierarchical modeling. The batting successes $y_i$ of 18 major league baseball players ($i = 1, ..., 18$) in their first 45 at bats of the 1970 season are given in the first column of Table E.1. The batting average defined as the ratio of a player's hits $y_i$ to his at bats (45 in this example), is one of the best acknowledged of all baseball statistics. They are given in the second column of Table E.1 and play the role of estimates $\hat{\mu}_i = \frac{y_i}{45}$ for the *true* $\mu_i$, *i.e.*, the unknown skill of player $i$ and the empirical score $\hat{\mu}_i$ has for long been considered as the classical best bet for the underlying true averages $\mu_i$. In addition, the two last columns provide a 90% Bayesian posterior credible interval for $\mu_i$ using a noninformative Uniform prior for $\mu_i$ with the Binomial model $Y_i \sim Binomial(\mu_i, 45)$ showing that a large uncertainty is attached to these estimates.

Now, before revealing the *true* $\mu_i$ in the next section, can we propose any better betting procedure than the empirical score $\hat{\mu}_i$?

[1]This supplementary material can be used as an introduction to Chapters 9 and 10.

| Players | First 45 at bats $y_i$ | Score $\hat{\mu}_i$ | IC 5% | IC 95% |
|---|---|---|---|---|
| Clemente | 18 | 0.400 | 0.254 | 0.546 |
| Robinson | 17 | 0.378 | 0.233 | 0.523 |
| Howard | 16 | 0.356 | 0.213 | 0.499 |
| Johnstone | 15 | 0.333 | 0.192 | 0.474 |
| Berry | 14 | 0.311 | 0.173 | 0.449 |
| Spencer | 14 | 0.311 | 0.173 | 0.449 |
| Kessinger | 13 | 0.289 | 0.154 | 0.424 |
| Alvarado | 12 | 0.267 | 0.135 | 0.399 |
| Santo | 11 | 0.244 | 0.116 | 0.372 |
| Swoboda | 11 | 0.244 | 0.116 | 0.372 |
| Unser | 10 | 0.222 | 0.098 | 0.346 |
| Williams | 10 | 0.222 | 0.098 | 0.346 |
| Scott | 10 | 0.222 | 0.098 | 0.346 |
| Petrocelli | 10 | 0.222 | 0.098 | 0.346 |
| Rodriguez | 10 | 0.222 | 0.098 | 0.346 |
| Campaneris | 9 | 0.200 | 0.081 | 0.319 |
| Munson | 8 | 0.178 | 0.064 | 0.292 |
| Alvis | 7 | 0.156 | 0.048 | 0.264 |

**TABLE E.1**: 18 major league baseball players in their first 45 at bats of the 1970 season. The prior for all scores is taken Uniform on [0,1].

## E.2 Borrowing strength from neighbors

### E.2.1 Estimate Williams' skills using all players' performances

Take for instance the case of player Williams ($k = 12$) with $y_k = 10$ successes out of 45 trials. The scores of the remaining 17 other players in the first column of Table E.1 can be considered as sample of batting averages of baseball professionals and this piece of information will be exploited to derive a prior distribution for $\mu_k$. To benefit from the Beta-Binomial conjugate properties as already seen in Chapter 2, we pick a Beta distribution with coefficients $a$ and $b$ to be estimated. The empirical mean and variance are respectively 0.268 and 0.005, and we can fit the two coefficients of the Beta pdf by the method of moments:

$$\begin{cases} \dfrac{a}{a+b} = 0.2679 \\ \dfrac{a}{a+b} \times \dfrac{b}{a+b} \times \dfrac{1}{a+b+1} = 0.005 \end{cases}$$

yielding $a = 10.18$, $b = 27.80$.

Keeping with the interpretation of Beta coefficients as prior distribution for Binomial trials, $a$ is to be interpreted as a prior number of successes while $b$ turns to be equivalent to a prior number of failures. In other words, we learn from the fellow players of Williams that a professional baseball player can a priori obtain around 10 successes at bats out of $10+28 = 38$ trials. Following the Beta-Binomial model from Eq. (2.8) with such a prior, a Bayesian predictor $\tilde{\mu}_k$ would be the posterior expectation of $\mu_k$, *i.e.*,

$$\begin{aligned} \tilde{\mu}_k &= \frac{y_k + a}{45 + a + b} \\ \tilde{\mu}_k &\approx \frac{10 + 10.18}{45 + 10.18 + 27.80} = 0.24 \end{aligned} \tag{E.1}$$

To sum it up, after transferring some information from the colleagues of Williams, we would rather propose to bet on $\tilde{\mu}_k = 0.24$ than on the empirical score $\hat{\mu}_k = 0.22$ computed from the sole individual results of Williams.

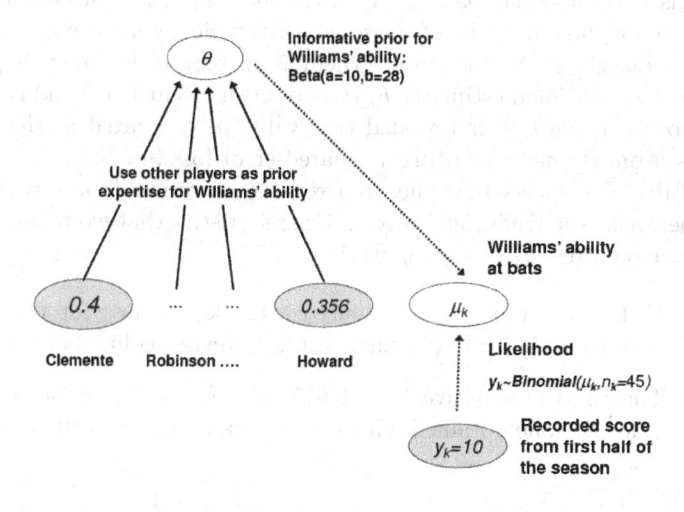


**FIGURE E.1**: Bayesian approach to assess Williams' ability at bat. Borrowing strength from the other players to estimate the skill $\mu_k$ for the player $k$. Note that this figure does not follow the conventions of Directed Acyclic Graphs as arrows indicate the flow of information rather than probabilistic conditioning.

Figure E.1 sums up the Bayesian association of the two sources of information that were combined to provide an estimation of William's ability at bats:

- The other players as a prior beta pdf;
- Its own recorded performance during the first 45 at bats.

### E.2.2 Winning bets

We could repeat the previous procedure for every player to obtain the estimation $\tilde{\mu}_i$, of each player's average ($i = 1, ..., 18$) for the remainder of the season. As a crude approximation, we assume here the same Beta prior for all players ($a = 10.18, b = 27.8$). Using a simple spreadsheet, Table E.2 gives estimates $\tilde{\mu}_i$ that we call Bayesian estimates in as much as they follow the Bayesian spirit of combining the two sources of information.

In most problems, the value of the unknown remain unrevealed allowing statisticians to elaborate on competing theories and endlessly discuss their merits around a cup of tea or some other stronger beverage. Yet, in this particular baseball example, the nice thing is that each player's average for the remainder of the season will typically involve several hundred more at bats so we can use them as surrogates for the true $\mu_i$ and compare traditional estimates $\hat{\mu}_i$ (based on the 45 at bats) and Bayesian proposals $\tilde{\mu}_i$ with their revealed true value $\mu_i$ estimated at the end of the season, for instance using a squared error loss function.

Table E.2 shows that the approximate Bayesian procedure clearly outperforms the traditional way of doing statistics that would only trust the personal results of each player:

- 15 times out of 18 the squared error loss is less for the former procedure (indicated by an asterisk in the last column of the table),
- The total cumulative loss (0.075) for the standard procedure is twice the one obtained with the Bayesian approach (0.032)!

## E.3 Fully exchangeable model

Figure E.1 can be transformed into a fully symmetric structure called a *hierarchical model*. The joint pdf for all quantities involved in the model can be read from the hierarchical structure and the conditional independence of the directed acyclic graph of Fig. E.2. This structure

| Players $i$ | True skill $\mu_i$ | Empirical score $\hat{\mu}_i$ | Loss $(\mu_i - \hat{\mu}_i)^2$ | Bayesian estimate $\tilde{\mu}_i$ | Loss $(\mu_i - \tilde{\mu}_i)^2$ |
|---|---|---|---|---|---|
| Clemente | 0.346 | 0.400 | 0.003 | 0.340 | 0.0000* |
| Robinson | 0.298 | 0.378 | 0.006 | 0.328 | 0.0009* |
| Howard | 0.276 | 0.356 | 0.006 | 0.315 | 0.0016* |
| Johnstone | 0.222 | 0.333 | 0.012 | 0.303 | 0.0066* |
| Berry | 0.273 | 0.311 | 0.001 | 0.291 | 0.0003* |
| Spencer | 0.270 | 0.311 | 0.002 | 0.291 | 0.0005* |
| Kessinger | 0.263 | 0.289 | 0.001 | 0.279 | 0.0003* |
| Alvarado | 0.210 | 0.267 | 0.003 | 0.267 | 0.0033 |
| Santo | 0.269 | 0.244 | 0.001 | 0.255 | 0.0002* |
| Swoboda | 0.230 | 0.244 | 0.000 | 0.255 | 0.0006 |
| Unser | 0.264 | 0.222 | 0.002 | 0.243 | 0.0004* |
| Williams | 0.256 | 0.222 | 0.001 | 0.243 | 0.0002* |
| Scott | 0.303 | 0.222 | 0.007 | 0.243 | 0.0036* |
| Petrocelli | 0.264 | 0.222 | 0.002 | 0.243 | 0.0004* |
| Rodriguez | 0.226 | 0.222 | 0.000 | 0.243 | 0.0003 |
| Campaneris | 0.285 | 0.200 | 0.007 | 0.231 | 0.0029* |
| Munson | 0.316 | 0.178 | 0.019 | 0.219 | 0.0094* |
| Alvis | 0.200 | 0.156 | 0.002 | 0.207 | 0.0000* |

**TABLE E.2**: *True* batting average calculated at the end of the season ($\mu_i$), average of the 45 first events (traditional estimate $\hat{\mu}_i$) and Bayesian estimate ($\tilde{\mu}_i$) for the 18 professional Baseball players in 1970.

involves two types of unknowns (latent variable and parameter) and one layer of observables:

- The *latent variable* $\mu_i$ here is the key concept that corresponds to the unobserved skill of player $i$;
- The latent variables are linked to the data $y_i$ by a *process of observations* (here a Binomial experiment);
- The latent variables are also linked together because they can be considered as stochastic draws from an *urn of resemblance* that would say what the standard baseball player looks like. This urn (we chose a Beta pdf with *population parameters* $\theta = (a, b)$) mimics the variations among players;
- In addition, external information can be encoded as a prior pdf for the parameter $\theta$ in a Bayesian setting.

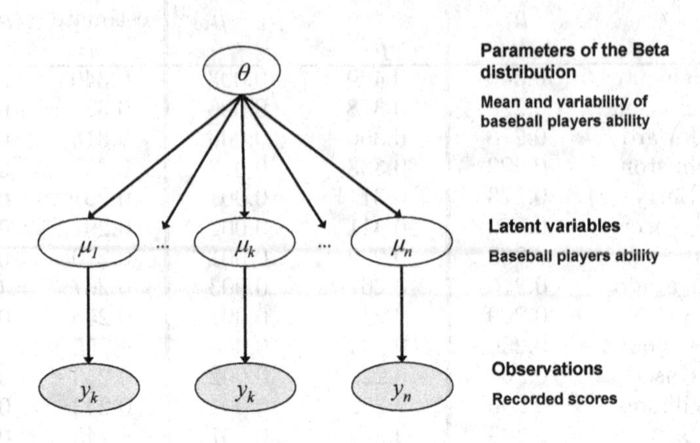


**FIGURE E.2**: Hierarchical Bayesian model for the 1970 baseball players.

## E.3.1 Prior distributions

### E.3.1.1 Prior on population parameters $\theta = (a, b)$

In this application, we propose to design the prior pdf $[a, b]$ by assigning independent distributions to $\bar{\mu} = \frac{a}{a+b}$ and $\nu = a + b$. Parameter $\bar{\mu}$ is the expected score of a professional baseball player and a convenient noninformative prior for $\bar{\mu}$ is a Uniform distribution on $[0, 1]$. Gamblers aware of baseball performances may argue that an informative Beta distribution will certainly be preferable: For baseball fans a batting average of 0.3 tends to be regarded as exceptionally good while a 0.4 average seems to be almost out of reach except for some of the greatest players in baseball history. Parameter $\nu$ is equivalent to a virtual number of prior baseball trials. We chose 1 virtual trial as a lower bound, constraining $a$ and $b$ to be both greater than $\frac{1}{2}$ in the Beta pdf. An upper bound $\nu_{\max}$ has also to be fixed. The bigger $\nu$, the more influential will be the prior information so taking $\nu_{\max} = 45$ is a reasonable choice to put no more weight on the prior than on one player's results. As $\nu$ is scaling the prior precision, we suggest to take $log(\nu)$ uniformly distributed on the interval $[log(1), log(\nu_{\max})]$ so as to remain invariant under a change of scale for the units of $\nu$. This choice will conveniently put more prior weight on small values for $\nu$. Writing for infinitesimal increments $d\bar{\mu}, dlog(\nu), da, db$

the conservation of probability mass under a change of variables

$$[\bar{\mu}] \times [log(\nu)] : d\bar{\mu} : dlog(\nu) = [a, b] : da : db.$$

Because the prior densities $[\bar{\mu}]$ and $[log(\nu)]$ are proportional to constants ($\sim$Uniform), this prior construction yields:

$$[a, b] \propto \left\| \frac{\Delta(\bar{\mu}, \log(\nu))}{\Delta(a, b)} \right\|$$

with the Jacobian $\left\| \frac{\Delta(m, \log \nu)}{\Delta(a,b)} \right\|$ such that

$$\left\| \frac{\Delta(\bar{\mu}, \log(\nu))}{\Delta(a, b)} \right\| = \det \begin{pmatrix} \frac{b}{(a+b)^2} & \frac{1}{(a+b)} \\ -\frac{a}{(a+b)^2} & \frac{1}{(a+b)} \end{pmatrix}.$$

Therefore, with such assumptions, the prior on $[a, b]$ is:

$$[a, b] \propto \frac{1}{(a+b)^2} \times 1_{a > \frac{1}{2}} \times 1_{b > \frac{1}{2}} \times 1_{a+b < \nu_{\max}} \tag{E.2}$$

#### E.3.1.2 Priors structure for the latent layer: Exchangeability

Given $(a, b)$, the $\mu_i$'s ($i = 1, ..., n$ with $n = 18$ in our example) are a priori independent. They are conditionally distributed as Beta random variables with parameters $(a, b)$:

$$[\mu_{1:n} | a, b] = \prod_{i=1}^{n} [\mu_i | a, b] \tag{E.3}$$

Unconditionally to $(a, b)$, they are correlated and their joint distribution is written as:

$$[\mu_{1:n}] = \oint_{a,b} \prod_{i=1}^{k} [\mu_i | a, b] \times [a, b] : da : db \tag{E.4}$$

As the pdf given in Eq. (E.4) is left unchanged by any permutation of the players' indices, the latent $\mu_i$'s are said to be *exchangeable* random variables.

### E.3.2 Posterior distributions and borrowing strength

#### E.3.2.1 Posterior distributions

The probabilistic link between all $\mu_i$'s which is established a priori by Eq. (E.4) allows for borrowing strength from the different units $i$ when updating the $\mu_i$'s by all data $y_{1:n}$.

For a Bayesian analysis, the joint distribution of all probabilistic quantities (*i.e.*, parameters, latent variables and observables) sums up the whole model structure that factorizes as:

$$\begin{aligned}&[Y_{1:n} = y_{1:n}, \mu_{1:n}, a, b] \\ &= [a, b] \times [\mu_{1:n}|a, b] \times [Y_{1:n} = y_{1:n}|\mu_{1:n}]\end{aligned} \tag{E.5}$$

Given $(a, b)$, the $\mu_i$'s are independent as shown in Fig. E.2. Given the $\mu_i$'s, the observables $Y_i$'s are independent and are conditionally distributed as Binomial random variables:

$$[Y_{1:n} = y_{1:n}|\mu_{1:n}] = \prod_{i=1}^{n} [Y_i = y_i|\mu_i]$$

The full joint distribution (Eq. (E.5)) then writes

$$\begin{aligned}&[Y_{1:n} = y_{1:n}, \mu_{1:n}, a, b] \\ &= \prod_{i=1}^{n} [Y_i = y_i|\mu_i] \times \prod_{i=1}^{n} [\mu_i|a, b] \times [a, b]\end{aligned} \tag{E.6}$$

The joint likelihood $[Y_{1:n} = y_{1:n}|a, b]$ is obtained by integrating out the latent variables:

$$\begin{aligned}&[Y_{1:n} = y_{1:n}|a, b] \\ &= \oint_{\mu_{1:n}} \prod_{i=1}^{n} [Y_i = y_i|\mu_i] \times \prod_{i=1}^{n} [\mu_i|a, b]\, d\mu_{1:n}\end{aligned} \tag{E.7}$$

The joint posterior distribution of parameters and latent variables is straightfully derived:

$$\begin{aligned}&[\mu_{1:n}, a, b|y_{1:n}] \\ &= \frac{1}{K(y_{1:n})} \times [a, b] \times \prod_{i=1}^{n} [Y_i = y_i|\mu_i] \times \prod_{i=1}^{n} [\mu_i|a, b]\end{aligned} \tag{E.8}$$

with the constant of integration $K(y_{1:n})$

$$K(y_{1:n}) = \oint_{a,b,\mu_{1:n}} \prod_{i=1}^{n} [Y_i = y_i|\mu_i] \times \prod_{i=1}^{n} [\mu_i|a, b] \times [a, b]\, da\, db\, d\mu_{1:n}$$

Therefore the marginal posterior distribution of parameters $(a, b)$ is obtained by integrating out the full joint posterior in Eq. (E.8) over all values of $(\mu_{1:n})$:

$$[a,b|y_{1:n}] = \frac{[a,b]}{K(y_{1:n})} \times \oint_{\mu_{1:n}} \prod_{i=1}^{k}[Y_i = y_i|\mu_i] \times \prod_{i=1}^{k}[\mu_i|a,b]\, d\mu_{1:n} \tag{E.9}$$

Similarly, the marginal posterior distribution of the latent variables $(\mu_1, ..., \mu_n)$ is:

$$[\mu_{1:n}|y_{1:n}] = \oint_{a,b} [\mu_{1:n}, a, b|y_{1:n}]\, da\, db \tag{E.10}$$

#### E.3.2.2 Borrowing strength

The *borrowing strength* concept (also nicely called the "Robin Hood" approach by Punt *et al.* [242]) is nicely illustrated by looking at the marginal posterior distribution of $\mu_k$ for one particular player $k$. Because all units $i$ are linked together through the hierarchical structure (a priori as in Eq. (E.4) and a posteriori as in Eq. (E.10)), the marginal posterior distribution of $\mu_k$ is conditioned by the data of all units $i$. It is given by integrating out Eq. (E.10) over all $\mu_i$'s except $\mu_k$:

$$[\mu_k|y_{1:n}] = \frac{1}{K(y_{1:n})} \times \oint_{a,b} [a,b] \times \oint_{\mu_i \neq \mu_k} \prod_{i \neq k}[Y_i = y_i|\mu_i] \times \prod_{i \neq k}[\mu_i|a,b]\, d\mu_{i \neq k} \times [Y_k = y_k|\mu_k] \times [\mu_k|a,b]\, da\, db \tag{E.11}$$

The previous formula (Eq. (E.11)) points out that the posterior knowledge about the skill of player $k$ is influenced by its own performance $[Y_k = y_k|\mu_k]$ (in the right-hand term of Eq. (E.11) which plays the role of the likelihood) but also by his fellow players via the left-hand term under the integral sign

$$[a,b] \times \oint_{\mu_i \neq \mu_k} \prod_{i \neq k}[Y_i = y_i|\mu_i] \times \prod_{i \neq k}[\mu_i|a,b] : d\mu_{i \neq k}.$$

This latter term can be seen as an update of the prior $[a, b]$ by all data except $y_k$ so that one should draw attention to the following pdf $[a, b|y_{i \neq k}]$ which is the marginal of the posterior of $(a, b)$ with all data except $y_i$ :

$$[a,b|y_{i \neq k}] = \frac{1}{K'(y_{i \neq k})} \times \oint_{\mu_i \neq \mu_k} [a,b] \times \prod_{i \neq k}[Y_i = y_i|\mu_i] \times \prod_{i \neq k}[\mu_i|a,b]\, d\mu_{i \neq k} \tag{E.12}$$

with the constant of integration

$$K'(y_{i\neq k}) = \oint_{a,b} \left( [a,b] \times \oint_{\mu_i \neq \mu_k} \prod_{i\neq k} [Y_i|\mu_i] \times \prod_{i \neq k} [\mu_i|a,b] \, d\mu_{i\neq k} \right) da \, db$$

Then in Eq. (E.11), the left-hand side which is closed to the posterior (Eq. (E.12)), does in fact play the role of a prior for $(a, b)$ but that already integrates the information conveyed by the data $y_{i\neq k}$. Equation E.11 can then be written more explicitly using the posterior $[a, b|y_{i\neq k}]$:

$$[\mu_k|y_{1:n}] = \frac{K'(y_{i\neq k})}{K(y_{1:n})} \oint_{a,b} [a,b|y_{i\neq k}] \times [Y_k = y_k|\mu_k] \times [\mu_k|a,b] \, da \, db \tag{E.13}$$

Equation E.13 clearly illustrates how the hierarchical structure borrows strength from the data $y_{i\neq k}$ to update the skill of the player $k$ through the population parameters $(a, b)$.

---

## E.4 Shrinkage effect in the exchangeable hierarchical structure

The method we suggested in Section E.2 to assess the $\mu_i$'s is the *empirical Bayes* procedure ([47];[48]). It is the weird marriage of:

- A frequentist point of view (such as the method of moments that we used) is taken to assess the parameters $\theta = (a, b)$ from the estimates $\hat{\mu}_i$ of the latent variables;
- A standard Bayesian updating technique that provides empirical posterior distributions of the latent $\mu_i$'s.

This empirical Bayes procedure is an approximation of the rigorous Bayesian solution given by Eq. (E.8) (which is developed with Eqs. (E.9) and (E.13)).

A major effect occurs for the exchangeable hierarchical model: For a player $k$ like Williams, taking into account the results of the other players makes a compromise between the *average* player with $a$ succes and $b$ failures and the observed first 45 at bats with respectively $y_k$ success

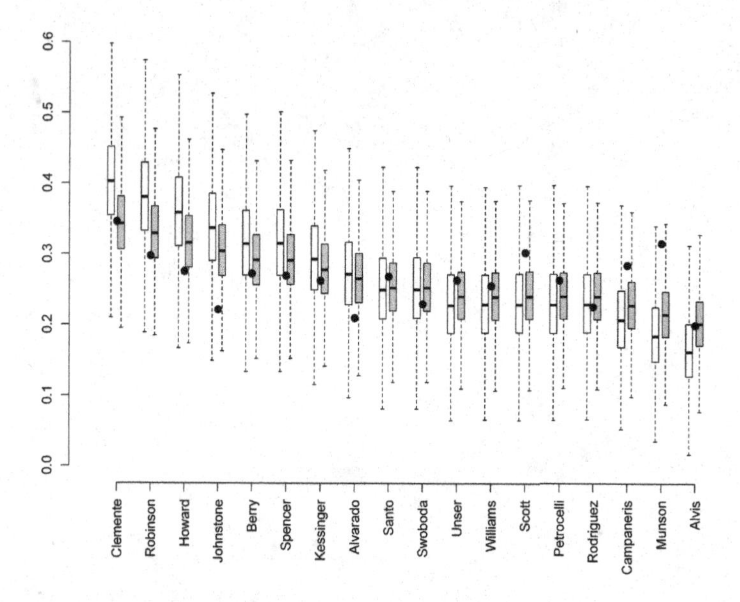


**FIGURE E.3**: Posterior pdf of baseball players' performance in 1970, computed under models assuming independence (white boxes) and hierarchically structured (greyed boxes). The dots are performances at the end of the season (surrogates for true values $\mu_i$).

and $45 - y_k$ failures. This compromise is clearly given by Eq. (E.1) for the empirical Bayes procedure and by the more elaborate Eq. (E.13) for the complete Bayesian solution. If the present score of player $k$ is lower than the average man $\frac{a}{a+b}$, his predictive score for the remainder of the season will be pushed up (as it is the case for Williams) because this first poor performance may happen due to bad luck. In the reverse case, the predictive bet will be diminished. This effect (see Fig. E.3 for the baseball example) is known as the *shrinkage effect* in the Bayesian literature.

# Bibliography


[1] C. Abraham and B. Cadre. Asymptotic global robustness in Bayesian decision theory. *Annals of Statistics*, 32:1341–1366, 2004.

[2] J.H. Albert. *Bayesian Computation with R.* Springer-Verlag, New York, NY, 2007.

[3] J.H. Albert and S. Chib. Bayesian analysis of binary and polychotomous response data. *Journal of the American Statistical Association*, 88(422):669–679, 1993.

[4] M. Allais. Le comportement de l'homme rationnel devant le risque: critique des postulats et axiomes de l'école américaine. *Econometrica*, 21(4), 1953.

[5] M. Allais. *Expected Utility Hypotheses and the Allais Paradox*, chapter 12: The so-called Allais paradox and rational decisions under uncertainty, pages 437–681. Dordrecht/Boston, Reidel, 1979.

[6] J.D. Armstrong, P.S. Kemp, G.J.A. Kennedy, M. Ladle, and N.J. Milner. Habitat requirements of Atlantic salmon and brown trout in rivers and streams. *Fisheries Research*, 62:143–170, 2003.

[7] P.R. Armsworth and J.E. Roughgarden. The economic value of ecological stability. *Proceedings of the National Academy of Science*, 100(12):7147–7151, 2003.

[8] B.C. Arnolds, E. Castillo, and J.M. Sarabia. *Conditional Specification of Statistical Models.* Springer-Verlag, New York, 1999.

[9] Y. Atchadé and J.S. Rosenthal. On adaptive Markov chain Monte Carlo algorithms. Technical report, Université de Montréal, 2003.

[10] J.-P. Aubin. *Viability Theory.* Modern Birkhauser Classics, Boston, 1985.

[11] J.-P. Aubin. *Dynamical Economic Theory: a Viability Approach.* Springer, Heidelberg, 1996.

[12] R. Axelrod. *The Evolution of Cooperation.* Basic Books, 1984.

[13] A. Baddeley, P. Gregori, J. Mateu, R. Stoica, and D. Stoyan, editors. *Case Studies in Spatial Point Process Modeling.* Lecture Notes in Statistics. Springer, New York, NY, 2005.

[14] J. Baglinière, P. Bomassi, B. Bousquet, F. Chancerel, H. DePontual, J. Dumas, G. Euzenat, G. Fontenelle, F. Fournel, F. Gayou, J. Luquet, G. Maisse, J. Martin Ventura, A. Marty, A. Nihouarn, J. Porcher, E. Prévost, P. Prouzet, G. Pustelnik, A. Richard, and H. Troadec. La détermination de l'age par scalimétrie chez le saumon atlantique (*Salmo salar*) dans son aire de répartition méridionale: utilisation pratique et difficultés de la méthode. *Bulletin Franais de la Pêche Pisciculture*, 298:69–105, 1985.

[15] J.L. Baglinière and A. Champigneulle. Population estimates of juveniles Atlantic salmon (*Salmo salar*) as indices of smolts production in the river scorff, Brittany. *Journal of Fish Biology*, 29:467–482, 1986.

[16] J.L. Baglinière, G. Maisse, and A. Nihouarn. Atlantic salmon (*Salmo salar*) wild smolt production. In R.J. Gibson and R.E. Cutting., editors, *Production of Juvenile Atlantic Salmon (Salmo salar) in Natural Waters*, volume 118, pages 189–201. Canadian Special Publication of Fisheries and Aquatic Sciences, 1993.

[17] J.L. Baglinière, F. Marchand, and V. Vauclin. Interrannual changes in recruitment of the Atlantic salmon (*Salmo salar*) population in the river Oir (lower Normandy France): Relationships with spawner and in-stream habitat. *ICES Journal of Marine Science*, 62:695–707, 2005.

[18] S. Banerjee, B. P. Carlin, and A.E. Gelfand. *Hierarchical Modeling and Analysis of Spatial Data.* Monographs on Statistics and Applied Probability. Chapman et Hall/CRC, Boca Raton, FL, 2004.

[19] A. Bardonnet and J.L. Baglinière. Freshwater habitat of Atlantic salmon. *Canadian Journal of Fisheries and Aquatic Sciences*, 57:497–506, 2000.

[20] T. Bayes. Essay towards solving a problem in the doctrine of chances. *Philosophical Transactions of the Royal Society of London*, 53:370–418, 1763. Reprinted in Biometrika, vol. 45, pp. 293-315, 1958.

[21] M.A. Beaumont, W. Zhang, and D.J. Balding. Approximate Bayesian computation in population genetics. *Genetics*, 62:2025–2035, 2002.

[22] J.R. Beddington, D.J. Agnew, and C.W. Clark. Current problems in the management of marine fisheries. *Science*, 316:1713–1716, 2007.

[23] R. Bellman. *Dynamic Programming.* Princeton University Press, 1957.

[24] E. Beltrami. *Mathematics for Dynamic Modeling.* Academic Press, 1987.

[25] C. Béné and L. Doyen. Storage and viability of a fishery with resource and market dephased seasonalities. *Environmental and Resource Economics*, 15:51–26, 2000.

[26] J.O. Berger. *Statistical Decision Theory and Bayesian Analysis.* In Statistics. Springer-Verlag, New York, 1985.

[27] J.O. Berger and D. Rios Insua. Recent developments in Bayesian inference with applications in hydrology. In E. Parent, P. Hubert, B. Bobée, and J. Miquel, editors, *Bayesian Methods in Hydrological Sciences*, pages 43–62. UNESCO Publishing, 1998.

[28] J.O. Berger and L.R. Pericchi. The intrinsic Bayes factor for model selection and prediction. *Journal of the American Statistical Association*, 91(43):109–122, 1996.

[29] J.M. Bernardo. Reference posterior distributions for Bayesian inference. *Journal of the Royal Statistical Society: Series B (Statistical Methodology)*, 41:113–147 (with discussion), 1979.

[30] J.M. Bernardo and A.F.M. Smith. *Bayesian Theory.* Wiley & Sons, Series in Probability and Mathematical Statistics, London, 1994.

[31] Donald A. Berry. *Statistics: A Bayesian Perspective.* Duxbury Press, Belmont, California, 1995.

[32] T. Bohlin and I.G. Cowx. Implication of unequal probability of capture by electric fishing on the estimation of population size. In I.G. Cowx., editor, *Development in Electric Fishing*, pages 145–155. Fishing News Books, Blackwell Scientific Publications Ltd, Oxford, 1990.

[33] T. Bohlin, S. Hamrin, T.G. Heggberget, G. Rasmussen, and J.S. Svein. Electrofishing theory and practice with special emphasis on salmonids. *Hydrobiologia*, 173:9–43, 1989.

[34] M.E. Borsuck, P. Reichert, A. Peter, E. Schager, and P. Burkhardt-Holm. Assessing the decline of brown trout (salmo trutta) in Swiss rivers using a Bayesian probability network. *Ecological Modelling*, 192:224–244, 2006.

[35] G.E.P. Box. Sampling and Bayes inference in scientific modelling and robustness (with discussion). *Journal of the Royal Statistical Society: Series A (Statistics in Society)*, 143:383–430, 1980.

[36] G.E.P. Box and G.T. Tiao. *Bayesian Inference in Statistical Analysis*. Addison-Wesley, Reading, 1973.

[37] S.P. Brooks. Markov chain Monte Carlo methods and its application. *The Statistician*, 47, part1:69–100, 1998.

[38] S.P. Brooks. Bayesian computation: A statistical revolution. *Philosophical Transactions of the Royal Society of London, Series A*, 15:2681–2697, 2003.

[39] M. Brun, C. Abraham, M. Jarry, J. Dumas, F. Lange, and E. Prévost. Estimating an homogeneous series of a population abundance indicator despite changes in data collection procedure: A hierarchical Bayesian modelling approach. *Ecological Modelling*, 222(5):1069–1079, 2011.

[40] S.T. Buckland, K.B. Newman, L. Thomas, and N.B. Koesters. State-space models for the dynamics of wild animal populations. *Ecological Modelling*, 171(1-2):157–175, 2004.

[41] S.T. Buckland, K.B. Newmann, C. Fernandez, L. Thomas, and J. Harwood. Embeding population dynamics models in inference. *Statistical Science*, 22(1):44–58, 2007.

[42] M. Buoro, E. Prévost, and O. Gimenez. Investigating evolutionary trade-offs in wild populations of Atlantic salmon (salmo salar): Incorporating detection probabilities and individual heterogeneity. *Evolution*, 64(9):2629–2642, 2010.

[43] K.P. Burnham and D.R. Anderson. *Model Selection and Multimodel Inference. A Practical Information-Theoretic Approach*. Springer-Verlag, New York, 2nd edition, 2002.

[44] O. Cappé, A. Guilin, Marin J.M., and C.P. Robert. Population Monte Carlo. *Journal of Computational and Graphical Statistics*, 13(4), 2004.

[45] F.L. Carle and M.R. Strube. A new method for estimating population size from removal data. *Biometrics*, 34:621–630, 1978.

[46] B.P. Carlin and S. Chib. Bayesian model choice via Markov chain Monte Carlo methods. *Journal of the Royal Statistical Society: Series B (Statistical Methodology)*, 57(3):473–484, 1995.

[47] B.P. Carlin and A.T. Louis. *Bayes and Empirical Bayes Methods for Data Analysis.* Chapman & Hall/CRC, 2nd edition, 2000.

[48] G. Casella. An introduction to empirical Bayes data analysis. *American Statistician*, 39(2):83–87, 1985.

[49] G. Casella and E. George. Explaining the Gibbs sampler. *Am. Stat.*, 46:167–174, 1992.

[50] H. Caswell. *Matrix Population Models Construction, Analysis and Interpretation.* Sinauer Associates Inc. Publishers, 2nd edition, 2001.

[51] H. Caswell, R.J. Naiman, and R. Morin. Evaluating the consequence of reproduction in complex salmonid life cycles. *Aquaculture (Elsevier Science Publishers)*, 43:123–134, 1984.

[52] G. Celeux, F. Forbes, C.P. Robert, and D.M. Titterington. Deviance information criteria for missing data. *Bayesian Analysis*, 1(4):651–674, 2006.

[53] M. Chaloupka and G. Balazs. Using Bayesian state-space modelling to assess the recovery and harvest potential of the Hawaian green sea turtle stock. *Ecological Modelling*, 205(1-2):93–109, 2007.

[54] G. Chaput, J. Allard, F. Caron, J.B. Dempson, C.C. Mullins, and M.F. O'Connell. River-specific target spawning requirements for Atlantic salmon (salmo salar) based on a generalized smolt production model. *Canadian Journal of Fisheries and Aquatic Sciences*, 55:246–261, 1998.

[55] S. Chib. Marginal likelihood from the Gibbs outputs. *Journal of the American Statistical Association*, 90(432):1313–1321, 1995.

[56] S. Chib and E. Greenberg. Understanding the Metropolis-Hastings algorithm. *The American Statistician*, 49(4):327–335, 1995.

[57] S. Chib and I. Jeliazkov. Marginal likelihood from the Metropolis-Hastings output. *Journal of the American Statistical Association*, 96(453):270–281, 2001.

[58] V. Christensen, S. GuÃnette, J. J. Heymans, C. J. Walters, R. Watson, D. Zeller, and D. Pauly. Hundred-year decline of north atlantic predatory fishes. *Fish and Fisheries*, 4(1):1–24, 2003.

[59] C.W. Clark. *Biometric Modelling and Fisheries Management.* Wiley-Interscience, New-York, 1985.

[60] C.W. Clark. *Mathematical Biometrics: The Optimal Management of Renewable Resources.* Wiley, 1990.

[61] C.W. Clark, A. Charles, J.R. Beddington, and M. Mangel. Optimal capacity decisions in a developping fisheries. *Marine Resource Economics*, 1:25–54, 1985.

[62] C.W. Clark and M. Mangel. *Dynamic State Variable Models in Ecology. Methods and Applications.* Oxford Series in Ecology and Evolution. Oxford University Press, 2000.

[63] J.S. Clark. Uncertainty and variability in demography and population growth; a hierarchical approach. *Ecology*, 84:1370–1381, 2003.

[64] J.S. Clark. Why environmental scientists are becoming Bayesians? *Ecology Letters*, 8:2–14, 2005.

[65] J.S. Clark. *Models for Ecological Data. An introduction.* Princeton University Press, 2007.

[66] J.S. Clark, S.R. Carpenter, M. Barber, S. Collins, A. Dobson, J.A. Foley, D.M. Lodge, M. Pascual, R. Pielke, W. Pizer, C. Pringle, W.V. Reid, K.A. Rose, O. Sala, W.H. Schlesinger, D.H. Wall, and D. Wear. Ecological forecasts: An emerging imperative. *Science*, 293(5530):657–660, 2001.

[67] J.S. Clark and A. Gelfand. *Applications of Computational Statistics in the Environmental Sciences: Hierarchical Bayes and MCMC Methods.* Oxford University Press, 2006.

[68] J.S. Clark and A.E. Gelfand. A future for models and data in environmental sciences. *Trends in Ecology & Evolution*, 21(7):375–380, 2006.

[69] J. Clobert, J.D. Lebreton, and D. Allaine. A general approach to survival rate estimation by recapture or resightings of marked birds. *Ardea*, 75:133–142, 1987.

[70] D. Collett. *Modelling Binary Data.* Chapman & Hall, 2nd edition, 2003.

[71] P. Congdon. *Bayesian Statistical Modelling.* John Wiley, 2001.

[72] R. Costanza, R. d'Arge, R. de Groot, S. Farber, M. Grasso, B. Hannon, K. Limburg, S. Naeem, R.V. O'Neill, J. Paruelo, R.G. Raskin, P. Sutton, and M. van den Belt. The value of the world's ecosystem services and natural capital. *Nature*, 387(6630):253–260, 1997.

[73] R.G. Cowell, A.P. Dawid, S.L. Spiegelhalter, and D.J. Lauritzen. *Probabilistic Network and Expert Systems.* Springer-Verlag, New York, NY, 1999.

[74] P.F. Craigmile, C.A. Calder, H. Li, R. Paul, and N. Cressie. Hierarchical model building, fitting, and checking: A behind-the-scenes look at a Bayesian analysis of arsenic exposure pathways. *Bayesian Analysis*, 4(01):1–36, 2009.

[75] N.A.C. Cressie, C.A. Calder, J.S. Clark, J.M. Ver Hoeff, and C.K. Wikle. Accounting for uncertainty in ecological analysis: the strenghts and limitations of hierarchical statitical modelling. *Ecological Applications*, 19(3):553–570, 2009.

[76] N.A.C. Cressie and C.K. Wikle. *Statistics for Spatio-Temporal Data.* Wiley Series in Probability and Statistics. John Wiley & Sons, Hoboken NJ, 2011.

[77] W.W. Crozier, E.C.E. Potter, E. Prévost, P.-J. Schön, and O. Maoiléidigh. A coordinated approach towards the development of a scientific basis for management of wild Atlantic salmon in the North-East Atlantic. Technical report, Concerted Action QLK5-CT1999e01546 (SALMODEL), 2003.

[78] R. Curtin and R. Prellezo. Understanding marine ecosystem based management: A literature review. *Marine Policy*, 34(5):821–830, 2010.

[79] P.M. Cury and Y. Miserey. *Une mer sans poissons.* Calmann-Levy, 2008.

[80] P.M. Cury, C. Mullon, S.M. Garcia, and L.J. Shannon. Viability theory for an ecosystem approach to fisheries. *ICES Journal of Marine Science*, 62:577–584, 2005.

[81] G. Dauphin, E. Prévost, C.E. Adams, and P. Boylan. A Bayesian approach to estimating Atlantic salmon fry densities using a rapid sampling technique. *Fisheries Management and Ecology*, 16(5):399–408, 2009.

[82] B. de Finetti. *La Prévision: ses Lois Logiques, ses Sources Subjectives.* Institut Henri Poincaré, Paris, 1937.

[83] M. De Lara and L. Doyen. *Sustainable Management of Natural Resources: Mathematical Models and Methods.* Springer-Verlag, Berlin, 2008.

[84] M. De Lara, L. Doyen, T. Guilbaud, and M.-J. Rochet. Is a management framework based on spawning-stock biomass indicators sustainable? *ICES Journal of Marine Science*, 64(4):761–767, 2007.

[85] M. De Lara and V. Martinet. Multi-criteria dynamic decision under uncertainty: a stochastic viability analysis and an application to sustainable fishery management. *Mathematical Biosciences*, 217(2):118–124, 2009.

[86] E. De Oliveira, N. Bez, and E. Prévost. Impact of local pollution on fish abundance using geostatistical simulations. In P. Monestiez, D. Allard, and R. Froidevaux., editors, *Geo ENV III - Geostatistics for Environmental Applications*, pages 181–191. Kluwer Academic Publishers, 2001.

[87] P. De Valpine. Shared challenges and common ground for Bayesian and classical analysis of hierarchical statistical models. *Ecological applications*, 19(3):558–588, 2009.

[88] B. Dennis. Discussion: Should ecologists become Bayesians? *Ecological Applications*, 6:1095–1102, 1996.

[89] D. Dey, S. K. Ghosh, and B.K Mallick. *Generalized Linear Models: A Bayesian Perspective.* CRC Press, Boca Raton, FL, 2000.

[90] J. Diamond. *Collapse, How Societies Choose to Fail or Succeed.* Penguin Books, 2005.

[91] P. Diggle and P.J. Ribeiro. *Model-based Geostatistics.* Springer, New York, NY, 2007.

[92] P. Dixon and Aaron M. E. Introduction: Ecological applications of Bayesian inference. *Ecological Applications*, 6(4):1034–1035, 1996.

[93] C.R. Dolan and L.E. Miranda. Immobilization thresholds of electrofishing relative to fish size. *Transactions of the American Fisheries Society*, 132:969–976, 2003.

[94] R.M. Dorazio and F.A. Johnson. Bayesian inference and decision theory. A framework for decision making in natural resource management. *Ecological Applications*, 13(2):556–563, 2003.

[95] A. Doucet, N. de Freitas, and N. Gordon. *Sequential Monte Carlo Methods in Practice.* Springer-Verlag, New York, 2001.

[96] N. R. Draper and H. Smith. *Applied Regression Analysis.* Probability and Statistics. John Wiley & Sons, Inc., New York, 1998.

[97] J. Dumas and P. Prouzet. Variability of demographic parameters and population dynamics of Atlantic salmon (salmo salar l.) in a SouthWest French river. *ICES Journal of Marine Science*, 60:356–370, 2003.

[98] B. Efron and C. Morris. Data analysis using Stein's estimator and its generalizations. *Journal of the American Statistical Association*, pages 311–319, 2004.

[99] J. Elliott. The relative role of density in the stock-recruitment relationship of salmonids. In E. Prévost and G. Chaput, editors, *Stock, Recruitment, and Reference Point. Assessment and Managment of Atlantic Salmon.*, pages 25–55. INRA editions, Paris, 2001.

[100] S. Engen, O. Bakke, and A. Islam. Demographic and environmental stochasticity-concepts and definitions. *Biometrics*, 54:840–846, 1998.

[101] L. Fahrmeir and G. Tutz. *Multivariate Statistical Modelling Based on Generalized Linear models.* Springer-Verlag, New York, NY, 1994.

[102] W. Feller. *An Introduction to Probability Theory and Its Applications*, volume 1. Wiley, New York, 3rd edition, 1968.

[103] C. Fernàndez, S. Cerviño, N. Pérez, and E. Jardim. Stock assessment and projections incorporating discard estimates in some years: an application to the hake stock in ICES divisions VIIIc and IXa. *ICES Journal of Marine Science*, 67(6):1185–1197, 2010.

[104] C. Fernandez, E. Ley, and M. Steel. Bayesian modelling of catch in a North-West Atlantic fishery. *Journal of the Royal Statistical Society: Series C (Applied Statistics)*, 51(3):257–280, 2002.

[105] C. Fernandez, E. Ley, and M.F.J. Steel. Benchmark priors for Bayesian model averaging. *Journal of Econometrics*, 100:381–427, 2001.

[106] K. R. Foster, P. Vecchia, and M. H. Repacholi. Science and the precautionary principle. *Science*, 288:979–981, 2000.

[107] W. Fox. An exponential surplus-yield model for optimizing exploited fish populations. *Transactions of the American Fisheries Society*, 99:80–88, 1970.

[108] B. Friedland. *Control System Design.* McGraw-Hill, New York, NY, 1986.

[109] D. Gamerman. *Markov chain Monte Carlo: Stochastic Simulation for Bayesian Inference.* Chapman & Hall/CRC Press, Boca Raton, FL, 1997.

[110] S. Garcia. The ecosystem approach to fisheries: issues, terminology, principles, insitutional foundations, implementation and outlook. *Food and Agriculture Organisation of the United Nations, Rome*, 2003.

[111] D. Gascuel, M. Barry, M. Laurans, and Sidibé A. *Evaluation des stocks démersaux en Afrique Nord-Ouest. Travaux du groupe "Analyses monospécifiques" du projet SIAP.* Number 65 in COPACE/PACE. FAO, Rome, 2003.

[112] A.E. Gelfand and D.K. Dey. Bayesian model choice: asymptotics and exact calculations. *Journal of the Royal Statistical Society: Series B (Statistical Methodology)*, 56(3):501–514, 1994.

[113] A.E. Gelfand, D.K. Dey, and H. Chang. Model determination using predictive distributions with implementation via sampling-based methods. *In Bayesian Statistics 4, Edited by J.M. Bernardo, J.O. Berger, A.P. Dawid and A.F.M. Smith, Oxford University Press, 1992.*, pages 147–167, 1992.

[114] A.E. Gelfand, P.J. Diggle, Montserrat F., and P. Guttorp, editors. *Handbooks of Spatial Statistics.* Handbooks of Modern Statistical Methods. Chapman & Hall/CRC Press, Boca Raton, FL, 2010.

[115] A.E. Gelfand and A.F.M. Smith. Sampling based approach to calculating marginal densities. *Journal of the American Statistical Association*, 85:398–409, 1990.

[116] A. Gelman. Prior distributions for variance parameters in hierarchical models. *Bayesian analysis*, 1(3):515–534, 2006.

[117] A. Gelman, J.B. Carlin, H.S. Stern, and D.B. Rubin. *Bayesian Data Analysis. Second Edition.* Col. Texts in Statistical Science. Chapman & Hall/CRC Press, New York, 2004.

[118] S. Geman and D. Geman. Stochastic relaxation, Gibbs distributions and the Bayesian restoration of image. *IEEE Trans. Pattern Analysis and Machine Intelligence*, 6:721–741, 1984.

[119] E. George and R. McCulloch. Variable selection via Gibbs sampling. *Journal of the American Statistical Association*, 88(423):881–889, 1993.

[120] W.R. Gilks, S. Richardson, and D. Spiegelhalter. *Markov chain Monte Carlo in Practice.* Chapman & Hall, London, 1996.

[121] O. Gimenez, S.J. Bonner, R. King, R.A. Parker, S.P. Brooks, L.E. Jamieson, V. Grosbois, B.J.T. Morgan, and L. Thomas. Winbugs for population ecologists: Bayesian modeling using Markov chain Monte Carlo methods. In D.L. Thomson, E.G. Cooch, and M.J. Conroy, editors, *Modeling Demographic Processes In Marked Populations*, volume 3 of *Environmental and Ecological Statistics*, pages 883–915. Springer, 2009.

[122] V. Ginot, C. Le Page, and S. Souissi. A multi-agent architecture to enhance end-user individual-based modelling. *Ecological Modelling*, 157:23–41, 2002.

[123] M. Goldstein and D. Wooff. *Bayes Linear Statistics, Theory & Methods.* John Wiley & Sons, Ltd, 2007.

[124] P.J. Green. Reversible jump MCMC computation and Bayesian model determination. Technical report, University of Bristol, 1994.

[125] V. Grimm and S. F. Railsback. *Individual-Based Modeling and Ecology.* Princeton University Press, Princeton NJ, 2005.

[126] J. Halley and P. Inchausti. Lognormality of ecological time series. *Oikos*, 99:518–530, 2002.

[127] T.R. Hammond. A recipe for Bayesian network driven stock assessment. *Canadian Journal of Fisheries and Aquatic Sciences*, 61:1647–1657, 2004.

[128] T.R. Hammond and V. Trenkel. Censored catch data in fisheries stock assessment. *ICES Journal of Marine Science*, 62:1118–1130, 2005.

[129] C. Han and B.P. Carlin. Markov chain Monte Carlo methods for computing Bayes factors: A comparative review. *Journal of the American Statistical Association*, 96:1122–1132, 2001.

[130] D.G. Hankin. Multistage sampling designs in fisheries research: Applications in small streams. *Canadian Journal of Aquatic and Fisheries Sciences*, 41:1575–1591, 1984.

[131] G. Hardin. Extensions of "the tragedy of the commons". *Science*, 280(5364):682–683, 1998.

[132] S.J. Harley and R.A. Myers. Hierarchical Bayesian models of length-specific catchability of research trawl surveys. *Canadian Journal of Fisheries and Aquatic Sciences*, 58:1569–1584, 2001.

[133] J. Harwood and K. Stokes. Coping with uncertainty in ecological advice: lessons from fisheries. *Trends in Ecology & Evolution*, 18(12):617–622, 2003.

[134] E. Hewitt and L.J. Savage. Symmetric measures on Cartesian products. *Transactions of the American Mathematical Society*, 80:470–501, 1955.

[135] R. Hilborn and M. Liermann. Standing on the shoulders of giants: Learning from experience in fisheries. *Reviews in Fish Biology and Fisheries*, 8:273–283, 1998.

[136] R. Hilborn and M. Mangel. *The Ecological Detective Confronting Models with Data*, volume 28 of *Monographs in population biology*. Princeton University Press, Princeton, New Jersey, 1997.

[137] R. Hilborn and C.J. Walters. *Quantitative Fisheries Stocks Assessment: Choice, Dynamics & Uncertainty*. Col. Natural Resources. Chapman & Hall, New York, 1992.

[138] N. T. Hobbs. New tools for insight from ecological models and data. *Ecological applications*, 19(3):551–552, 2009.

[139] J.A. Hoeting, D. Madigan, A. Raftery, and C.T. Volinsky. Bayesian model averaging: a tutorial. *Statistical Science*, 14:382–417, 1999.

[140] P.D. Hoff. *A First Course in Bayesian Statistical Methods*. Springer, New York, 2009.

[141] M.B. Hooten, C.K. Wikle, R.M. Dorazio, and J.A. Royle. Hierarchical spatiotemporal matrix models for characterizing invasions. *Biometrics*, 63(2):558–567, 2007.

[142] J.A. Hutchings. Collapse and recovery of marine fishes. *Nature*, 406(6798):882–885, 2000.

[143] J.A. Hutchings and M.E.B. Jones. Life history variation and growth rate thresholds for maturity in Atlantic salmon, salmo salar. *Canadian Journal of Fisheries and Aquatic Sciences*, 55(Suppl. 1):22–47, 1998.

[144] ICSEAF. Historical series data selected for cape hake assessment. Technical report, ICSEAF, 1989.

[145] G. Iooss. *Bifurcations of maps and applications.* Lecture Notes, Mathematical Studies. North-Holland, Amsterdam, 1979.

[146] D. Jacobson and D. Mayne. *Differential Dynamic Programming.* American Elsevier Publishing Company, 1970.

[147] W.H. Jeffreys and J.O. Berger. Ockham's razor and Bayesian analysis. *Am. Sci. (American Scientist)*, 80:64–72, 1992.

[148] J.B. Johnson and K.S. Omland. Model selection in ecology and evolution. *Trends in Ecology & Evolution*, 19(2):101–108, 2004.

[149] I.D. Jonsen, J.M. Flemming, and R.A. Myers. Robust state-space modeling of animal movement data. *Ecology*, 86(11):2874–2880, 2005.

[150] N. Jonsson, B. Jonsson, and L.P. Hansen. The relative role of density-dependent and density-independant survival in the life cycle of Atlantic salmon salmo salar. *Journal of Animal Ecology*, 67:751–762, 1998.

[151] I. Jordaan. *Decisions Under Uncertainty: Probabilistic Analysis For Engineering Decisions.* Cambridge University Press, 2005.

[152] M.I. Jordan. *Learning in Graphical Models.* MIT Press, Cambridge, MA, 1999.

[153] J.B. Kadane. *Principles of Uncertainty.* Texts in Statistical Science. Chapman & Hall/ CRC, Boca Raton, FL, 2011.

[154] J.B. Kadane and N.A. Lazar. Methods and criteria for model selection. *Journal of the American Statistical Association*, 99:279–290, 2004.

[155] J.B. Kadane, L.J. Wolson, A. O'Hagan, and K. Craig. Papers on elicitation with discussions. *The Statistician*, pages 3–53, 1998.

[156] R.E. Kass, B.P. Carlin, A. Gelman, and R.M. Neal. Markov chain Monte Carlo in practice: a roundtable discussion. *The American Statistician*, 52(2):93–100, 1998.

[157] R.E. Kass and A.E. Raftery. Bayes factors. *Journal of the American Statistical Association*, 90(430):773–795, 1995.

[158] R.L. Keeney. A utility function for examining policy affecting salmon in Skeena river. *Journal of Fisheries Research Board Canada*, 39:49–63, 1976.

[159] G.J.A. Kennedy and W.W. Crozier. Factors affecting recruitment success in salmonids. In Harper D.M. and A.J.D. Fergusson, editors, *The Ecological Basis for River Management.*, Chichester, 1995. Wiley & Sons.

[160] M. Kery. *An Introduction to WinBUGS for Ecologists.* Academic Press, Elsevier Inc., 2010.

[161] M. Kery and M. Schaub. *Bayesian Population Analysis using WinBUGS/OpenBUGS: a Hierarchical Perspective.* Academic Press, Burlington, 2012.

[162] D.K. Kimura, J.W. Balsiger, and D.H. Ito. Kalman filtering the delay-difference equation: practical approaches and simulations. *Fishery Bulletin*, 94:678–691, 1996.

[163] R. King and S.P. Brooks. On the Bayesian analysis of population size. *Biometrika*, 88(2):317–336, 2001.

[164] R. King, B.J.T. Morgan, O. Gimenez, and S.P. Brooks. *Bayesian Analysis for Population Ecology.* Chapman & Hall/CRC, Boca Raton, FL, 2010.

[165] M. Kitti, M. Lindros, and V. Kaitala. Optimal harvesting of the Norvegian spring-spawning herring: Variable versus fied harvesting strategies. *Environmental Modeling and Assessment*, 7:47–55, 2002.

[166] P.M. Kuhnert, Martin T.G., and Griffiths S.P. A guide to eliciting and using expert knowledge in Bayesian ecological models. *Ecology Letters*, 13(7):900–914, 2010.

[167] S. Kuikka, M. Hilden, H. Gislason, S. Hansson, H. Sparholt, and G. Varis. Modelling environmentally driven uncertainties in Baltic cod (gadus morhua) management by Bayesian influence diagrams. *Canadian Journal of Fisheries and Aquatic Sciences*, 56:629–641, 1999.

[168] M. Kyung, J. Gilly, M. Ghoshz, and G. Casella. Penalized regression, standard errors, and Bayesian Lassos. *Bayesian Analysis*, 5(2):369–412, 2010.

[169] M. Laurans. *Evaluation des ressources halieutiques en Afrique de l'ouest: dynamique des populations et variabilité écologique.* Thèse de doctorat, ENSAR - Mention Halieutique, 2005.

[170] S.L. Lauritzen and D.J. Spiegelhalter. Local computations with probabilities on graphical structures and their application to expert systems (with discussion). *Journal of the Royal Statistical Society: Series B (Statistical Methodology)*, 50(2):157–224, 1988.

[171] D.C. Lee and B. Rieman. Population viability assessment of salmonids by using a probabilistic network. *North American Journal of Fisheries Management*, 17:1144–1157, 1997.

[172] P. M. Lee. *Bayesian Statistics: An Introduction.* Wiley, Chichester, 1997.

[173] S. Lek. Uncertainty in ecological models. *Ecology Model. (Ecological Modelling)*, 207:1–2, 2007.

[174] S. R. Lele and B. Dennis. Bayesian methods for hierarchical models: Are ecologists making a Faustian bargain? *Ecological Applications*, 19(3):581–584, 1999.

[175] WC. Lewin, J. Freyhof, V. Huckstorf, T. Mehner, and C. Wolter. When no catches matter: Coping with zeros in environmental assessments. *Ecological Indicators*, 10:572–583, 2010.

[176] P. Lewy and A. Nielsen. Modelling stochastic fish stock dynamics using Markov chain Monte Carlo. *ICES Journal of Marine Science*, 60:743–752, 2003.

[177] D.V. Lindley. A statistical paradox. *Biometrika*, pages 187–192, 1965.

[178] D.V. Lindley. *Making Decision.* Wiley, New York, NY, 2nd edition, 1985.

[179] H. Linhart and W. Zucchini. *Model selection.* John Wiley & Sons, New York, NY, 1986.

[180] M.R. Link and R.M. Peterman. Estimating the value of in-season estimates of abundance of sockeye salmon (oncorhynchus nerka). *Canadian Journal of Fisheries and Aquatic Sciences*, 55(6):1408–1418, 1998.

[181] W. Link and R.J. Barker. *Bayesian Inference with Ecological Applications.* Academic Press, Boston, MA, 2010.

[182] W. Link and R.J. Barker. Model weights and the foundation of multimodel inference. *Ecology*, 87(10):2626–2635, 2010.

[183] W.A. Link, E. Cam, J.D. Nichols, and E.G. Cooch. Of bugs and birds: Markov chain Monte Carlo for hierarchical modeling in wildlife research. *Journal of Wildlife Management*, 66(2):277–291, 2002.

[184] D. Ludwig, R. Hilborn, and C.J. Walters. Uncertainty, resources exploitation and conservation: lessons from history. *Science*, 260:17–36, 1993.

[185] D.J. Lunn, A. Thomas, N. Best, and D. Spiegelhalter. Winbugs - a Bayesian modelling framework: Concepts, structure, and extensibility. *Statistics and Computing*, 10(4):325–337, 2000.

[186] M. Machina. Expected utility without the independence axiom. *Econometrica*, 50, 1982.

[187] S. Mäntyniemi, A. Romakkaniemi, and E. Arjas. Bayesian removal estimation of population size under unequal catchability. *Canadian Journal of Fisheries and Aquatic Sciences*, 62:291–300, 2005.

[188] B.G. Marcot, R.S. Holthaousen, M.G. Raphael, M.M. Rowl, and M.J. Wisdom. Using Bayesian belief networks to evaluate fish and wildlife population viability under land management altenatives from an environmental impact statement. *Forest Ecology and Management*, 153:29–42, 2001.

[189] J.M. Marin and C. P. Robert. *Bayesian Core.* Springer, New York, NY, 2007.

[190] T.G. Martin, B.A. Wintle, J.R. Rhodes, P.M. Kuhnert, S.A. Field, S.J. Low-Choy, A.J. Tyre, and H.P. Possingham. Zero tolerance ecology: improving ecological inference by modelling the source of zero observations. *Ecology Letters*, 8:1235–1248, 2005.

[191] V. Martinet, L. Doyen, and Thébaud. Defining viable recovery paths toward sustainable fisheries. *Ecological Economics*, 64(2):441–422, 2007.

[192] R.M. May. Biological population with nonoverlapping generations: stable points, stable cycles and chaos. *Science*, 6:645–647, 1974.

[193] R.M. May. Simple mathematical models with very complicated dynamics. *Nature*, 261:459–467, 1976.

[194] M.K McAllister, S.L. Hill, D.J. Agnew, G.P. Kirkwood, and J.R. Beddington. A Bayesian hierarchical formulation of the DeLury stock assessment model for abundance estimation of Falkland islands' squid (Loligo gahi). *Canadian Journal of Fisheries and Aquatic Sciences*, 61(6):1048–1059, 2004.

[195] M.K. McAllister and C.H. Kirchner. Accounting for structural uncertainty to facilitate precautionary fishery management: illustration with Namibian orange roughy. *In "Targets, Thresholds, and the Burden of Proof in Fisheries Management" Mangel, M. ed. Bull. of Mar. Sci.*, 70(2):499–540, 2002.

[196] M.K. McAllister and G.P. Kirkwood. Bayesian stock assessment: A review and example application using the logistic model. *ICES Journal of Marine Science*, 55:1031–1060, 1998.

[197] M.K. McAllister, E.K. Pikitch, A.E. Punt, and A. Hilborn. A Bayesian approach to stock assessment and harvest decisions using the sampling/importance resampling algorithm. *Canadian Journal of Fisheries and Aquatic Sciences*, 51:2673–2687, 1994.

[198] M.A. McCarthy and P. Masters. Profiting from prior information in Bayesian analyses of ecological data. *J. Appl. Ecology (Journal of Applied Ecology)*, 42:1012–1019, 2005.

[199] P. McCullagh and J.A. Nelder. *Generalized Linear Models.* Number 37 in Monographs on Statistics and Applied Probability. Chapman & Hall/CRC, Boca Raton, FL, 1989.

[200] C.E. McCulloch, S.R. Searle, and J.M. Neuhaus. *Generalized, Linear and Mixed Models.* Wiley, New York, NY, 2008.

[201] S. B. McGrayne. *The Theory That Would Not Die.* Yale University Press, New Haven, CT, 2011.

[202] R. Meyer and R.B. Millar. Bayesian stock assessment using a state-space implementation of the delay difference model. *Canadian Journal of Fisheries and Aquatic Sciences*, 56:37–52, 1999.

[203] R. Meyer and R.B. Millar. BUGS in Bayesian stock assessments. *Canadian Journal of Fisheries and Aquatic Sciences*, 56:1078–1086, 1999.

[204] C.G.J. Michielsen and M.K. McAllister. A Bayesian hierarchical analysis of stock-recruit data: Quantifying structural parameter uncertainty. *Canadian Journal of Fisheries and Aquatic Sciences*, 61:1032–1047, 2004.

[205] R.B. Millar. Reference priors for Bayesian fisheries models. *Canadian Journal of Fisheries and Aquatic Sciences*, 59:1492–1502, 2002.

[206] R.B. Millar and R. Meyer. Bayesian state-space modeling of age-structured data: fitting a model is just the beginning. *Canadian Journal of Fisheries and Aquatic Sciences*, 57:43–50, 2000.

[207] N.J. Milner, J.M. Elliott, J.D. Armstrong, R. Gardiner, J.S. Welton, and M. Ladle. The natural control of salmon and trout populations in streams. *Fisheries Research*, 62:111–125, 2003.

[208] T.J. Mitchell and J.J. Beauchamp. Bayesian variable selection in linear regression. *Journal of the American Statistical Association*, 83:1023–1036, 1998.

[209] M.G. Mitro and A.V. Zale. Predicting fish abundance using single-pass removal sampling. *Canadian Journal of Fisheries and Aquatic Sciences*, 57:951–961, 2000.

[210] J.M. Morales, D.T. Haydon, J. Frair, K.E. Holsinger, and J.M. Fryxell. Extracting more out of relocation data: building movement models as mixtures of random walks. *Ecology*, 85(9):2436–2445, 2004.

[211] S.B. Munch, A. Kottas, and M. Mangel. Bayesian nonparametric analysis of stock-recruitment relationships. *Canadian Journal of Fisheries and Aquatic Sciences*, 62:1808–1821, 2005.

[212] B. Munier. *Risk, Decision and Rationality.* Dordrecht, Reidel Publishing Company, 2008.

[213] C. M. Muntshinda, R. B. O'Hara, and I. P. Woivod. A multispecies perspective on ecological impacts of climating forcing. *Journal of Animal Ecology*, 80:101–107, 2011.

[214] R. A. Myers and B. Worm. Rapid worldwide depletion of predatory fish communities. *Nature*, 423:280–283, 2003.

[215] S. Nadkarni and P.S. Shenoy. A causal mapping approach to constructing Bayesian networks. *Decision Support Systems*, 38:259–281, 2004.

[216] C.L. Needle. Recruitment models: diagnosis and prognosis. *Reviews in Fish Biology and Fisheries*, 11:95–111, 2002.

[217] J. Von Neumann and O. Morgenstein. *Theory of Games and Economic Behaviour.* Princeton University Press, New Jersey, 1953.

[218] K.B. Newman. State-space modeling of animal movement and mortality with application to salmon. *Biometrics*, 54:1290–1314, 1998.

[219] K.B. Newman, C. Fernandez, L. Thomas, and S.T. Buckland. Monte Carlo inference for state-space models of wild animal populations. *Biometrics*, 65(2):572–583, 2009.

[220] A.G. Nicieza, F. Brana, and M.M. Toledo. Development of length-bimodality and smolting in wild Atlantic salmon, salmo salar l, under different growth conditions. *Journal of Fish Biology*, 38:509–523, 1991.

[221] K. Ogle. Hierarchical Bayesian statistics: merging experimental and modeling approaches in ecology. *Ecological Applications*, 19(3):577–581, 2009.

[222] A. O'Hagan. Fractional Bayes factors for model comparison. *Journal of the Royal Statistical Society: Series B (Statistical Methodology)*, 57(1):99–138, 1995.

[223] A. OHagan, C.E. Buck, A. Daneshkhah, J.R. Eiser, P.H. Garthwaite, D.J. Jenkinson, J.E. Oakley, and T. Rakow. *Uncertain Judgements: Eliciting Experts Probabilities.* Wiley, Chichester, UK, 2006.

[224] E. Parent and J. Bernier. *Le Raisonnement Bayésien: Modélisation et Inférence.* Springer France, Paris, 2007.

[225] E. Parent and E. Prévost. Inférences Bayèsiennes de la taille d'une population de saumons par utilisation de sources multiples d'information. *Revue de Statistique Appliquée*, LI(3):5–38, 2003.

[226] T. Park and G. Casella. The Bayesian Lasso. *Journal of the American Statistical Association*, 103(482):681–686, 2008.

[227] T. A. Patterson, L. Thomas, C. Wilcox, O. Ovaskainen, and J. Matthiopoulos. State-space models of individual animal movement. *Trends in Ecology & Evolution*, 23(2):87–94, 2008.

[228] D. Pauly, V. Christiensen, J. Dalsgaard, R. Freese, and F. Torres. Fishing down marine food webs. *Science*, 279:860–863, 1998.

[229] J.T. Peterson, R.F. Thurow, and J.W. Gusevich. An evaluation of multipass electrofishing for estimating the abundance of stream dwelling salmonids. *Transactions of the American Fisheries Society*, 133:462–475, 2004.

[230] L.I. Pettit and K.D.S. Young. Measuring the effect of observations on Bayes factors. *Biometrika*, 77(3):455–66, 1990.

[231] M. Piattelli-Palmarini. *La Réforme du Jugement: Ou Comment Ne Plus Se Tromper.* Odile Jacobs, 1995.

[232] E.K. Pikitch, C. Santora, E.A. Babcock, A. Bakun, R. Bonfil, D.O. Conover, P. Dayton, P. Doukakis, D. Fluharty, B. Heneman, E.D. Houde, J. Link, P.A. Livingston, M. Mangel, M.K. McAllister, J. Pope, and K.J. Sainsbury. Ecosystem-based fishery management. *Science*, 305(5682):346–347, 2004.

[233] Martyn Plummer. Penalized loss functions for bayesian model comparison. *Biostatistics*, 9(3):523–539, 2008.

[234] E.C.E. Potter, J.C. MacLean, R.J. Wyatt, and R.N.B. Campbell. Managing the exploitation of migratory salmonids. *Fisheries Research*, 62:127–142, 2003.

[235] J.W. Pratt, H. Raiffa, and R. Schlaifer. The foundations of decision under uncertainty: an elementary exposition. *Journal of the American Statistical Association*, 59(306):353–375, 1964.

[236] S. J. Press and J.M. Tanur. *The Subjectivity of Scientists and the Bayesian Approach.* John Wiley, New York, NY, 2001.

[237] E. Prévost, J.L. Baglinière, G. Maisse, and A. Nihouarn. Premiers éléments d'une relation stock / recrutement chez le saumon atlantique (salmo salar) en France. *Cybium*, 20(3):7–26, 1996.

[238] E. Prévost, G. Chaput, and (Ed.). *Stock, Recruitment and Reference Points Assessment and Management of Atlantic salmon.* INRA Editions, Paris, 2001.

[239] E. Prévost, E. Parent, W. Crozier, I. Davidson, J. Dumas, G. Gudbergsson, K. Hindar, P. McGinnity, J. MacLean, and L.M. Sattem. Setting biological reference points for Atlantic salmon stocks: Transfer of information from data-rich to sparse-data situations by Bayesian hierarchical modelling. *ICES Journal of Marine Science*, 60:1177–1193, 2003.

[240] A.E. Punt. Extending production models to include process error in the population dynamics. *Canadian Journal of Fisheries and Aquatic Sciences*, 60:1217–1228, 2003.

[241] A.E. Punt and R. Hilborn. Fisheries stock assessment and decision analysis: the Bayesian approach. *Reviews in Fish Biology and Fisheries*, 7:35–63, 1997.

[242] A.E. Punt, D.C. Smith, and A.D. M. Smith. Among-stock comparisons for improving stock assessments of data-poor stocks: the "Robin Hood" approach. *ICES Journal of Marine Science*, 68(5):972 -981, 2011.

[243] A. Quiggin. A theory of anticipated utility. *Journal of Economic Behavior and Organization*, 3:323–343, 1964.

[244] J.J. Quinn, I and R.B. Deriso. *Quantitative Fish Dynamics*, chapter 3: Stock and Recruitment. Biological resource management. Oxford University Press, Oxford, New York, 1999.

[245] J.M.Newton M.A.and Krivitsky P.N. Raftery, A.E.Satagopan. Estimating the integrated likelihood via posterior simulation using the harmonic mean identity. *Bayesian statistics 8 (Eds) J.M. Bernardo, M.J. Bayarri, J.O. Berger, A.P. David, Heckerman D., Smith A.F.M. and West M. Oxford University Press.*, pp. 1-45:–, 2007.

[246] H. Raiffa. *Decision Analysis - Introductory lectures on choices under uncertainty.* Addison Wesley, Reading, MA, 1968. traduction française: Analyse de la décision: introduction aux choix en avenir incertain, Dunod, 1973.

[247] H. Raiffa and R. Schlaifer. *Applied Statistical Decision Theory.* Harvard University Press, Harvard, 1961.

[248] R.G. Randall. Effect of water temperature, depth, conductivity and survey area on the catchability of juvenile Atlantic salmon by electric fishing in new brunswick streams. In I.G. Cowx., editor, *Developments in Electric Fishing*, pages 79–90. Fishing News Books, Blackwell Scientific Publications, 1990.

[249] C.R. Rao and H. Toutenburg. *Linear Models: Least Squares and Alternatives.* Springer, New York, NY, 1999.

[250] W.J. Reed. The steady state of a stochastic harvesting model. *Mathematical Biosciences*, 41:273–307, 1978.

[251] W.J. Reed. Optimal escapement levels in stochastic and deterministic harvesting models. *Journal of Environnemental Economics and Management*, 6:350–363, 1979.

[252] H. Regan, M. Colyvan, and M.A. Burgman. A taxonomy and treatment of uncertainty for ecology and conservation biology. *Ecological Applications*, 12(2):618–628, 2002.

[253] D. Revuz. *Markov Chains*. North Holland, Amsterdam, 1984.

[254] S.C. Riley, R.L. Haedrich, and R.J. Gibson. Negative bias removal estimates of Atlantic salmon parr relative to stream size. *Journal of Freshwater Ecology*, 8:97–101, 1993.

[255] E. Rivot and E. Prévost. Hierarchical Bayesian analysis of capture-mark-recapture data. *Canadian Journal of Fisheries and Aquatic Sciences*, 59:1768–1784, 2002.

[256] E. Rivot and E. Prévost. *Concevoir et construire la décision*, chapter 13: Aide à la décision pour la régulation de l' exploitation des populations naturelles de saumon atlantique (Salmo salar). Quae: Sciences et Technologies, 2009.

[257] E. Rivot, E. Prévost, A. Cuzol, J.L. Baglinière, and E. Parent. Hierarchial Bayesian modelling with habitat and time covariates for estimating riverine fish population size by successive removal method. *Canadian Journal of Fisheries and Aquatic Sciences*, 65:117–133, 2008.

[258] E. Rivot, E. Prévost, and E. Parent. How robust are Bayesian posterior inferences based on a Ricker model with regards to measurement errors and prior assumptions about parameters? *Canadian Journal of Fisheries and Aquatic Sciences*, 58:2284–2297, 2001.

[259] E. Rivot, E. Prévost, E. Parent, and J.L. Baglinière. A Bayesian state-space modelling framework for fitting a salmon stage-structured population dynamic model to multiple time series of field data. *Ecological Modelling*, 179:463–485, 2004.

[260] C.P. Robert and G. Casella. *Monte Carlo Statistical Methods*. Springer-Verlag, New York, NY, 1998.

[261] C.P. Robert and G. Casella. *Introducing Monte Carlo Methods with R*. Springer-Verlag, New York, NY, 2010.

[262] M. Robert, A. Faraj, M. K. McAllister, and E. Rivot. Bayesian state-space modelling of the DeLury depletion model: strengths

and limitations of the method, and application to the Moroccan octopus fishery. *ICES Journal of Marine Science*, 67:1272–1290, 2010.

[263] J.F. Rodgers, M.F. Solazzi, S.L. Johnson, and M.A. Buckman. Comparison of three techniques to estimate juvenile coho salmon populations in small streams. *North American Journal of Fisheries Management*, 12:79–86, 1992.

[264] A.E. Rosenberg and J.B. Duham. Validation of abundance estimates from mark-recapture and removal techniques for rainbow trout captured by electrofishing in small streams. *Canadian Journal of Fisheries and Aquatic Sciences*, 25:1395–1410, 2005.

[265] J. A. Royle and R. M. Dorazio. *Hierarchical Modeling and Inference in Ecology: the Analysis of Data from Populations, Metapopulations and Communities.* Academic Press, New York, NY, 2008.

[266] H. Rue and L. Held. *Gaussian Markov Random Fields: Theory and Applications.* Chapman & Hall, CRC, Boca Raton, FL, 2005.

[267] D. Ruelle. *Chaotic evolution and strange attractors.* Cambridge University Press, Cambridge, UK, 1989.

[268] L. J. Savage. *The Foundations of Statistics.* Dover Publications, New York, 1954.

[269] M. B. Schaefer. Some aspects of the dynamics of populations important to the management of marine fisheries. *Journal of the Fisheries Research Board of Canada*, 14:26–56, 1954.

[270] H. Scheffé. *The Analysis of Variance.* John Wiley, New York, NY, 1959.

[271] J. Schnute. A new approach to estimating populations by the removal method. *Canadian Journal of Fisheries and Aquatic Sciences*, 40:2153–2169, 1983.

[272] J.T. Schnute. Data uncertainty, model ambiguity and model identification. *Natural Resources Modeling*, 2(2):159–212, 1987.

[273] J.T. Schnute, A. Cass, and L. Richards. A Bayesian decision analysis to set escapment goals for Frazer river Sockeye salmon (onchorhynchus nerka) fishery. *Canadian Journal of Fisheries and Aquatic Sciences*, 57:962–979, 2000.

[274] J.T. Schnute and A.R. Kronlund. A management oriented approach to stock recruitment analysis. *Canadian Journal of Fisheries and Aquatic Sciences*, 53:1281–1293, 1996.

[275] J.T. Schnute and A.R. Kronlund. Estimating salmon stock-recruitement relationships from catch and escapement data. *Canadian Journal of Fisheries and Aquatic Sciences*, 59:433–449, 2002.

[276] J.T. Schnute and L.J. Richards. The influence of error on population estimates from catch-age models. *Canadian Journal of Fisheries and Aquatic Sciences*, 52:2063–2077, 1995.

[277] J.T. Schnute and L.J. Richards. Use and abuse of fishery models. *Canadian Journal of Fisheries and Aquatic Sciences*, 58:1–8, 2001.

[278] G.A.F. Seber. *The Estimation of Animal Abundance and Related Parameters.* Charles Griffin & Compagny Ltd., London and High Wycombe, 2nd edition, 1982.

[279] E.J. Simmonds, E. Portilla, D. Skagen, D. Beare, and D.G. Reid. Investigating agreement between different data sources using Bayesian state-space models: an application to estimating NE Atlantic mackerel catch and stock abundance. *ICES Journal of Marine Science*, 67(6):1138–1153, 2010.

[280] D.S. Sivia. *Data Analysis: A Bayesian Tutorial.* Clarendon Press, Oxford, 1996.

[281] D. Sorensen and D. Gianola. *Likelihood, Bayesian, and MCMC Methods in Quantitative Genetics.* Springer, New York, NY, 2002.

[282] D.W. Speas, C.J. Walters, D.L. Ward, and R.S. Rogers. Effects of intraspecific density and environmental variables on electrofishing catchability of brown and rainbow trout in the colorado river. *North American Journal of Fisheries Management*, 24:586–596, 2004.

[283] D.J. Spiegelhalter, N.G. Best, B.P. Carlin, and A. Van der Linde. Bayesian measures of model complexity and fit. *J. R. Statist. So.*, 64(3):1–34, 2002.

[284] D.J. Spiegelhalter, A. Thomas, and N. Best. Winbugs version 14 user manual. *MRC and Imperial College of Science, Technology and Medicine*, (available at: http://www.mrc-bsu.cam.ac.uk/bugs), 2003.

[285] D.J. Spiegelhalter, A. Thomas, and N.G. Best. Computation on graphical Bayesian models. In J.M. Bernardo, J.O. Berger, A.P. Dawid, and A.F.M. Smith, editors, *Bayesian Statistics 5*, pages 407–425. Oxford University Press, 1996.

[286] C. Stein. Inadmissibility of the usual estimator for the mean of a multivariate distribution. *Proceedings of the Third Berkeley Symposium on Mathematical and Statistical Probability*, 1:197–206, 1956.

[287] P.A. Stephens, S.W. Buskirk, and C. M. del Rio. Inference in ecology and evolution. *Trends in Ecology & Evolution*, 22(4):192–197, 2007.

[288] S. Sturtz, U. Ligges, and A. Gelman. R2winbugs: A package for running winbugs. *Journal of Statistical Software*, 12(3):1–16, 2005.

[289] Z. Su and M.D. Adikson. Optimal in season management of pink salmon (*Onchorhynchus gorbusha*) given uncertain run sizes and seasonal changes in economic values. *Canadian Journal of Fisheries and Aquatic Sciences*, 59:1648–1659, 2002.

[290] D.P. Swain, I.D. Jonsen, J.E. Simon, and R.A. Myers. Assessing threats to species at risk using stage-structured state-space models: mortality trends in skate populations. *Ecological Applications*, 19(5):1347–1364, 2009.

[291] C.S. Tapiero. *Applied Stochastic Models and Control in Management.* North Holland, Amsterdam, NL, 1988.

[292] A. Thomas, B. O'Hara, U. Ligges, and S. Sturtz. Making BUGS open. *Journal of Statistical Software*, 6(1):12–17, 2006.

[293] L. Thomas, S.T. Buckland, K.B. Newman, and J. Harwood. A unified framework for modelling wildlife population dynamics. *Australian and New Zeland Journal of Statistics*, 47(1):19–34, 2005.

[294] J.E. Thorpe, M. Mangel, M. Metcalfe, and F.A. Huntingford. Modelling the proximate basis of salmonids life-history variation, with application to Atlantic salmon, salmo salar l. *Evolutionary Ecology*, 12:581–599, 1998.

[295] R. Tibshirani. Regression shrinkage and selection via the Lasso. *Journal of the Royal Statistical Society, Series B*, 58:267–288, 1996.

[296] L. Tierney. Markov chains for exploring posterior distributions (with discussion). *The Annals of Statistics*, 22:1701–1762, 1994.

[297] L. Tierney. *Markov chain Monte Carlo in Practice*, chapter Introduction to General State-Space Markov Chain Theory, pages 59–74. Chapman & Hall, London, UK, 1996.

[298] V.M. Trenkel, D.A. Elston, and S.T. Buckland. Fitting population dynamic model to count and cull data using sequential importance sampling. *Journal of the American Statistical Association*, 95(450):363–374, 2000.

[299] M. Tribus. *Rational, Descriptions, Decisions and Designs.* Pergamon Press, New York, 1969.

[300] S. Tuljapurkar and H. Caswell. *Structured-population Models in Marine, Terrestrial and Freshwater Systems.* Chapman & Hall, New York, NY, 1997.

[301] A. Tvesky and D. Kahneman. Advances in prospect theory: Cumulative representation of uncertainty. *Journal of Risk and Uncertainty*, 5:297–321, 1992.

[302] C.G. Utrilla and J. Lobon-Cervilla. Life history patterns in a southern population of Atlantic salmon. *Journal of Fish Biology*, 55:68–83, 1999.

[303] O. Varis and S. Kuikka. Joint use of multiple environmental assessments models by a Bayesian meta-model: the Baltic salmon case. *Ecological Modelling*, 102:341–351, 1997.

[304] O. Varis and S. Kuikka. Learning Bayesian decision analysis by doing: lessons from environmental and natural resources management. *Ecological Modelling*, 119:177–195, 1999.

[305] G. Verbeke and G. Molenberghs. *Linear Mixed Models for Longitudinal Data.* Springer, New York, NY, 2000.

[306] P. F. Verhulst. Notice sur la loi que la population suit dans son accroissement. *Correspondance Mathématique et Physique*, 10:113–121, 1838.

[307] Y. Vermard, E. Rivot, S. Mahëvas, P. Marchal, and D. Gascuel. Identifying fishing trip behaviour and estimating fishing effort from VMS data using Bayesian hidden Markov models. *Ecological Modelling*, 221(15):1757–1769, 2010.

[308] C.J. Walters. Optimal harvest strategies for salmon in relation to environmental variability and uncertain production parameters. *Journal of Fisheries Research Board Canada*, 32:1777–1784, 1975.

[309] Y.G. Wang and N.R. Loneragan. An extravariation model for improving confidence intervals of population size estimates from removal data. *Canadian Journal of Fisheries and Aquatic Sciences*, 53:2533–2539, 1996.

[310] E.J. Ward. A review and comparison of four commonly used Bayesian and maximum likelihood model selection tools. *Ecological Modelling*, 211:1–10, 2008.

[311] L. Wasserman. *All of Statistics: a Concice Course in Statistical Inference*. Springer, New York, NY, 2004.

[312] C.K. Wikle. Hierarchical Bayesian models for predicting the spread of ecological processes. *Ecology*, 84(6):1382–1394, 2003.

[313] C.K. Wikle. Hierarchical models in environmental science. *International Statistical Review*, 71(2):181–199, 2003.

[314] B. Worm, E.B. Barbier, N. Beaumont, J.E. Duffy, C. Folke, B.S. Halpern, J.B.C. Jackson, H.K. Lotze, F. Micheli, S.R. Palumbi, E. Sala, K.A. Selkoe, J.J. Stachowicz, and R. Watson. Impacts of biodiversity loss on ocean ecosystem services. *Science*, 314(5800):787–790, 2006.

[315] B. Worm, R. Hilborn, J.K. Baum, T.A. Branch, J.S. Collie, C. Costello, M.J. Fogarty, E.A. Fulton, J.A. Hutchings, S. Jennings, O.P. Jensen, H.K. Lotze, P.M. Mace, T.R. McClanahan, C. Minto, S.R. Palumbi, A.M. Parma, D. Ricard, A.A. Rosenberg, R. Watson, and D. Zeller. Rebuilding global fisheries. *Science*, 325(5940):578–585, 2009.

[316] R.J. Wyatt. Estimating riverine fish population size from single and multiple-pass removal sampling using a hierarchical model. *Canadian Journal of Fisheries and Aquatic Sciences*, 59:695–706, 2002.

[317] R.J. Wyatt. Mapping the abundance of riverine fish populations: Integrating hierarchical Bayesian models with a geographic information system (GIS). *Canadian Journal of Fisheries and Aquatic Sciences*, 60:997–1006, 2003.

[318] K. Yamamura, H. Matsuda, H. Yokomizo, K. Kaji, H. Uno, K. Tamada, T. Kurumada, T. Saitoh, and H. Hirakawa. Harvest-based Bayesian estimation of sika deer populations using state-space models. *Population Ecology*, 50(2):131–144, 2008.

[319] A. Zellner. *An Introduction to Bayesian Inference in Econometrics.* Wiley series in probability and mathematical statistics. Wiley, New York, NY, 1971.

[320] A. Zellner. *Bayesian inference and decision techniques: Essays in Honor of Bruno De Finetti*, volume 6 of *Studies in Bayesian econometrics and statistics*, chapter On assessing Prior Distributions and Bayesian Regression analysis with g-prior distribution regression using Bayesian variable selection, pages 233–243. North-Holland, Amsterdam, NL, 1986.

[321] A. F. Zuur, E. N. Iono, and J. N. Walker. *Mixed Effect Models and Extension in Ecology with R.* Springer, New York, NY, 2009.

# Index

Printed in the United States
by Baker & Taylor Publisher Services

Printed in the United States
by Baker & Taylor Publisher Services